Rethinking
Resource Management

Rethinking Resource Management offers students and practitioners a sophisticated framework for rethinking the dominant approaches to resource management in a complex world. Drawing on a deep understanding of relationships between resource projects and indigenous peoples, the book argues that current resource management practices consider important human values irrelevant and invisible.

The book uses case studies to argue that professional resource managers do not take responsibility for the social and environmental consequences of their decisions on the often powerless and vulnerable indigenous communities they effect. It offers an approach to social impact assessment methods that are more participatory and empowering than many alternative technical approaches. It discusses the invisibility of indigenous peoples' values and knowledge in the dominant paradigms of resource management. By drawing on contemporary social philosophy it offers a relational framework for thinking about interaction and change in resource management systems. This philosophical discussion is followed by a critical evaluation of case study methods and looks at case studies from Australia, North America and Norway.

Finally, *Rethinking Resource Management* investigates methodological issues of social impact assessment, policy development, applied research and the relevance of geographical perspectives and ethics to professional practice. In advocating more just, equitable and sustainable professional practice, the book explores new ways of seeing and thinking as a foundation for new practices. *Rethinking Resource Management* is empirically informed, theoretically sophisticated and ethically engaged in a way that will force resource managers at any point in their career to reassess what they think resource management is, should be and could be about.

Richard Howitt is an Associate Professor in Human Geography at Macquarie University, Australia. He received the Australian Award for University Teaching in Social Sciences in 1999.

The front cover illustration shows 'Dialectical Dynamics' by Fiona Cross (© 1993), who produced this image as a response to studying Resource Management with Richard Howitt in 1992. She wrote of the piece: 'Dialectical dynamic of roots and leaves, sky and dirt! At the moment I feel like the plant is me!'

Frontispiece 'Journey' by John Robinson, sculptor (© Macquarie University, photographer Mario Bianchino). This sculpture is displayed in the Macquarie University Sculpture Park, Sydney. The stainless steel shape captures, distorts and reflects the surrounding environment and its constant changes. The fantastic shape includes a small opening which provides a 'window' through which the viewer can see part of the environment without distortion. Like so many aspects of resource management, the acts of seeing, interpreting and responding represented in this work are given meaning by their context rather than having any unambiguous prior meaning.

Rethinking Resource Management

Justice, sustainability and indigenous peoples

Richard Howitt

London and New York

First published 2001
by Routledge
11 New Fetter Lane, London EC4P 4EE

Simultaneously published in the USA and Canada
by Routledge
29 West 35th Street, New York, NY 10001

Routledge is an imprint of the Taylor & Francis Group

© 2001 Richard Howitt

Typeset in Galliard by Bookcraft Ltd, Stroud, Gloucestershire
Printed and bound in Great Britain by TJ International Ltd, Padstow,
Cornwall

British Library Cataloguing in Publication Data
A catalogue record for this book is available from the
British Library

Library of Congress Cataloging in Publication Data
Howitt, Richard.
    Rethinking resource management: justice, sustainability and
    indigenous peoples/Richard Howitt.
        p. cm.
    Includes bibliographical references and index.
    1. Natural resources–Management–Case studies. 2. Sustainable
    development–Case studies. 3. Economic justice–Case studies.
    4. Indigenous peoples–Case studies. I. Title

    HC59.15.H69 2001
    333.7–dc21   00–067372

ISBN 0–415–12332–1 (hbk)
ISBN 0–415–12333–X (pbk)

Contents

Figures

Maps

Plates

Tables

Preface

Across the world, indigenous peoples have faced displacement, dispossession, cultural and physical genocide and exposure to great risk from all manner of activities that have been justified in terms of their contributions to industrialisation, development and somebody else's national (or even international) interest. As the governments of the old world order's three worlds of development pursued their goals, indigenous peoples remained an anomalous fourth world – they resisted development. Somehow they (sometimes) survived.

In the final decade of the twentieth century, amid contested assertions of a 'New World Order', the United Nations agreed to a decade dedicated to the world's indigenous peoples. Five hundred years after Columbus' voyage of 'discovery' transformed the diverse self-governing worlds of the Americas into a single 'new world' for Europe to exploit, to govern and to transform, the persistent presence of indigenous peoples continues to challenge many of the assumptions underlying developmentalism.

Nowhere is the power of this challenge clearer than in the realm of resource management. Indigenous rights and concerns are implicated in many resource-based development projects. At the turning of the century, they have intruded into the policies and practices of many international agencies, transnational resource companies and inter-governmental and non-government bodies. Indigenous rights have also rapidly emerged as central in the industrialisation of biodiversity. In most nation states, even the concept of indigenous *rights* is controversial. Why should indigenous people be given rights unavailable to other citizens? This question is raised over and over again as a basis for restricting 'concessions' to 'special interest groups'. A commitment to equality becomes the basis for imposing conditions on indigenous citizenship of and participation in national society. This process was clearly seen in Australia in the late 1990s, where the conservative Liberal–National coalition government substantially amended the Native Title Act 1993. Among the amendments were changes to the Act's 'right to negotiate' provisions. The government's defence was that this 'right' was unavailable to other property holders, it was not an inherent element of native title and it was a concession to indigenous people made by a previous government that it was not bound to retain. This vision of equality turns upside down the notion of indigenous rights. Neither indigenous Australians nor the international indigenous rights movement generally claim new rights: they aim to preserve existing ones – rights that they 'already had before they were subjected to some colonising state' (Brösted 1987: 156). It is often the case that resource projects are at the front line of relations between nation states and indigenous peoples. This places a heavy responsibility on resource management professionals.

The recent expansion of employment opportunities for resource management specialists in many fields of business and government activity has produced something of a boom in student enrolments in resource and environmental management courses at universities and colleges around the world. For readers seeking professional employment in the diverse fields of resource management, the issues raised in this book have probably been pushed aside in an effort to demonstrate technical excellence, or a detailed, if fragmented, understanding of specific physical, ecological and biological processes affected by various aspects of professional, scientific resource management. The importance of the social, political, cultural and ethical contexts of resource management practices, however, cannot be avoided in the Realpolitik of professional practice. Literacy in the complex geopolitics of resources is an essential part of contemporary resource managers' fundamental conceptual toolkit.

Most readers of this book will inevitably be irrevocably dependent on the dominant national and international systems of resource management that deliver the means of everyday survival. For most, the diverse world of traditional, indigenous, local-scale resource management systems will be so unfamiliar as to be invisible. In many cases these unfamiliar resource management systems may also seem so unproductive, inefficient or so incomprehensible as to be worthless.

This is certainly how it has appeared to many professional resource managers, to mining company executives, forest economists, energy ministers, fisheries experts and countless others when they are faced with subsistence economies based on commercially valuable resources, or located in areas containing potentially commercial resources not used by the subsistence sector. These professionals make decisions which have dramatic and far-reaching consequences for these unfamiliar, incomprehensible and largely invisible other worlds.

This book aims to render visible much that is conventionally left invisible in resource management education. It examines professional resource managers' decisions and the systems that make them possible in the light of indigenous peoples' experience. It is both a critique and a reconstruction; simultaneously a challenge and a guide, for students and professionals in various areas of resource management. In the process, the book seeks to confront readers with the need to rethink the field of resource management.

The book's core argument is quite simple: we must rethink resource management in order to make resource management decisions more accountable to critical human values such as social justice, ecological sustainability, economic equity and cultural diversity. Contemporary industrial resource management systems have the power to turn upside down the taken-for-granted worlds of the communities they affect. Those responsible for making key decisions about resources require a professional literacy that equips them to read and respond humanely to the complex situations in which they are inevitably immersed.

In advancing its argument, this volume focuses on indigenous peoples' experience. The same basic argument, however, and similar conclusions could be reached from many other vantage points. For example, focusing on questions of inter-generational equity, women's rights, the experience of workers in resource industries, issues of environmental quality and so on could equally lead one to the conclusion that there is an urgent need to radically rethink resource management practices in the industrial world. So the focus here on indigenous peoples should be seen not only as a substantial focus in its own right, but also as a case study of the reasons for rethinking currently dominant resource management practices.

In broad terms, the field of resource management is currently dominated by a regime that is utilitarian, reductionist, technocentric and market driven. This book argues that, despite its spectacular commercial successes, this dominant paradigm needs to be rethought because it fails to meet human needs in several important areas. Specifically, it treats critically important issues such as justice, sustainability and human rights as externalities – as someone else's problems. The dominant paradigm claims to deal with these externalities with ostensibly objective, authoritative and dispassionate market tools. This book demonstrates that naïve market-based solutions to issues of justice, sustainability, equity and diversity are inadequate and unsupportable.

In developing criteria for evaluating successful resource management, I want to propose that we rethink industrial resource management systems from the vantage points of these core values: justice, sustainability, equity and diversity. A practical agenda for change will be presented. In supporting such changes, I seek to displace narrowly economistic notions of value and accountability with wider, more coherently and complexly contextualised notions of human landscapes in which resource management decisions are held accountable to a wider range of human values and experience. The intention, therefore, is not to be 'objective', but to challenge the underlying notion of objectivity; not to be prescriptive, but to open lines of debate; not to be authoritative, but to challenge the foundations of authority; and not to be dispassionate, but to deal openly with passionate human issues. As a text, then, the purpose is deliberately subversive. Current industrial resource management paradigms have failed. They need to be rethought, reshaped and restructured towards more humane goals. This book seeks to provoke contributions to this process.

Acknowledgements

This book has had a long gestation. It was conceived when I was appointed to teach Resource Management at Macquarie University in 1992, and since then I have accumulated many debts to people who have contributed to my education in the new geopolitics of resource management, sustainability and indigenous rights. Students in my undergraduate courses at Macquarie University have helped to clarify ideas and direction over many offerings of these courses. My extraordinarily talented group of honours and postgraduate students have all influenced my thinking, and in the case of Sandie Suchet, Rochelle Braaf, Jan Turner, Marcia Langton and Sue Jackson, have made very direct input to the thinking reflected in this work. Fiona Cross, who studied with me as an undergraduate and has gone on to a career that combines her passions for justice and visual arts, kindly allowed me to use her artistic response to the course on the front cover. Sandie Suchet, Rochelle Braaf, Stephen Hodge, Richard Cook, Robyn Dowling, Libby Ellis, Graeme Aplin, Peter Mitchell, Lyn Hudson and Bob Fagan have shared the teaching in these units and I have appreciated their support and contributions more than I can possibly say. The work has also benefited from the contributions of many research assistants over the years, including Bronwyn Parry, Beth Norris, Erica Klimpsch, Toney Hallahan, Leah Gibbs, Venessa Kealy, Andria Durney, Jo Fox and my wonderful colleague Ian Bryson. I have also benefited greatly from shared fieldwork on major projects with some wonderful colleagues over the years, particularly Ciaran O'Faircheallaigh, Annie Holden and Sue Jackson.

My colleagues in Human Geography, Aboriginal Studies and the Resource and Environmental Management programmes at Macquarie University have offered a critical, constructive and supportive context for teaching, research and writing over a sustained period. Colleagues at Cambridge and Oxford Universities, Aristotle University of Thessaloniki and Université du Québec à Chicoutimi have also offered support at important times. Staff at libraries at Macquarie University, the Scott Polar Research Institute, Cambridge and the Université du Québec à Chicoutimi also provided valuable assistance.

Among my Aboriginal colleagues there are several whose patience in teaching me and working with me over many years needs to be properly acknowledged, although I think there are some debts that one is unable to repay. I hope this book goes some way toward reflecting their efforts. Bella Savo and Sandy Callope in Weipa, Freeda Archibald in Muswellbrook, and Parry Agius in South Australia have spent many hours as co-researchers and colleagues in the field. Marcia Langton, Patrick Dodson, Mick Dodson, Tracker Tilmouth, Noel Pearson, Olga Havenan, Peter Yu and Banduk Marika in particular have all offered support, guidance and ideas that have shaped the arguments developed here. The Aboriginal rights movement in Australia greatly benefits from their contributions and I acknowledge their influence here.

My teachers, particularly Mary Hall, Bill Jonas and Frank Williamson laid firm foundations for the work reflected here and I thank them. Many colleagues, particularly Bob Fagan, Bertell Ollman, Dick Bryan and Ron Horvath have continued to shape these foundations through constructive debate over many years. At Routledge a succession of commissioning editors, commencing with Tristan Palmer and ending with the very patient Ann Michael, have waited patiently each August for a promised manuscript. I am indebted for their patience and trust. And many friends have tolerated my absences and vagueness arising from the gestation of this book for far too many years.

In addition to these professional debts, there are other debts that are beyond repayment. There is a great cost in a book like this that is inevitably borne by one's family. I am sure that there have been many moments when my wife Anastasia has wished to see no more of this manuscript. I am grateful she voiced that wish so rarely. My son Alexei has lived with versions of this manuscript for almost his entire life! His patience with a father who has been absent in the classroom and the field too often is treasured. I hope the products of these absences have perhaps enriched his life as well. I must also express my immense gratitude to my parents-in-law, Helen and Chris Toliopoulos. Without their unending support, much of the work underpinning this book could simply not have been done. And from my own parents, I have drawn strongly on the foundations they offered in shaping the arguments presented here.

So much support, love, friendship and trust is impossible to repay, and my gratitude impossible to express. To the many friends and colleagues whose names I've not mentioned, but whose contributions, however small, have shaped this work, I thank you and hope that this modest contribution to debate reciprocates your generosity.

In addition to these personal acknowledgments, I am grateful for the following permissions to use copyright materials in this book: A.P. Watt Ltd on behalf of Michael B. Yeats and Simon & Schuster for rights to reproduce the poem 'The Second Coming' by W.B.Yeats; Edward Arnold for permission to reproduce Table 3.2; The Australian National Gallery and Lyn Williams for the right to reproduce the images 'Forest' and 'Red Trees' by Fred Williams (Figure 2.1); Buffy Sainte-Marie and Creative Native for permission to reproduce the lyrics of 'My Country 'Tis of Thy People You're Dying'; Canadian Press Picture Archive for permission to reproduce the image 'Face to Face'(Plate 1.1); Cordon Art for the right to reproduce the image 'Day and night' by M.C.Escher (Figure 2.3); Deborah Bird Rose for permission to reproduce her works; Fiona Cross for permission to reproduce the image 'Dialectical dynamics' on the front cover; Glenn Welker for permission to reproduce the text of the 'Declaration of Indigenous Peoples' from the text published on his website: http://www.whitestareagle.com/natlit/declare.htm; IWGIA for permission to reproduce text and illustrations from Paine (1982); Jim Birckhead for permission to reproduce Figure 1.5; Leon Rosselson for permission to reproduce the lyrics of 'The World Turned Upside Down'; Emily Benedek and Random House for permission to use extracts from her work; Mario Bianchino and Macquarie University for permission to reproduce the image of the sculpture 'Journey' by John Robinson; Michael Gallagher for permission to reproduce his photographs of Noonkanbah (Plates 8.3–8.5); Rolf Gerritsen and Martha MacIntyre for permission to reproduce Figure 12.5; The Office of Economic Development of the Navajo Nation for permission to reproduce material from their statistical records (Table 9.1); The Speaker of the House of Representatives of the Parliament of Australia who, after consultation with the traditional authorities in Yirrkala, granted permission to reproduce the image and text of the Yirrkala Bark Petition (Plate 8.1).

Part I

Introduction
(and disorientation)

The World Turned Upside Down

In sixteen forty nine to St George's Hill
A ragged band they called the Diggers came to show the people's will.
They defied the landlords, they defied the laws,
They were the dispossessed reclaiming what was theirs.

We come in peace, they said, to dig and sow.
We come to work the land in common and to make the waste ground grow
This earth divided, we will make whole
So it can be a common treasury for all.

The sin of property we do disdain.
No-one has any right to buy and sell this earth for private gain.
By theft and murder they took the land
Now everywhere the walls rise up at their command.

They make the laws to chain us well.
The clergy dazzle us with heaven or they damn us into hell.
We will not worship the God they serve,
The god of greed who feeds the rich while poor folk starve.

We work, we eat together, we need no swords.
We will not bow to masters or pay rent to the lords.
Still we are free, though we are poor.
You Diggers all stand up for glory, stand up now.

From the men of property the orders came.
They sent their hired men and troopers to wipe out the Diggers' claim.
Tear down their cottages, destroy their corn.
They were dispersed, but still the vision lingers on.

You poor take courage, you rich take care.
This earth was made a common treasury for everyone to share.
All things in common. All people one.
We come in peace. The orders came to cut them down.

Song lyric by Leon Rosselson,
recorded on 'Rosselsongs', Fuse Records

1 Worlds turned upside down

The sin of property we do disdain.
No-one has any right to buy or sell this earth for private gain.
By theft and murder they took the land
Now everywhere the walls rise up at their command

Leon Rosselson[1]

Resources, politics and people

Conflicts over resources are an important and influential element of political, social and economic processes throughout the world. Resources and their management have long been central in all political processes. As political scientist Adrian Leftwich puts it:

> Politics consists of all the activities of and conflict, within and between societies, whereby the human species goes about obtaining, using, producing and distributing resources in the production and reproduction of its social and biological life.
>
> (Leftwich 1983: 11)

Resources themselves need to be understood not as pre-existing substances or things, but in terms of functions and relationships. This approach to defining resources as simultaneously economic, cultural and physical in character, although crucial for the argument presented here, is hardly new. In 1956, for example, Spoehr observed:

> It is doubtful that many other societies ... think about natural resources in the same way we do. It is probable that the term itself ... is primarily a product of our own industrial civilization.
>
> (Spoehr 1956: 93)

Spoehr went on to examine the ways in which different peoples' definitions of 'resources' reflected the specific technology available, the social relations within the particular cultural group and the society's interpretation of ecological circumstances. Spoehr's 1950s approach, with his 'bearded figure of Darwin watching quietly from the shadows' (1956: 101), may seem dated. Unlike some more recent texts, however, it takes the interplay of culture, environment, economy and technology into account, and does not try to reduce the complex task of resource management to a technical task.

Even in 1933, Zimmerman's influential *World Resources and Industries* sought to provide a 'new synthesis between cultural geography and economics' (Zimmerman 1964: vii). For Zimmerman, culture was central in creating even those resources popularly seen as 'natural'. Resources were not pre-existing substances, but:

> living phenomena, expanding and contracting in response to human effort and behaviour ... . To a large extent, they are man's [sic] own creation. *Man's own wisdom is his premier resource – the key that unlocks the universe.*
>
> (ibid.: 7, emphasis in original)

He went on to provide a definition of 'resources' which is worth considering at some depth, even after more than sixty years:

> The word 'resource' *does not refer to a thing nor a substance but to a function which a thing or a substance may perform or to an operation in which it may take part,* namely, the function or operation of attaining a given end such as satisfying a want. In other words, the word 'resource' is an abstraction reflecting human appraisal and relating to a function or operation. As such, it is akin to words such as food, property, or capital, but much wider in its sweep than any of these.
>
> (ibid.: 8, emphasis in original)

In other words, just as Leftwich's definition of 'politics' emphasises the centrality of 'resources' in human politics, Zimmerman's definition of resources reminds us that resources are fundamentally a matter of relationships not things. They do not exist outside the complex relationships between societies, technologies, cultures, economics and environments in some pre-ordained form, waiting to be discovered. They are created by these relationships. The geopolitics of resources, therefore, is not simply about access to and trade in pre-existing 'things' called resources. Rather, it is about fundamental transactions of power, wealth and privilege.

This book grows out of experience at a critical location within resource geopolitics – the interface between resource-based development and indigenous peoples. As a researcher, as a teacher and simply as a human being, I have become increasingly convinced of the urgent need for those involved in resource management systems to be more literate in the complexities of socio-political processes than they currently are. This need is demonstrated most urgently in the troubled relations between resource management and indigenous peoples.

Industrialisation and development: core goals of the Cold War world

In the rapidly changing world of the early 1950s and 1960s, superpower tensions over resources escalated rapidly. Western governments actively repressed Communist influences in economic, cultural and political spheres; governments in the Soviet bloc pursued industrialisation at a breakneck speed, regardless of the human and environmental costs. Newly independent governments in the former colonial empires sought to escape the legacies of European imperialism and American neo-colonialism; and nascent social movements demanding civil rights, women's rights, human rights and a range of fundamental freedoms began to challenge the previously unchallenged verities of everyday life in many places.

In the wake of post-war austerities, booming industrial economies raised hopes for an improved quality of life throughout the old world order's First, Second and Third Worlds. In each of these imagined places governments, communities, opposition movements and international agencies adopted industrialisation and development as core societal goals. There was a widespread, optimistic faith in the power of science, technology and good governance. In the context of Cold War geopolitics, however, it often seemed that 'development' of one of the old world order's imagined worlds could only be achieved if development of the other was suppressed. In particular, viewed from the West, development and industrialisation of the Soviet bloc was constructed as the key threat to development of the First World, and its ostensibly generous paternalistic approach to development in the Third World.

Industrialisation and development, however, are demanding masters for all their disciples. Both have huge appetites for resources. They hunger for energy, minerals, timber, land, food, labour, information and consumers; they require the raw materials with which to make things to sell; and they demand (and produce) the raw materials to build economic, political and social power. For much of their history, optimistic disciples have fed the appetites of industrialisation and development with little regard for long-term environmental costs, and with scant recognition of the complex and often contradictory social processes they set in motion. Management of these appetites is a matter of enormous importance for human societies. Despite this, the systematic assumptions underlying the practices of resource managers are rarely subjected to critical evaluation outside the contingencies of specific cases. Instead resource management is increasingly defined as a specialist field for qualified and neutral experts and objective scientific precision.

The myth of objective and neutral resource management has brought many human communities and the environments on which their lives have been built to, and sometimes beyond, the brink of catastrophe. In the landscapes of the poor and marginalised, the iconoclastic promises of resource-based development have been used to justify all manner of clever schemes – extraction, submergence, division, plantation, clearance and so on. Among the many tragic stories of resource mismanagement that could be told, it is the experience of indigenous peoples around the world that exemplifies most starkly the need for change. It is their homelands, their lives, their cultures and their rights that have too often become the Ground Zero for testing and implementing the theories and practices of scientific resource management.

This book deconstructs the hegemonic ideologies of scientific resource management. It offers a reconstruction of the field with wider and more functional professional literacy that encompasses the social, cultural, political, economic and environmental issues raised, as well as the technical expertise required of resource managers. In reconstructing the field, it is argued that resource managers must develop the knowledge, skills and sensitivities to deal with the moral, ethical and political domains of resource management as well as the technical domain. As we move into the twenty-first century, it is a dangerous and unjustified folly not to do so.

The risk of failure, of course, is that we demonstrate that, at the planetary scale, we all live at Ground Zero.

Ground Zero: Emu Test Site, Australia

In 1953 Great Britain and Australia detonated three atomic bombs over the desert homelands of the Yankatjara and Pitjantjatjara people in South Australia. The governments named the site 'Emu', and the tests 'Totem'.

The flightless emu is one of the ancestral characters who created the desert landscapes of the region in the creation stories of the Yankatjara and Pitjantjatjara people. Along with the kangaroo, it was also adopted as an icon of the Australian nation state as part of the Commonwealth coat of arms. Selection of the name 'Emu' for the atomic testing site, then, had considerable symbolic significance.

The tests were part of Britain's nuclear weapons development programme intended to arm the West against the nuclear might of the Soviet Union and its Communist allies. In 1953, indigenous Australians had no status as citizens of the nation, and no recognition as its prior owners. When selecting a test site, no one considered asking for permission from the Yankatjara or Pitjantjatjara because, to all intents and purposes, Australia was treated as *terra nullius*. It was still a loyal post-colonial daughter of the empire. These remote desert lands were the emptiest of lands in the continent that the colonisers and their descendants asserted no one owned. These lands were a strategic resource that was free for the taking in the governments' eyes.

So the three Totem tests were undertaken after a minimal effort to move Aboriginal people (Anangu) from the area, and signposting the area with warnings in English – for a population who had no access to literacy education. Totem 1, 2 and 3 represent three traumatic events in the life of the Maralinga Tjarutja Lands, and the lives of the Anangu who called it home in the 1950s (see for example Milliken 1986; McClellan *et al.* 1985; Toyne and Vachon 1984).

Thirty-five years later, in 1988, in the arid Mulga woodland country of the Maralinga Tjarutja lands, over 150 people, mostly Anangu from the surrounding settlements and communities, came together to discuss Aboriginal involvement in and empowerment through land management and conservation issues at the Emu site (Kean *et al.* 1988) – Ground Zero of the Totem tests.

For Australians concerned with social justice and environmental protection, the Emu test site, contaminated by plutonium and symbolic of the nation's mistreatment of its indigenous citizens and environment as well as its subservience to Britain, is indeed a totem. It is a lasting monument to non-Aboriginal environmental vandalism in Australia. Among the Anangu whose lands were literally blasted into oblivion, its choice as the site for debating questions of resource, environmental management and community empowerment had great symbolic power – of all places, this was a location where the abject failure of non-Aboriginal stewardship of the land was clear for all to see, the need for Aboriginal involvement in conservation and land management could hardly need justification.

Professional literacy in a changing world

The field of resource management covers a great diversity of human endeavour. It includes the technically sophisticated work of exploration geologists, project engineers, foresters, resource economists and marine ecologists, as well as the support, planning and regulatory work of government employees, and the hard physical work of production. It involves not only some of the world's largest market-driven capitalist enterprises, but also diverse small enterprises and myriad small-scale producers in artisanal and other non-capitalist modes of production. It also involves not only wage labourers in organisations of varying economic efficiency, but also peasant farmers, hunters, fishers, pastoralists and myriad others in their communities in the management of everyday lives by acting to ensure continuity of the means of survival. It covers not only production of physical resources, but also the management of conservation areas, tourism sites, cultural materials, information and services. As Leftwich's useful definition of politics reminds us, not only are all human communities involved in politics, but they are also all involved in resource management. Resource management systems are also political systems. They not only produce resource-commodities, but also produce power.

Cultural differences between peoples construct different understandings about what constitutes both 'resources' and 'power'. Consequently, many cultural (and ecological) consequences of resource management decisions simply become invisible because of the way that the cultural construction of knowledge constructs one's understanding of resources themselves. For many resource managers, it is easy for the resources they manage to become 'naturalised' – to appear as if they are substances or things created (and therefore manageable) outside any cultural context. It is easy to see how management of 'natural' resources is reconstructed as a technical and professional task for experts, whose vision of the cultural domain is limited.

In constantly globalising markets for resource commodities such as food, timber, minerals, energy, tourism destinations, agricultural lands, urban lands, waste disposal sites and so on, resource management has emerged quickly as an important field of professional expertise connecting diverse places (and peoples) to global marketplaces. In the process, those places and their peoples are transformed – often irrevocably and often at great social, ecological and cultural cost.

For many of the professionals involved in the management of such change, however, these transformations are both invisible and unimportant. Protection of social and biophysical environments, if it is considered at all, is widely seen in professional circles as a matter for government specialists, not for operational managers. Many professional resource managers see their duty in terms of efficiency – minimising costs, maximising profits, guaranteeing outputs, maintaining supplies and so on. Professional literacy, then, is generally seen as being about using the best available techniques, understanding the technical literature and reading in a specialised field – whether it be aerial geomagnetic surveying, futures markets for gold or aluminium, or ecological aspects of the life cycle of a commercially exploited fish or fowl. This book is not about that sort of professional literacy, except as a target for transformation. To nurture this professional literacy (and subsequently improved outcomes 'on the ground'), it is argued that three basic steps need to be taken (Figure 1.1). We first need to develop new ways of 'seeing' the field of resource management in ways that make visible the complex consequences of resource management decisions. Second, we need to develop new ways of 'thinking' that accept the contextual complexities of resource decision making. And finally, we need to develop new ways of 'doing' resource management.

Instead, this book addresses the transformational politics that are constructed by or might be possible around and within industrial resource management systems. It is about understanding and responding to the new geographies produced by resource management practices. Perhaps most significantly, it is also about putting in place a vision of resource management practice that not only opposes reproduction of the social and ecological catastrophes of the past, but also actively contributes to sustainable and just human futures. Unlike many textbooks about 'the geography of resource management' (for example Mitchell 1989; Castillon 1992), this is not a manual of techniques. It does tackle some technical and methodological issues, and it certainly aims to be 'applied' and 'practical', but its approach to questions of method, technique, practice and application is always in terms of the broader process of professional literacy which is being targeted.

The approach developed here does not involve advocacy of some particular approach as a universal best practice. Rather, it advocates some basic principles and perspectives for creating better practices. Dogmatic adherence to a particular methodology or theoretical approach because it is pre-defined by so-called experts as best practice is part of the problem under examination here, not least because the best practice of one generation is the obsolete myopia of

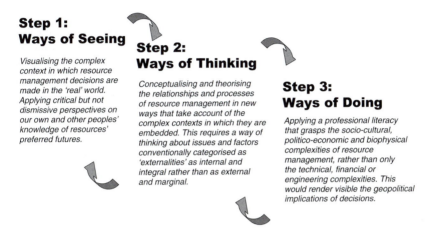

The Challenge

to develop a more diverse professional literacy in the language of cultural and social diversity, the language of landscape, and the language of values as they interact with resource management systems

Step 1:
Ways of Seeing

Visualising the complex context in which resource management decisions are made in the 'real' world. Applying critical but not dismissive perspectives on our own and other peoples' knowledge of resources' preferred futures.

Step 2:
Ways of Thinking

Conceptualising and theorising the relationships and processes of resource management in new ways that take account of the complex contexts in which they are embedded. This requires a way of thinking about issues and factors conventionally categorised as 'externalities' as internal and integral rather than as external and marginal.

Step 3:
Ways of Doing

Applying a professional literacy that grasps the socio-cultural, politico-economic and biophysical complexities of resource management, rather than only the technical, financial or engineering complexities. This would render visible the geopolitical implications of decisions.

Figure 1.1   Steps towards literacy in resource geopolitics

the next. In the rapidly changing world of resource geopolitics, having the best current technique is only a small part of the story. Using examples of indigenous peoples' experience of resource-based development processes, this book provides a framework for future generations (and hopefully some of this generation) of resource managers to do a better job in husbanding the planet's resources and nurturing the human and non-human communities that rely upon them.

Core values for resource management

As a field of academic study and professional practice, resource management has unquestionably been dominated by a concern with technical sophistication as a source of credibility and social relevance. For example, it remains common for resource managers' professional education to emphasise technical skills and methodological matters above (typically way above) the 'soft' skills of social, cultural and political literacy. Yet just what do these sophisticated techniques really achieve? How do they achieve 'better' resource management outcomes?

The answer to such questions depends, of course, on what is defined as 'better'. It is here that many professionals retreat into their politically edged shell of 'value-free' science. For them 'better' is not a value-laden term. For them, 'better' is an objective issue – more is better, and 'more' can be easily measured. Sophisticated exploration techniques allow geologists to identify 'more' oil, 'more' gas, 'more' minerals. Sophisticated forest management techniques allow foresters to squeeze 'more' timber from forests. Sophisticated project management or systems engineering techniques allow investors to construct 'more' efficient processing plants that produce 'more' material for every dollar invested and every unit of raw material input.

Having reduced the untidily complex and value-laden term 'better' to the neatly quantifiable 'more', the market alchemists' work has really only just begun. In the language of the marketplace, wealth is reduced to money; resources become commodities; and value becomes price. In changing complex realities into simplified models, these 'experts' develop some highly sophisticated stupidities. In Papua New Guinea, for example, where some of humanity's oldest sustainable agricultural systems have been in place for hundreds of generations, national economic figures do not include subsistence economic activity. In the process, the livelihoods (and cultural life) of a substantial proportion of the population have simply disappeared – replaced with the miraculous growth (and spectacular busts) conventionally associated with resource-based economies. As wealth is no longer measured in terms of a community's ability to feed itself, to undertake cultural obligations and to live, sophisticated and 'objectively' measurable economic indicators such as Gross Domestic Product can become the main measure of wealth.

In the process, these sophisticated stupidities succeed in hiding the most basic of issues in producing 'better' resource management – the question of goals; the question of 'why' rather than 'how' we might manage our resources. In rendering the complex simple and the value-laden objective, the dominant paradigms of resource management have lost sight of the underlying purpose of managing resources.

In this book, the issue of underlying purpose – the why of resource management – is addressed in terms of four core values. 'Successful' resource management achieves sustainable improvements in human lives in terms of

- social justice;
- ecological sustainability;
- economic equity;
- cultural diversity.

Like all human values, these are not universal. They reflect the particular context in which I have operated. For me, they have developed in the crucible of multicultural and indigenous politics in the Australian mining industry, and in the debates within radical geography since the 1970s. For others who share my commitment to these as core values for their professional behaviour, the particular inequities of other situations, or the emergence of the green political movements and other new social movements since the 1960s have been the catalyst. For all of us, external pressures in the form of unsustainable environmental practices, critical issues in human rights abuses, and the disabling inequities in the modern economy have demanded a broader focus than is provided by reductionist, objectivist, scientistic sophistication.

Recent experience has demonstrated the power of the currently dominant professional practice of resource managers to change forever and irreversibly patterns of daily life, patterns of social and cultural meaning for people affected by or involved in resource industries – the power to turn worlds upside down. The development of integrated, global-scale human systems means that modern resource management systems can generate situations where 'costs' and 'benefits', 'crises' and 'solutions' are no longer contained within systems at a single scale. 'Culprits' and 'victims' are often no longer contained within systems reflecting a common society, worldview or system of regulation (Lipietz 1996).

The language of resource management: new words/new worlds?

The language used in resource management is a significant issue. In my teaching, for example, students often find issues of language (not just terminology, but the deeper issues of the relationship between words, meaning and power) the most troubling ones. Language reflects, shapes and limits the way we articulate and understand the world around us. It not only provides the building blocks from which we construct our way of seeing complex realities. It also constructs the limits of our vision. Language reflects and constructs power. *Our* language renders invisible many things given importance by *other* people. And in the contemporary world of industrial resource management, the invisible is generally considered unimportant. Dominant economistic and scientistic epistemologies, or patterns of thinking about the world, thus render the concerns and aspirations of many people both invisible *and* unimportant. In the process of managing resources, ostensibly for the betterment of humanity, resource managers quite literally turn the world upside down. The means for survival are no longer under the control of human communities, but subject to the vagaries of the marketplace.

The Cold War confrontation between capitalism and Communism has collapsed, and the free market rules the world, or at least that is what we are often expected to believe. Almost anything can be traded in commodity markets. Culture, finance and markets have been globalised. But serious questions need to be asked about what is rendered invisible and unimportant by the markets of the world's resource industries.

The language of economic models is the language most often used to describe and explain market processes. Many geographers have observed that economic models typically render geography – the complex and dynamic characteristics of and relationships between people and real places – invisible and unimportant. Yet even a superficial knowledge of commodity markets and resource industries is enough to confirm that geopolitical dynamics – the real (and very complex) geographies of oil in the Middle East; metals in Japan, the former Soviet Union, the USA and Europe; timber in Malaysia, Papua New Guinea and Brazil; tuna and other fisheries in international waters – are fundamental to the operation of commodity markets.

At the intersection of geography, politics and environmental processes, a new geopolitics of resources is being forged by complex and dynamic processes and the relationships between overlapping and competing interests of many sorts – buyers and sellers; owners and managers; workers and bosses; producers and consumers; lobbyists, advocates and regulators. We need to explore new ways of thinking about these issues and relationships. We need to develop a new way of talking about them.

The language of the market is simply incapable of encompassing in its vision many of the crucial non-market elements that influence contemporary resource management systems. By excluding them, this language – and the models, behaviour, political structures and theories it reflects and constructs – becomes dangerous because it renders invisible and unimportant very real processes and relationships that need to be addressed in the understanding of and participating in the complex landscapes of resource management. Many of these processes are integral to the resource management systems in which we operate.

Guidance on how to envision complexity, how to capture these complex interactions between society, economy, politics and environment, is not easily found. We need to be able not only to envisage existing complexities, but also to envisage new worlds – new ways of approaching the tasks of resource management consistent with the core values of social justice, ecological sustainability, economic equity and cultural diversity. One way of understanding this task is to consider how writers of fiction approach the daunting task of writing new worlds. Rushdie, for example, speaks of the 'imaginary homelands' created by writers exiled from their real homelands:

> It may be that writers in my position, exiles or emigrants or expatriates, are haunted by some sense of loss, some urge to reclaim, to look back, even at the risk of being mutated into pillars of salt. But if we do look back, we must also do so in the knowledge ... that our physical alienation from India [or any other inaccessible homeland] almost inevitably means that we will not be capable of reclaiming precisely the thing that was lost; that we will, in short, create fictions, not actual cities or villages, but invisible ones, imaginary homelands, Indias of the mind ... .
>
> These are of course political questions, and must be answered at least in political terms. I must say first of all that description is itself a political act. The black American writer Richard Wright once wrote that black and white Americans were engaged in a war over the nature of descriptions. Their descriptions were incompatible. So it is clear that redescribing a world is the necessary first step to changing it.
>
> (Rushdie 1992: 10, 13–14)

Le Guin also reminds us of the importance of imagination in the politicised work of 'making the world different' which is, she notes, a task requiring 'political imagination' (1989: 46). Western approaches to this task have been dramatically captured by the imaginary of Columbus' 'New World'. The collective Western obsession with newness and an ethnocentric sense of discovery has often blinded us to perspectives that value things other than newness. The interplay between the imaginaries and realities of colonial and post-colonial dispossession and marginalisation, between 'fact' and 'fiction', between the privileged discourses of power and the imaginary homelands of alternative futures, requires more than technically sophisticated research. It also requires impassioned imagination. And it requires us to be self-consciously aware of our own place in the world, our own 'metaphorical location as participants in social transformation' (Howitt 1993a: 7). As Ruiz puts it:

> To be located is to rediscover the specificity and plurality of experience ... One's critical consciousness is inextricably related to one's location, although it is not determined by it.
>
> (Ruiz 1988: 162)

The work of political imagination to which Le Guin refers is thus both political and epistemological in nature. And it is not limited to the production of fiction. In addressing the experience of indigenous peoples in industrial resource management systems, we need to construct a way of seeing that rejects the notion of a single, privileged centre or a single way of representing 'truth'.

Industrial resource management and global crisis

The inadequacies of the dominant paradigms in industrial resource management are simultaneously exemplars of and contributors to a wider problem. Contemporary resource management systems are pivotal in both the constitution of the current global crisis, and also as a focus for action to overcome it. Ekins (1992: 1–2), for example, identifies four elements in his description of the global problematic that requires 'a patchwork of overlapping approaches' for resolution (Table 1.1).

Elements of these inter-related crises can be found at all scales and in many of the place-based conflicts over resources that generate and reflect struggles for wider change. The relationships between the local and the global have been a central concern of both human geography and the green political movement throughout the late 1970s and 1980s. For both, the processes of change set in train by the 1973 oil crisis represented a significant challenge. For human geographers, many of the core topics of their discipline (industrial location decisions, urban forms, spatial patterns in transport and trade and so on) were clearly shown to be integrated into complex global systems. In economic geography, for example, attention was turned to the strategies of global

Table 1.1 The global problematic

The environmental crisis	Environmental pollution and ecosystem and species destruction at such a rate and on such a scale that the very biospheric processes of organic regeneration are under threat
The military machine	The existence and spread of nuclear and other weapons of mass destruction and the overall level of military expenditure
The holocaust of poverty	The affliction with hunger and absolute poverty of some 20 per cent of the human race, mainly in what is misleadingly called the Third World
The denial of human rights	Intensifying human repression resulting from the increasing denial by governments of the most fundamental human rights and the inability of increasing numbers of people to develop even a small part of their human potential.

Source: Based on Ekins 1992.

corporations, and the local manifestations of global power. Systemic models aimed at explaining the global system were in vogue, including Wallerstein's World Systems Theory and Marxist theories of capitalism and imperialism, particularly through the work of David Harvey (1973, 1982, 1985; also Taylor 1982, 1993). For the environmental activists, the world scale constraints on local action revealed in the 1970s produced rapid recognition of the need to pay attention to both scales – to think globally and act locally (for example Gardner and Roseland 1989a, b).

For our purposes here, the 1973 oil crisis represents something of an 'aha experience' for the field of resource management. An 'aha experience' is something that enables, even forces one to say 'Aha. I understand things differently now'. In this case, the actions of OPEC dramatically changed the balance of power in trade relations between the industrialised 'West' and the oil-producing countries. This made it clear that international resource systems and national political economies were not independent of each other. Coming as it did in a period that saw the publication of *Limits to Growth* (Meadows *et al.* 1972) and increasing sensitivity to the real meaning of the metaphor of spaceship earth created by the extraordinary view of earth from space provided by the Apollo expeditions of the late 1960s, OPEC reinforced the idea that social and political forces and processes were as much a constraint on economic growth as the constraints of physical resources and technologies to develop them.

In other words, the OPEC-related oil crisis served to signal that not only are resource management systems embedded in particular locations and particular economic contexts, but they are also simultaneously embedded in

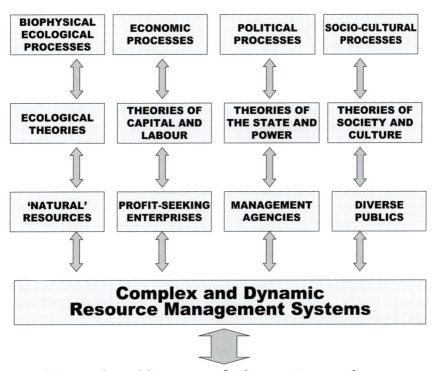

| BIOPHYSICAL ECOLOGICAL PROCESSES | ECONOMIC PROCESSES | POLITICAL PROCESSES | SOCIO-CULTURAL PROCESSES |

| ECOLOGICAL THEORIES | THEORIES OF CAPITAL AND LABOUR | THEORIES OF THE STATE AND POWER | THEORIES OF SOCIETY AND CULTURE |

| 'NATURAL' RESOURCES | PROFIT-SEEKING ENTERPRISES | MANAGEMENT AGENCIES | DIVERSE PUBLICS |

Complex and Dynamic Resource Management Systems

intersecting with a range of other systems and process

Figure 1.2　Each element in a resource management system is complex and dynamic in its own right

both social and political structures at various geographical scales and also a range of cultural and epistemological systems. Such crises emphasise the importance of understanding how resource management decisions reflect and affect the social, cultural and political settings that in turn themselves constitute the resource management systems which professional resource managers live and work in. Each element in a resource management system is complex and dynamic (Figure 1.2). Not only does each element present its own challenges for resource management. It also interacts with other processes and elements in the system to constitute unique sets of relationships and circumstances.

While it is a diagrammatic convenience to separate certain features, the notional separation of categories in a diagram should not be mistaken for a fixed relationship, or a simple 'categorical' separation in reality. Nor should any particular set of identified processes be given, *a priori*, greater causal power, higher explanatory standing or more epistemological privilege. In these terms, human geography's development, particularly in its recent radical permutations, as a synthesising, 'self-consciously decentred' discipline (Graham

1992: 153), provides a powerful foundation for developing an effective critique and reconstruction of resource management.

While most disciplinary and epistemological positions highlight specific core determining processes, geography can, and sometimes does, reasonably comfortably span the complexity in which I seek to contextualise resource management. In many cases, disciplinary positions render completely invisible many of the things that this book argues are integral elements of the resource management systems themselves. Disciplinary blinkers and subsequent partiality seems to favour the interests of those who are, generally speaking, beneficiaries of the existing systems and modes of thinking – those who are enriched and empowered by them.

Professional education which relies on developing technical skills of resource managers in isolation from an understanding of the social, cultural, political, economic and ecological contexts in which resource management decisions are made is, by definition, unable to equip students of resource management with a professional literacy which enables them to understand the human consequences of the advice they might give, the decisions they might make, or the responses of other people to their decisions. This makes their decisions and advice vulnerable to the potentially showstopping effects of human (and environmental) responses to these consequences. Such education not only reinforces resource management as part of the old order of 'top-down' approaches to planning and practice in industrialisation and development, but it also makes it inevitable that well-educated professionals will continue to be unable to see potential catastrophes (ecological, economic, cultural, social) before it is too late to avoid them. Thus, the dominant paradigms not only produce inadequate resource managers, but they also block the development of 'bottom-up' approaches to key issues, and effective accommodation of the best from both 'top-down' and 'bottom-up' approaches in specific circumstances.

Towards a new world order: two vantage points for rethinking resource management

The persistent tension between bottom-up and top-down approaches to questions of resource management is a central theme in the history and current configurations of resource geopolitics. In general terms, these two approaches to resource management systems provide very different perspectives on even the most basic questions of goals and purposes. But characterising this tension is no easy task once one begins to accept and try to work with the complexity that exists in the Realpolitik of resource management systems. The sort of questions asked, and the answers constructed, for example, depend considerably on where you think the 'top' and 'bottom' of the system are located. For instance, if the 'top' is the arena of national government, then 'top-down' policies, regulation and facilitation of resource-based industries might consist of a range of economic, environmental, legal and health and safety statutes and regulations. In contrast, if the 'top' is seen as the global institutions such as the World Bank and major global resource corporations, then the power of even many nation states to impose 'top-down' plans of their own is extremely constrained.

Despite recent rumblings about the emergence of an ostensibly New World Order in the wake of the Gulf War and the collapse of doctrinaire Communism in Eastern Europe, the tension and conflict between bottom-up and top-down approaches to resource-based development is still likely to be resolved in favour of beneficiaries of the existing order, or of the already privileged, empowered and enabled. Even where the pattern shifts, as in the emergence, for example, of the so-called 'tiger' economies of Southeast Asia, the new formulation reproduces many of the structural patterns of the old – similar patterns of uneven development, marginalisation of key groups (such as indigenous people, women, young people, aged people, ethnic, religious or other minorities) from economic and political power; entrenched patterns of state power and élite privilege, and so on. For indigenous peoples, the shift from colonial to post-colonial administration has rarely changed entrenched marginalisation. In any system, however, the power of the already rich and powerful is never left unchallenged; it is always under challenge from many sources: competitors seeking to wrest for themselves the trappings and benefits of wealth and power; 'ordinary people' in search of a better future; and an amazing array of lunatics, desperados and visionaries always trying to transform the basis for privilege at its source. Inevitably, this means that there is a wide range of views (and underlying rationalities) that can be characterised as either 'top-down' or 'bottom-up'.

In the case of the interlocking crises Ekins labels a global problematic, the idea of the need for a new world order (a fundamental transformation of the structural logic of international political relations) can be constructed (and politically justified) in entirely different ways from vantage points at the top and bottom of the system (see Figure 1.3).

In characterising an all-encompassing global crisis as a 'top-down' problem, for example, the constituent elements of crisis become enmeshed as problems of such scale and magnitude that they can only be addressed globally. The issue of poverty requires a powerful World Bank to look after the generation (but not the distribution) of wealth; the changing patterns of global climates related to the greenhouse effect for example, require international treaties; issues of widespread deforestation require an International Tropical Forestry Action Plan; problems in trade and international economic relations call for a World Trade Organisation as arbiter in trade-related disputes; and the protection of biodiversity needs an international treaty that commodifies and values genetic information and indigenous knowledge in new ways that make it worth conserving. And this is endorsed in terms of human-centred rhetoric. For example, at the end of the Gulf War former US President George Bush outlined his vision of the 'new world order' as an era of unprecedented peace and stability in a world dominated by democratic institutions and fair markets and where:

Figure 1.3   'Top-down' and 'bottom-up' perspectives on world order

diverse nations are drawn together in common cause to achieve the universal aspirations of mankind [sic] – peace and security, freedom and the rule of law.

(Bush, quoted in O'Tuathail 1993: 123)

In constructing a singular crisis of global scale, the New World Order (with capital letters) emerges as a logical and desirable outcome. In this vision of new world order, it is necessary to impose new order from the top down in order to address the global crisis. The idea of a new world order emerges as one in which the already successful are empowered to dictate the terms of settlement on those who are not so privileged: the USA becomes the world's military policeman, the World Trade Organisation reduces all international relations (for example environmental protection legislation; workers' health, safety and wage rates; child labour concerns; women's rights and so on) to issues of 'free' trade.

In contrast, if the global problematic is conceived as a series of interlocking, interacting and overlapping crises within particular localities and regions and nations, and as vulnerable to a myriad of partial and even sometimes contradictory 'solutions' in different places at different times and places, with different goals and priorities, then the new world order (decidedly without capital letters) that emerges to counteract these crises is very different from the vision promulgated by the first President Bush.

In this book, tension between local scale, bottom-up and non-local, top-down resource management is conceived and critiqued as a critical driving force in the dynamics of resource geopolitics at all geographical scales. Top-down solutions and proposals, generally oriented to the whims of external (non-local) commodity markets, aim at maximising benefits (usually economic) to vested interests or else some fortuitously defined national interest that excludes the interests of those people displaced, dispossessed or distressed by any particular mine or dam or forestry project. In contrast, almost without exception, bottom-up challenges to the technically or politically preferred solutions which come from central governments, global and national resource corporations and external consultants can be conveniently dismissed by them as parochial vested interests undermining the wider public interest, national development aspirations and community welfare. In doing so, the tendency is for 'top-down' approaches to obliterate, belittle and invalidate the 'bottom-up'. From the perspective of a resource manager at the 'top' – wherever that is thought to be – this gives licence to do almost anything in the pursuit of industrialisation and development, and in the process of exercising this licence, the myths of the dominant paradigm are exposed. This is not the implementation of 'objective', 'scientifically-determined' best practices, but the reinforcing of privilege that is constructed and renewed socially. It is by analysing this tension, principally in the context of relations between industrial resource management systems and indigenous peoples, that this book seeks to open dialogues about new approaches to the big issues of resource geopolitics at a variety of geographical scales.

People without geography: Indigenous peoples and resource management systems

> They make the laws to chain us well.
> The clergy dazzle us with heaven or they damn us into hell.
> We will not worship the God they serve,
> The god of greed who feeds the rich while poor folk starve.
>   Leon Rosselson[2]

Leon Rossleson's powerful song of the Diggers' struggle during the English Revolution is based on the words of a seventeenth-century visionary, Gerrard Winstanley. It refers to a revolutionary period of English history when:

> various groups of the common people (tried to) impose their own solutions to the problems of their time, in opposition to the wishes of their betters.
>   (Hill 1972: 11)

Plate 1.1   Stand-off at Oka, Montréal 1990. A protester at the Mohawk protest
camp faces a Canadian soldier

Source: CP Picture Archive (Shaney Komulainen).

These people lived in a period of unprecedented social and political turmoil –
a period Hill characterises as 'the world turned upside down'. Technological
and political change imposed almost incomprehensible pressures upon ordinary people's lives. Groups such as the Diggers responded to the emergent
new order of power and privilege with their existing values and understandings to assert an alternative to the chaos being created around them. They
based their actions on a vision of human society rooted in natural rights and
common property. Although often portrayed as destructive groups who
simplistically and hopelessly rejected change, the Diggers and other social
movements of the era such as the Levellers and Luddites sought to exercise
control over change, and to use it to bring about acceptable outcomes for the
people affected by it.

The Diggers' ill-fated challenge to the world order of the seventeenth century has many parallels with the concerns of this book. Despite the order to
cut the Diggers down, their vision lingers on. Like the Diggers, many groups
on the bottom rungs of the late twentieth-century world order have set in
train social movements for political and economic change based on visions of
social justice, environmental sustainability, economic equity and acceptance
of diversity and difference. And, like the Diggers, such visions face opposition

from powerfully entrenched beneficiaries of the existing order of things, who often seek to cut them down in the most barbaric and inhuman ways, even in mature modern democracies.

Ground Zero: Oka, Montréal, Canada

In July 1990 Canada and the world were confronted with the images of armed confrontation between Mohawk Indian warriors and the Sûreté du Québec at Oka in suburban Montréal, as the Indians sought to stop expansion of a local golf course and parking area into a disputed piece of land that included a Mohawk cemetery. During a raid to arrest the protesting Mohawks blockading the site, a provincial police officer was shot and killed. Over the next three months, the conflict escalated. Commuter traffic on one of the city's main road bridges was blockaded, and eventually the Canadian army confronted the Mohawks with tanks and guns.

For many Canadians, confrontation between troops and warriors at Oka (Plate 1.1) was an image which clashed incomprehensibly with their idea of modern Canada. For them, whatever injustices might have been done in the past, Indian grievances should be handled within the legal framework provided by the Canadian state. Armed conflict with warrior societies belonged on the early colonial frontier and not in modern Montréal. Yet the roots of this most disturbing conflict are to be found in the continuities that link modern Canada with those colonial frontiers.

The territory of the Iroquois confederacy, of which the Mohawk nation was a member, straddled the present borders of the USA and Canada. The sophisticated political traditions of the Iroquois influenced the drafters of the US Constitution (see Williams 1990). The Iroquois leaders had signed international agreements which were not treaties of settlement and conquest but treaties of international co-operation and recognition. The rights recognised in these treaties were confirmed in the Treaty of Paris of 1760 and King George III's Royal Proclamation of 1763. Regardless of the political sophistication of the Iroquois and the terms of these treaties, the disputed land within the Mohawk Reservation at Oka was 'granted' to a Catholic religious order by the French governor of New France in 1717.

Despite Mohawk opposition, parcels of the land in the area were sold to French settlers and a francophone community was established within the Mohawk Reservation. Throughout the nineteenth century, Mohawk protests (and arrests) continued, despite the simultaneous widening of internal divisions within the fragmented Mohawk nation and the Iroquois Confederacy. Protesters, including chiefs, were imprisoned, excommunicated and 'disappeared'.

In the 1950s, construction of the St Lawrence Seaway, the massive canal system that allows ocean-going vessels to pass beyond the rapids at Montréal, destroyed further Mohawk land and disrupted Mohawk community life. By this time the Mohawk nation's resistance was broken, and few protests were heard. There was little or no government effort to address any negative effects of the development on the community. The benefits of the seaway for Canada's industrialisation and development were self-evident, but the once powerful Mohawks had become all but invisible.

In the 1960s and 1970s, as part of a wider resurgence of Native American nationalism, there was a revitalisation of Iroquois and Mohawk cultural and political organisations. This included a re-emergence of the warrior societies with links to traditional religious practices, militant nationalism and an assertion of Mohawk and Iroquois sovereignty within Canada and the USA. Revitalisation of Mohawk nationalism was also linked to a revitalisation of culture across the Iroquois nations in the 1960s and 1970s, with a widespread revival of the longhouse, the traditional religious institutions of the confederacy, the assertion of land rights in Canada and the USA, and the emergence of Mohawk schools and political organisations.

The local government decision to 'develop' land at Oka as a golf course and car park catalysed Mohawk frustration and anger. For many whose parents had watched powerlessly as the seaway was pushed through their land, the land at Oka was an opportunity to take a stand against further alienation and disempowerment – to reassert sovereignty. In defending sovereignty by force of arms, the protesters at Oka faced a dilemma – if they tried to assert sovereignty peacefully, the land at Oka would be destroyed. If they used force to defend it, they would be criminalised by governments who claimed they were no longer sovereign. One of the young Canadian soldiers facing the warriors recognised this dilemma:

> These people are convinced that they're right … . They have a certain patriotism. Unfortunately, they are tossing aside the laws of our white governments. They're in a vicious circle. As long as we don't recognise them as a nation with their own protective force, we can't accept that they can bear military arms. But as long as they don't possess military arms, they will not be able to affirm their rights as a nation.
>
> (quoted in York and Pindera 1991: 314)

This is a dilemma that will be recognised by dispossessed and oppressed peoples throughout the world. It was certainly recognised by native communities across Canada, who took spontaneous action to support the protesting Mohawks in Montréal, including blockades of the major transcontinental rail link and damage to property and road blockades across the nation. Even Québec nationalists, who more recently have been particularly critical of indigenous arguments of sovereignty (see Drache and Perrin 1992; Trent *et al.* 1996), condemned the way in which the police and provincial government were dealing with the Mohawks.

> [At Oka] the state is once again criminalizing a valid social protest, it is trying to dismiss social demands, demands for sovereignty, as criminal activities.
>
> (Pierre Vallières, an early leader of the Québec nationalist movement, quoted in York and Pindera 1991: 415)

The planning dispute at Oka quickly became a potent symbol of the daily encroachment of outsiders onto Indian lands across Canada. Despite widespread disquiet about the implied violence of the Mohawk warriors' approach to defending the land, virtually every Indian experienced a 'shock of understanding', in which there were many lessons to be learnt.

It's what we see every day ... . We know we've never given up those mountains or forests, and yet they're being mined every day. We see those big trucks running by, taking logs out, and we know there's no benefit to our people. We know our treaties have been signed, but they are not fulfilled yet. Oka was an opportunity for people to remember the empty promises. When we saw people deciding to stand up and be counted, deciding to end this kind of abuse and non-recognition, there was a real outpouring of support.

(George Erasmus quoted in York and Pindera 1991: 274)

Other aboriginal leaders experienced the same shock of understanding when they saw the masked warriors on television. They had tried to follow the path of non-violence, they had tried to obey the rules of the game, but they had gotten nowhere. For the first time, they were beginning to suspect that the guns of the warriors were the only tactic that might bring justice to their people. 'Everything else has collapsed and failed', said Ethel Blondin, a Dene Indian and Liberal MP from the Northwest Territories. 'I could never denounce the warriors. They symbolize something I believe in – the struggle to defend our land and our rights'.

(York and Pindera 1991: 274)

In many ways, we all live in Oka – the voices of protest at Oka can be heard echoing in many places and many conflicts around the world. Wherever the assumption exists that the self-appointed disciples of developmentalism and industrialisation have an unassailable right to manage land, resources and people just as they see fit, the protests at Oka echo in the challenges from popular protest.

Indigenous land and primitive accumulation

Of all the resources for industrialisation, it is the land and its riches and potential that is often most central in establishing the pre-conditions for industrial development. The process which Karl Marx (1954 [1887]: 667–724) labelled primitive accumulation, sundered the relationship between people and their traditional estates and in the process created the means for producing and appropriating wealth in new ways. In Marx's view, it was not merely the land which was expropriated from the people, but also the people who were expropriated from the land and left with nothing but the sale of their labour power as a means of surviving. He concluded 'the history of this ... expropriation, is written in the annals of mankind in letters of blood and fire' (Marx 1954 [1887]: 669). His account of one episode of clearance in the Scottish Highlands (see Box below), holds echoes of events experienced by indigenous peoples around the world.

Access to resources for industrialisation and development has been an important motivation for both *in situ* intensification and geographical expansion of industrial economies. Both these processes have brought industrial societies into contact and conflict with tribal and indigenous peoples. The

Ground Zero: Scottish Highland clearances

As an example of the method obtaining in the nineteenth century, the 'clearance' made by the Duchess of Sutherland will suffice here. This person ... resolved ... to effect a radical cure, and to turn the whole country, whose population had already been ... reduced to 15 000 into a sheep-walk. From 1814 to 1820 these 15 000 inhabitants, about 3000 families, were systematically hunted and rooted out. All their villages were destroyed and burnt, all their fields turned into pasturage. British soldiers enforced this eviction, and came to blows with the inhabitants. One old woman was burnt to death in the flames of the hut, which she refused to leave. This fine lady appropriated 794 000 acres of land that had from time immemorial belonged to the clan. She assigned to the expelled inhabitants about 6000 acres on the seashore – 2 acres per family. The 6000 acres had until this time lain waste, and brought in no income to their owners. The Duchess, in the nobility of her heart, actually went so far as to let these at an average rent of 2s. 6d. per acre to the clansmen, who for centuries had shed their blood for her family. The whole of the stolen clanland she divided up into 29 great sheep farms, each inhabited by a single family, for the most part imported English farm-servants. In the year 1835 the 15 000 Gaels were already replaced by 131 000 sheep. The remnant of the aborigines flung on the sea shore, tried to live by catching fish. They became amphibious and lived, as an English author says, half on land and half on water, and withal only half on both.

(Marx 1954 [1887]: 682–3).

eighteenth- and nineteenth-century enclosures and clearances in Scotland and Ireland referred to by Marx laid the foundations of industrial capitalism and displaced people who became the optimistic settlers of new lands in their oppressors' colonies. Expanding industrial societies rapidly appropriated the lands, resources and even lives of tribal peoples at the frontiers. In places such as Siberia and the Russian Far East, we can see that the issue is emphatically not a product of only capitalist economies, but is common to industrial economies generally (see also Wolf 1982).

The process of primitive accumulation has generally been relegated in political economy to an historically interesting concept related only to the prehistory of capital accumulation. Yet in indigenous territories, the process of primitive accumulation described by Marx and developed by Luxemburg (1963), is constantly renewed. In the case of minerals, timber, wildlife and genetic material, the late twentieth century is a period of intense primitive accumulation in indigenous domains.

In previous generations, the destruction of indigenous and tribal cultures, the devastation of whole societies was dismissed as necessary for imperial success and justified by appeals to religious, racial and cultural superiority (See *inter alia* Wolf 1982; Chomsky 1993; Berger 1991; Stevenson 1992).[3] According to Joseph Conrad, the geographical expansion of empire – 'geography militant' (1955 [1926]) – was a primitive accumulation justified by 'an idea', within which lay a heart of darkness (see Box). Non-industrial societies were characterised as primitive, barbaric, inferior – doomed to extinction in the face of advanced humanity. Superiority became a blanket justification for barbaric behaviour by the civilised nations in a crude imperialist race for resources – land, minerals, labour, timber and other forest products, energy, food and other valuable commodities. Under the aegis of imperialism, the destruction of cultural diversity, of human life, was no more significant than the destruction of exotic environments and the biological diversity they contained.

From *Heart of Darkness*

'And this also', said Marlow suddenly, 'has been one of the dark places of the earth.' ...

   'Mind, none of us would feel exactly like this. What saves us is efficiency – the devotion to efficiency. But these chaps were not much account, really. They were no colonists ... They were conquerors, and for that you want only brute force – nothing to boast of, when you have it, since your strength is just an accident arising from the weakness of others. They grabbed what they could get for the sake of what was to be got. It was just robbery with violence, aggravated murder on a great scale, and men going at it blind – as is very proper for those who tackle a darkness. The conquest of the earth, which mostly means the taking it away from those who have a different complexion or slightly flatter noses than ourselves, is not a pretty thing when you look into it too much. What redeems it is the idea only. An idea at the back of it; not a sentimental pretence but an idea; and an unselfish belief in an idea – something you can set up, and bow down before, and offer a sacrifice to.'

(Conrad 1995 [1917]:18, 20)

In the late twentieth century, such barbaric excuses for the actions of the powerful are no longer politically acceptable, but racism, intolerance and ignorance continue to abound.[4] Despite the emergence of a new cultural politics of difference (West 1990; Bhaba 1994), paternalism, ignorance and misunderstanding continues to characterise intercultural relations between many indigenous and non-indigenous peoples. For example, environmentalists have found it easier to advocate protection of 'natural' environments and warm furry animals than to prioritise protection of the rights of indigenous peoples whose stewardship of habitats and use of many warm furry animals is harder to encapsulate as a bumper sticker. Environmentalists have often opposed indigenous use and occupation of (even access to) lands they classify as having high conservation values (Langton 1995). Animal rights have often been accorded priority over indigenous rights by Western campaigners (Gray 1991). In the case of increased global recognition of and commitment to ecological sustainability and the conservation of biodiversity (mainly for industrial purposes) since the 1992 Earth Summit in Rio de Janiero, political commitment to the environment has not been matched by a recognition of the need for strategies consistent with the notion of conservation of cultural diversity or the sustainability of social and cultural identity. Indeed, in many quarters, the quest for sustainability has been rapidly incorporated into the ideology of industrialisation and development (for example Howitt 1995; also Schmidheiny 1992). In the process, environmental protection becomes dependent on the financial resources made available by development (debt-for-wilderness swaps, for example), and protection of indigenous peoples becomes conditional on incorporation of their lands, communities and resources into the developmentalist project. In other words, primitive accumulation continues to dominate indigenous politics. Even where, as in Australia, there is some rhetorical commitment to reconciliation between indigenous and non-indigenous groups, ignorance, paternalism and racism limit the extent to which indigenous people are empowered to propose political, social and economic agendas rooted in their own traditions rather than subservient to the appetites of industrialisation and development.

Who are indigenous peoples?

For many professional resource managers, indigenous peoples are an unknown quantity. Within the dominant paradigm, they have been defined as outside the formal systems of resource management. In professional education systems for resource managers, where development of 'people skills' is generally undervalued, the more difficult areas of cross-cultural and intercultural communication, conflict resolution and indigenous rights are rarely dealt with at all. So it is hardly surprising that tensions between resource developers and indigenous peoples are so readily categorised as 'too hard' or 'somebody else's problem'. In many cases, the social myths and misunderstandings that characterise structural relations between indigenous peoples and the dominant society are reinforced and reinvented in local social relations between resource projects and local groups.

So who are 'indigenous peoples', and why should resource managers be expected to know anything about them or be prepared to deal with their concerns and experiences? It is in answering these questions that one can glimpse the complexity in which the work of resource managers is embedded out there in the 'real' world.

The figures available from the principal support and advocacy organisations for indigenous peoples suggest that there are currently around 5000 indigenous and tribal cultural groups in existence. The 200 million people in these groups comprise about 4 per cent of the global population, but account for 90–95 per cent of contemporary cultural diversity (Gray 1991; Maybury-Lewis 1992; Tauli-Corpuz 1993). Connell and Howitt (1991a: 3–4) emphasise that no single definition of indigenous peoples is possible. The process of identification as indigenous is historically contingent rather than categorical. In many places, government definitions of indigenous groups have emphasised what people are not – indigenous people are not literate, not healthy, not civilised, not 'us' (Dodson 1994a). The United Nations has talked about indigenous peoples, communities and nations as:

> those which, having a historical continuity with pre-invasion and post-colonial societies that developed on their territories, consider themselves distinct from other sectors of the societies now prevailing in those territories, or parts of them. They form at present non-dominant sectors of society and are determined to preserve, develop and transmit to future generations their ancestral territories, and ethnic identity, as the basis of their continued existence as peoples, in accordance with their own cultural patterns, social institutions and legal systems.
>
> (Cobo 1986: 1–4)

Howitt (1996: 11) point out, however, that even this definition, despite its strengths, fails to encompass some groups within the international indigenous peoples movement. The process of defining the nature and content of indigenous identity is itself a highly politicised act. What is encompassed within a particular definition, and its implications for practical processes such as resource management, will depend on who is doing the defining and why it is being done. Nation states seeking to exercise social control over indigenous minorities will define indigenous status in a different way to an autonomous tribal organisation seeking to limit control membership. It is significant, however, that self-determination and self-identification are so prominent in Cobo's definition. Issues of self-determination have been central in the international indigenous rights movement's dealings with the nation states. As Michael Dodson has observed, the nation states within the United Nations have strongly asserted:

> that justice for colonized peoples requires their freedom to assert the right to be their own rulers and be free from subjugation to alien masters … . We [indigenous peoples] meet all the same criteria in terms of being distinct peoples united by common territories, cultures, traditions, languages, institutions and beliefs. We share a sense of kinship and identity, a consciousness as distinct peoples and a political will to exist as distinct peoples … . [The position of the nation states in the UN,

however,] is that indigenous peoples do *not* qualify for the right to self-determination in international law … [because] it is feared that recognition of the right to self-determination would pose a threat to the principle of territorial integrity.

(Dodson 1994b: 69–70, emphasis in original)

This leaves indigenous peoples in a problematic situation in terms of human rights. In many nation states, assertion of indigenous identity – for example by using indigenous languages, practising traditional cultural, religious or even economic activities, acting to protect indigenous territories from unwanted intrusions by settlers, developers or state institutions, or taking legal action to establish rights to cultural or territorial autonomy – is treated as treasonable behaviour. State-sponsored suppression of indigenous identities, on the other hand, is protected from international intervention by categorising such matters as internal domestic matters.

Such criminalisation of indigenous practices has been long-entrenched in the United States and Canada (Institute for Natural Progress 1992; Tough 1993). Its contemporary seriousness as a repressive strategy has been reinforced in recent years in places such as Turkey, where the national government has repressed Kurdish nationalism with a ruthless military campaign, and Mexico, where the national government tried to use military force to satisfy the North American investment community that a coalition of Indians and peasants seeking social justice, economic equity, environmental protection and self-determination in the southern province of Chiapas would not threaten the prerogatives and privileges of capital in that country. Even in advanced democratic states such as Norway, Canada, Australia and the United States, state-sponsored attacks on indigenous identity and political activity have been commonplace. Such attacks are vigorously defended in terms of 'national cohesion', 'territorial integrity' and the 'right' of the national community or sovereign sub-national entities (states, provinces and territories) to pursue development via commercial exploitation of resources on 'national' territory – even if it is previously unwanted 'wasteland' under indigenous control.

Wasteland?

When it comes to broken election promises … [c]onsider this humdinger from the New Testament policy speech – 'The meek shall inherit the earth'. So far the best the meek have managed earth-wise is to have dirt kicked in their faces, or to be spattered with mud from the wheels of the rich man's carriage. Any earth the meek did inherit was No-Man's Land, the most blasted of heath. When it came to the meek, the privileged employed a scorched earth policy, shoving them into this corner of moonscape or that wretched 'Reservation'. If it was unproductive, exhausted, desolate, barren, god-forsaken or uninhabitable, the meek were welcome to it. Yet lately, with awesome irony, some of the wastelands of despised ... have turned out to conceal wealth beyond the dreams of avarice. No, that's wrong. There is NO wealth beyond avarice. Greed's appetite grows with every mouthful. So now the privileged want the wastelands back again and the only earth the meek are likely to inherit will be the clods tossed on their graves.

(Adams 1980: 24)

Historical circumstances have pushed many indigenous groups into a marginal existence on the peripheries of the mainstreams of social and political life of the dominant cultures in the nation states in which they live. For many, physical survival has been possible only through assimilation, often incomplete, resisted and resented, into the settler populations around them. For others, their existence on lands desired by others has been sufficient cause for genocidal attacks, sometimes sanctioned by the state. While the tenacious survival of indigenous cultures around the world is testimony to the strength of the human spirit, it is clear that for many groups, survival is not guaranteed into the future. Many groups face extremely serious crises; these crises come from a variety of sources, many of which relate directly to resource management (Table 1.2) (Burger 1990).

The historical contingencies of 'first contact' have left lasting legacies in all areas. The inconsistent approach of European powers in dealing with existing property rights in their colonial empires left a confusing and complex pattern of treaties, conquests, common law and ambiguity (Williams 1990). The colonial interplay of economic and religious zealotry fragmented, disoriented and disabled many indigenous groups. Similarly, the diverse forms of modern internal colonialism and post-colonialism with their continued religious,

Table 1.2 Sources of crisis for indigenous peoples

'first contact'

modern colonialism

frontier violence

forestry

dams

mines

militarisation

environmental collapse

cultural collapse

economic collapse

Source: Based on Burger 1990.

economic and state institutionalisation of indigenous peoples, the myriad forms of dependence, marginalisation and exclusion, and entrenched structures of racism and disadvantage reflected in the economy, education systems, legal systems, prison systems and so on, all contribute to current crises for indigenous survival.

In many parts of the world, the interface between indigenous and settler populations remains genuinely disputed territory, with high levels of direct violence, sometimes state-sanctioned and sometimes communal. In many parts of Latin America, this frontier violence continues to threaten indigenous peoples with genocide. During 1994 in Brazilian Amazonia, for example, illegal gold miners and agricultural settlers executed Yanomami Indian villagers. Refugees from such frontier violence inevitably find it extraordinarily difficult to maintain cultural identities intact. Similar stories of violence, displacement and loss of cultural identity and economic autonomy in refugee camps are repeated in Africa, Asia and the Middle East. Yet, as Sharp (1994) notes, this frontier is simultaneously a zone of engagement and accommodation – a place in which recognition and reconciliation can be pursued.

While both direct and structural violence against indigenous peoples from the outside world takes its toll, so does internal violence. Alienation and disorientation has seen many communities unable to effect change on the wider structures of oppression turn inwards in frustration, anger and despair. In many cultures, previously institutionalised or ritualised violence has had social controls removed and had more powerful weapons placed at its disposal. In Papua New Guinea, for example, inter-tribal tensions and ritualised warfare is escalating as a result of wider social contacts and the availability of guns and easier transportation. The realities behind violent images of Maori society portrayed in works such as *Once Were Warriors* (Duff 1990) and *The Bone People* (Hulme 1994) and the realities they reflect have been widely challenged within Maoritanga, but nevertheless impose a life of fear on many Maori. The terrible consequences of neglected communities, poor health, crises of personal and social identity, alcohol and other substance abuse, powerlessness and alienation have been well documented in many indigenous groups, including the massive report of the Australian Royal Commission into Aboriginal Deaths in Custody (Johnston 1991; Dodson 1991) and the more recent inquiry into Australian governments' genocidal policies of removing Aboriginal children from their families (Australia 1997; Tatz 1999). Similarly, the recent Canadian Royal Commission on Aboriginal Peoples documented the consequences of community violence for Canadian First Nations (Royal Commission on Aboriginal Peoples 1996). If even a fraction of the internalised violence occurring in these frustrated, disempowered 'communities' were directed against the wider population, it would probably be perceived as a state of warfare that would not be tolerated. The reaction of North American governments to the crises at Oka, Wounded Knee and Chiapas reminds us of the profound truth of this.

Specific threats to indigenous territories, and the ability of indigenous peoples to sustain their cultural relationships, their duties and customs involving their traditional lands, accompany these general processes threatening indigenous survival. Militarisation of indigenous territories is widespread. The remote areas and often sparse settlement patterns of many indigenous territories attract the attention of military planners seeking locations for weapons testing, military training, storage of weapons and wastes, and secure bases. In some cases, development of military facilities to protect interests encroaching on indigenous lands adds further insult to dispossession. On Cape York Peninsula in northern Queensland, for example, the strategic importance and vulnerability of one of Australia's major bauxite mines has justified development of a large military air base on Aboriginal land. In remote Canada, Innu people have faced terrible disruption to their personal and economic lives as a result of low-level military training flights in Québec and Labrador which disrupt hunting and community life. In Russia, Kazhakstan, China, Australia, France's Pacific territories and the USA, nuclear weapons testing has taken place on indigenous lands. In other places, indigenous peoples have suffered from the direct impacts of resource-based encroachments on indigenous autonomy. Exploitation of forests used by tribal peoples, development of large-scale mining projects, intrusions by small-scale, disorganised miners to exploit high-value minerals, the displacement of entire populations to make way for hydro-electricity and irrigation reservoirs and the sickening legacy of environmental destruction around the poorly planned and badly maintained resource projects all testify to the difficulties arising from resource-based development of indigenous territories. In the Russian Arctic, oil and gas exploration fuels new threats to indigenous peoples.

Ground Zero: a new frontier in Western Siberia

In the 1990s, with the transition from Communist Party rule in Russia and the break-up of the former Soviet Union, the West perceived new opportunities to acquire valuable mineral, energy and timber resources in Siberia and the Russian Far East. In the West, the vast and complex biophysical, socio-cultural and politico-economic geographies of 'Siberia' were long ago reduced to a simplistic icon of the worst features of the Soviet system – the deadly gulags whose characterisation by Solzhenitsyn (1974) had so gripped the political imagination of the West.

Below the vast plains of Western Siberia, between the Ob and Yenisey Rivers, the world's largest structural sedimentary basin has accumulated enormous hydrocarbon deposits. Exploited by the Soviet Union since the 1950s, this area of 'the Soviet Amazon' became a focus for international energy transnationals when new discoveries, and a new political and administrative environment heralded a new phase of development activity. At the same time, however, environmental damage from previous phases of poorly designed, poorly managed and poorly maintained oil and gas developments threatened catastrophe. The vast expanses of these fragile plains hold some of the largest reservoirs of fresh water and are integral to global ecological balances. They are also the enduring homelands of the Khanti, Mansi and Nenets. According to one scientific study:

> Every day since 1989 an average of four underground pipelines fractured, spilling seven million barrels of oil annually into the lakes and rivers that they traverse in their journey to refineries thousands of kilometres away. The concentration of oil in the water in the larger oil fields is up to 440 times above international safety standards. Over the past 20 years in the Khanti-Mansi region, 100 lakes and rivers, 17 million hectares of fish spawning grounds, and an area three-quarters the size of Great Britain in forests and grazing land have been irreversibly ruined.
>
> (quoted in Campbell 1991: 32)

As resource transnationals lined up to gain access to the mineral, energy and forest resources of this vast area, capitalist development of these resources offered some Siberians a hope of escaping from their harsh past. There was little mention among the key corporate or government decision makers of the need for resource and regional development strategies that gave priority to questions of justice, equity, sustainability or fostering diversity. For the Russian government, the foreign exchange value, contribution to industrial production and market leverage delivered by these resources was central to the very survival of the state. In 1991, for example, West Siberian production accounted for 64 per cent of oil and 71 per cent of gas output for all the former Soviet republics (Sagers and Kryokov 1993: 127):

> Ministry representative: 'The Ministry of Oil and Gas Construction is a construction ministry and we have to construct pipelines and not concern ourselves with the devil knows what. Preserving nature, saving reindeer – that's not our business ... . We must build. Time is passing. According to you, the workers just have to stand idle.'
>
> (quoted in and translated by Vitebsky 1990: 21)

> 'The policy practised towards the indigenous people (has been) one of genocide and ecocide. We needed to exploit their country so we expelled these people from their land. In fact, we cut them off from their roots. Now they are disappearing and nothing is being done to save them.'
>
> (V. Katasonov, economist, High Party School of Economics, Moscow, quoted in Campbell 1991: 32)

Away from the political confusion and market hype, however, criticism of the combination of ecological and social damage from the Soviet Union's industrialisation and development policies was gaining momentum under *perestroika* and *glasnost*. Indigenous peoples of the Russian Arctic spoke out about the heavy toll of decades of Communist development:

> Today our ancestors' land is crying for mercy. It has been invaded by industrial enterprises geared to maximum exploitation of natural resources. Gold, diamonds, and mica are extracted in our territory. During the last

ten years oil extraction has increased 2.1 times, gas extraction 4.8 times, and you cannot tell how many forests have been felled without mercy. Surveying the territory from a helicopter you will see how the dense taiga, that was still there yesterday, is gone today – barbourously taken away.

Under the pretext of fulfilling important state plans, ministries and local authorities are by means of truths and untruths financing these activities and are continuously building new industrial enterprises, railways, nuclear power stations, hydro-electric stations, and they are making plans for the extraction of oil and gas from new fields and for felling enormous stretches of forests. And not in one single case do we find scientifically or economically well-founded programmes which have been accepted by the indigenous peoples. And even if such programmes do exist nobody has thought of presenting them to the local population. In fact, the Northern peoples have become hostages in the hands of the industrial 'magnates' (ministries).

As a consequence the ecological situation is critical and conditions for hunting, fishing, and reindeer-herding have deteriorated drastically. ... In other words, the living conditions have been damaged for all the peoples in the area without exception.

> (Chuner Taksami, opening speech at the
> Congress of Small Indigenous Peoples of the Soviet North,
> Moscow, March 1989, in IWGIA 1990: 24–25)

Reindeer herding is the mainstay of economy and culture of many indigenous peoples in the Russian Arctic. For the Nentsy, the indigenous people of the Yamal Peninsula which was targeted for massive hydrocarbon-based development in the 1990s, 'there is no alternative occupation' (Vitebsky 1990: 21). The intimate socio-ecological relationship between reindeer and their herders was threatened with massive disruption from construction, production and accidents.

Mikhail Gorbachev's new policy approach held promise for a new approach to industrialisation and development. Indigenous peoples hoped that the period of 'revolutionary reconstruction and renewal of Soviet society' (Taksami in IWGIA 1990: 23) might provide an opportunity to address the tragic legacies of industrialisation and developmentalism Soviet-style. Indeed, Gorbachev himself spoke of:

> The situation of the small peoples of the North, Siberia and the Far East. The industrial development of the territory in which they live is being carried out without due consideration for their way of life or for the social and ecological consequences. These peoples need special protection and help from the state. It is essential to assign to the Councils (soviets) of Peoples' Deputies of these territories the exclusive right to their economic utilization, that is, to hunting grounds, pastures, inland waters, inshore water, forests, to the established reserve zones with the aim of restoring and preserving the homelands of [these] people.
>
> (Gorbachev 1989 quoted in Vitebsky 1990: 24)

The oil and gas industry burns off massive amounts of 'waste' and releases oil through fractured pipelines and spillages which take longer to break down in the Arctic temperatures. Trapped in Artic waters, the oil poses great threats to marine life and marine industries. With the wider scale of the threat of global warming endangering the delicate ecology of permafrost in areas such as the Yamal Peninsula, exploration and development are posing significant and immediate local-scale threats. Campbell (1991: 32 also 1990) quotes a regional ecology inspector as fearing that the West Siberian plain will 'cease to service its ecological function' by 2005. For the region's indigenous peoples, the consequences of genocidal and ecocidal industrialisation and development have been an everyday reality for many decades. Efforts to reverse both cultural and ecological threats continue, but the sheer scale of the problem is reproduced in many parts of the former Soviet Union. Despite the end of the Cold War, developmentalism and industrialisation continue to hunger for the precious resources of Siberia. It seems that the homelands of the Nenets, Evenki, Yakuts, Khants and other indigenous peoples of Russia's North are also a resource management Ground Zero.

The combined effect of these processes is an awful combination of economic and cultural collapse and environmental degradation that undermines the foundations of indigenous survival. It may seem that such crises must overwhelm indigenous groups, and that cultural survival faces hopeless odds. It is all too easy for observers whose lives are privileged by industrialisation and development to conclude, as did the beneficiaries of earlier periods of colonial dispossession, that indigenous groups are 'doomed'. Yet resistance, determination and even optimism continue among indigenous peoples. Within international forums such as the United Nations, the World Bank and the International Labour Organisation, indigenous peoples have established stronger grounds for securing their futures. Advances in specific jurisdictions, such as the recognition of native title at common law in Australia, the development of increased respect for treaty rights in Canada, the USA and New Zealand, the negotiation of settlements of comprehensive claims in Canada, and the development of national and international institutions involving indigenous peoples such as the four-nation Sami Parliament, the multinational Inuit Circumpolar Conference, the World Council of Indigenous Peoples and so on all provide the foundations for counter-tendencies to those who would predict (and have so long predicted) the demise of indigenous peoples. Within many indigenous groups, the closing years of the millennium have seen dramatic cultural revivals and a reassertion of indigenous sovereignty and identity (Maybury-Lewis 1992). There is a complex dialectical relationship between threat and resistance, knowledge and power, past, present and future in the struggle for indigenous recognition and survival. It is precisely these relationships that resource managers enter into when their activities affect indigenous interests. And it is precisely this complex of relationships that resource managers need to understand better in order to establish better resource management practices.

In the case of the struggles around traditional ecological knowledge, a shift away from a binarised distinction between indigenous and non-indigenous cultures and related binaries such as 'traditional'/non-traditional, pre-modern/modern, authentic/tainted and so on forces an engagement with the social, political and environmental relations of the people affected by their activities rather than the discursive imaginaries constructed around false notions of 'authentic', 'tribal' and 'rights'.

What is traditional ecological knowledge?

> Native knowledge about nature is firmly rooted in reality, in keen personal observation, interaction, and thought, sharpened by the daily rigours of uncertain survival.
>
> (Knudtson and Suzuki 1992: 16)

> The importance of Traditional Knowledge lies not in its understanding of environmental impacts but in an ability to extract money from government. Why else would aboriginal leaders concentrate so intensely on the astonishing claim that Traditional Knowledge is 'intellectual property' for which its holders must be paid?
>
> (Howard and Widdowson 1996: 36)

> It is part of our responsibility to be looking after our country. If you don't look after country, country won't look after you … . The country tells you when and where to burn. To carry out this task you must know your country. You wouldn't, you just would not attempt to burn someone else's country.
>
> (Bright 1994: 59)

The terms 'indigenous knowledge' and 'traditional ecological knowledge' have entered the resource management literature rapidly since the mid–1980s. As might be garnered from the quotes introducing this section, the increased recognition accorded to the traditional ecological knowledge of indigenous peoples, peasant farmers and even rural communities in advanced capitalist nations, is highly contested in some quarters. Even among the advocates of 'indigenous knowledge', there is considerable debate over its nature, content and utility (see Agrawal 1995a, b; Indigenous Knowledge and Development Monitor 1996a,b). In surveys of the literature (Mailhot 1994; Kuhn and Duerden 1997), it is acknowledged that traditional ecological knowledge is much more than different sorts of taxonomies of natural phenomena used by hunter-gatherer societies, which had been documented as 'ethnoscience' since the 1950s. Mailhot identifies studies of this sort in the fields of medicine, anatomy, colours, kinship, fauna, flora and even skin colouring and types of ice (Mailhot 1994: 4). Similarly, contemporary understanding of traditional ecological knowledge, and its application to a variety of tasks, also involves more than the technological and environmental determinism of early cultural ecology approaches to hunter-gatherer societies (Steward 1936, cited in Mailhot 1994: 9). Kuhn and Duerden emphasise the increased 'integration of TEK and Western knowledge in formal resource management decision-making structures' as an important element of the increased attention given to indigenous knowledge (1996: 76). Specifically, in their review of recent literature, they identify wildlife management, fisheries management, comanagement agreements for conservation areas, agricultural projects, mining projects, climate change studies, health, human settlement studies, and environmental and cumulative impact assessment as areas in which traditional ecological knowledge has gained prominence in resource management. In work for Canada's Royal Commission on Aboriginal Peoples, Brascoupé (1997) acknowledges the relevance of a quite extraordinarily wide range of disciplines to the study of indigenous knowledge (namely, ecology, soil science, veterinary medicine, forestry, human health, aquatic resource management, botany, zoology, agronomy, agricultural economics, rural sociology, mathematics, management science, agricultural education and extension, fisheries, range management, information science, wildlife management and water resource management).

After noting that the concept traditional ecological knowledge is 'relatively new and still evolving' (1994: 11), Mailhot defines the term as:

> The sum of the data and ideas acquired by a human group on its environment as a result of the group's use and occupation of a region over very many generations.
>
> (ibid.: 11)

She notes that this definition encompasses both practical or empirical aspects of traditional ecological knowledge and also its ideological aspects. In other words, traditional knowledge systems are not just information sets. They are also coherent, culturally contextualised ways of seeing, understanding and relating to the world (human, environmental and cosmological).

Traditional ecological knowledge is both information (specific knowledge and representations of environmental relations in particular places) and a way of knowing (an environmental ethic). Suzuki sees traditional ecological knowledge presenting a significant challenge to traditional scientific methods of addressing environmental information:

> [W]hile science yields powerful insights into isolated fragments of the world, the sum total of these insights is a disconnected, inadequate description of the whole ... As a practising scientist about fifteen years ago, I began to realize that if Western science really could deliver the promised benefits to humankind, then the quality of human life should have vastly improved during the 1960s and 1970s, as science grew explosively ... . Too often, most of us assume that 'they' – the scientists and

engineers – will do something to pull us through. But we are waking to the dangers of clinging to a faith that science and technology can forever resolve the problems they created in the first place.

Are there other perspectives from which to make our judgements and assessments, other ways of perceiving our place in the cosmos? I began to realize that other, profoundly different notions of our relationship with Nature do indeed exist when I became involved in the early 1980s in the battle to save the forests in the southern part of the Queen Charlotte Islands. This was the first in a series of experiences I had with different aboriginal peoples that opened me up to new possibilities and different, richer perspectives for understanding the world.

(Knudston and Suzuki 1992: xxii–xxv)

Suzuki's 'personal foreword' is an interesting reflection by a leading scientist on the power of the ideological dimension of traditional ecological knowledge. Yet here too we find another binary being constructed and resolved. In this case, it is the 'Native mind' and the 'Scientific mind' that are first held in tension and then resolved (see Knudtson and Suzuki 1992: 8–19).

The issue of the relationship between traditional ecological knowledge and Western science is one which has generated much debate. In the context of environmental research in the Canadian Arctic, Hobson (1992) argues emphatically that 'traditional knowledge *is* science' (his emphasis). Agrawal suggests that this perceived tension between 'Western' and 'indigenous' knowledges – often presented in quite unproblematic terms – is now deeply implicated in debates about development. Agrawal is ambivalent about the recognition development theorists are giving to indigenous knowledge:

Current formulations about indigenous knowledge ... recognize that derogatory characterizations of the knowledge of the poor and the marginalized populations may have been hasty and naïve. In reaction against Modernization Theorists and Marxists, advocates of indigenous knowledge systems underscore the promise it holds for agricultural production systems and sustainable development ...

The focus on indigenous knowledge and production systems heralds a long overdue move. It represents a shift from the preoccupation with the centralized, technically oriented solutions of the past decades that failed to alter life prospects for a majority of peasants and small farmers in the world.

(Agrawal 1995a: 413–414)

Reviewing the work of such advocates, whom he labels 'neo-indigenistas', Agrawal questions both the so-called divide between indigenous and Western scientific knowledges, and the approach to documentation and application of indigenous knowledge adopted by neo-indigenista development advocates. Elsewhere, he suggests that these debates need to be considered in terms of power:

The critical difference between indigenous and scientific knowledge is not at an epistemological level: rather it lies in their relationship to power … the question is one of power. Who has access to resources and can deploy them in order to disadvantage others? Clearly, it is not the holders of indigenous knowledge who exercise the power to marginalize. Indeed, no matter how you slice the cake, the criterion of power will triumph when local, traditional, or practical knowledge is contrasted with global, modern, or theoretical knowledge. To this extent, and only to this extent, the attention to 'indigenous', the adoption of the idiom of the 'indigenous', and the attempts to direct resources toward the 'indigenous' can and must be welcomed.

(Agrawal 1996a)

The risk is, of course, that the disciples of industrialisation and development simply appropriate indigenous peoples' (and other local communities') traditional ecological knowledge as another means of pursuing developmentalist agendas, regardless of their biophysical, political-economic or socio-cultural consequences: that indigenous knowledge is reduced to documented information resources which become accessible for others, particularly commercial interests 'to mine, manipulate, or plunder' (Knudston and Suzuki 1992: 19; see also Kuhn and Duerden 1997: 79). The documentation and application of traditional ecological knowledge is, therefore, a critical area for consideration. Mailhot (1994: 19) identifies three principal areas of practical application of traditional ecological knowledge – development projects, renewable resource management and impact studies. The extent to which resource management systems are implicated in each of these fields is considerable. In terms of indigenous empowerment, struggles for recognition of traditional ecological knowledge have wide implications. Yet, in terms of indigenous peoples' struggles for justice, they are not ultimately reducible in any useful way to the sorts of struggles that Fraser characterises as 'recognition struggles' (1995, 1997a; see also Young 1997: 158–9).

Struggles to recognise, protect and use traditional ecological knowledge

Since the acknowledgment of the relevance of traditional ecological knowledge to resource management decisions in Canada's Mackenzie Valley Pipeline Inquiry in the mid–1970s (Berger 1977, 1988), Canada has made greater progress towards entrenching traditional ecological knowledge in resource management systems than any other country (Kuhn and Duerden 1997). The changes have afforded both affirmation of First Nation identities in Canada and transformation of many dimensions of the underlying political, cultural and economic framework. In this case, struggles to secure environmental justice for First Nations have been co-equal with the struggle to secure remedies for economic and cultural injustices:

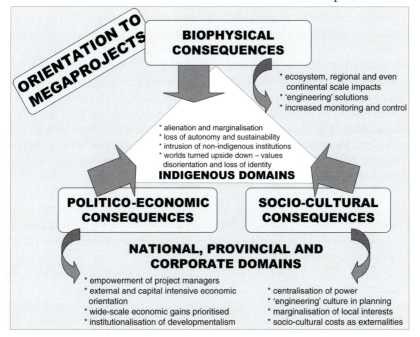

Figure 1.4 Implications of the dominant orientation towards megaprojects as the focus for development planning and resource management

As aboriginal issues have climbed the political agenda in Canada, land claim agreements have been reached, constitutional amendments in favour of the self-government of First Nations have been advocated, and court decisions are once again beginning to explore the nature and limits of aboriginal rights.

(Jacobs and Mulvihill 1995: 8)

In an earlier paper, Jacobs (1988: 55) suggested that research into traditional ecological knowledge and land use in the Canadian Arctic was part of a critical path to sustainable and equitable futures for the north. He suggested that such futures would require both a shift of paradigms in planning and decision making, and a shift in praxis. In part, this requires a shift away from unproblematic acceptance of industrialisation and development as the measure of success. Resource management and environmental decision making in Canada's north have been largely dominated by a commitment to resource-based megaprojects (Boothroyd *et al.* 1995; Boothroyd 1995; Maxwell *et al.* 1997). This orientation of national and provincial policy has influenced both the paradigm and praxis of planning and decision making in both the dominant domain of national, provincial and corporate planning, and also in indigenous domains in the north. The dominance of this megaproject orientation has interrelated consequences in biophysical, politico-economic and socio-cultural domains (Figure 1.4).

It is important to recognise, however, that the dominant megaproject orientation of society is both contested and ambiguous. It is not only indigenous interests that question and challenge the reliance on resource megaprojects as a key element of national social and economic policy. Environmental groups, small businesses, local governments and others are often opposed to the specific implications of megaproject proposals. In the wake of the 1992 United Nations Conference on Environment and Development in Rio de Janiero, systemic shifts towards 'sustainable' development have also established a basis for a wider scale challenge to the sort of developmentalist and industrialisation values reflected in orientation towards resource megaprojects. However, as Maxwell *et al.* noted, although megaprojects are:

> frequently sited in regions where they are least likely to encounter politically opposition – often communities composed of the most disadvantaged groups … [few of these groups] have the resources and political skill to prevent construction.
>
> (Maxwell *et al.* 1997: 34)

In some cases, however, resources and skills with which to challenge national, provincial and corporate developmentalism are available. Among indigenous groups, the long struggle for survival and deep engagement in struggles for recognition, redistribution and environmental protection has fostered development of formidable political and organisational skills, including significant skills in shifting scales in project-specific disputes to bring international pressure to bear on local issues (Jhappan 1990, 1992) and to shape local and national alliances (ibid. 1990). While affected communities often lack financial resources to mount major campaigns against megaprojects, some celebrated exceptions indicate the ambiguity of the developmentalist approach. The insight of Karl Marx, that social formations contain the seeds of their own transformation, seems to hold true in such cases.[5]

Canada's Berger Inquiry

In the case of communities affected by proposals to construct a gas pipeline along the Mackenzie River Valley to link oil and gas resources in Canada's Western Arctic to southern markets, the size of the project demanded a substantial inquiry into its technical, economic, environmental and social implications. While the energy companies involved were able to finance formidable legal and technical expertise to support and defend their proposals, local community groups faced substantial difficulties in matching their resources. The pipeline proponent's application in this case 'cost $60 million to prepare and weighed three tons' (Funk 1985: 122). The social impact analysis presented in support of the proposal reached conclusions which Funk suggested 'were based on the assumption that the pipeline would only speed a well-established process of evolution from a traditional to a modern way of life among native peoples' (ibid.:

125). The March 1974 appointment of Justice Thomas Berger to undertake an *ad hoc* inquiry into this proposal set in train a review process which changed the taken-for-granted privileging of wealth and power that characterised development planning and resource management in Canada. The proposal was of unprecedented scope and complexity, and although the terms of reference for Berger's inquiry required him to provide advice on terms and conditions for development, a 'full and fair inquiry' into all aspects of the economic, social and environmental consequences of proceeding was also required. As Berger himself noted:

> The Inquiry ... was unique in Canadian experience because, for the first time, we were to try to determine the impact of a large-scale frontier project before and not after the fact.
>
> (Berger 1977, vol II: 224)

Berger's Inquiry proceeded initially with preliminary hearings to discuss how the Inquiry should be conducted:

> The Inquiry did not start out with a prescribed set of procedures or a pre-conceived notion of what would transpire. Its form and content were established on the basis of testimony heard during the preliminary hearings ... Only by a thorough and balanced assessment could the sensitive areas be detected and examined ... .
>
> If there were any preconceptions about how the Inquiry should proceed, they lay in the direction of ensuring that it be thorough, fair, flexible, and accessible.
>
> (Gamble 1978: 948)

To achieve this, Berger concluded that representation of all interests affected by the proposal was fundamental to a 'fair and complete' inquiry. On Berger's recommendation:

> funding was provided by the Government of Canada to the 'native organizations, the environmental groups, northern municipalities, and northern business to allow them to participate on an equal footing ... with the pipeline companies – to enable them to support, challenge, or seek to modify the project.
>
> (Berger 1977, vol II: 225)

Funding guidelines required groups seeking funding to establish a clear interest that ought to be represented, a need for separate representation, a demonstrated commitment to the interest being represented, a lack of alternative funds to facilitate participation in the Inquiry, and a clear proposal for using and accounting for the funds. This innovation facilitated unprecedented levels of participation in the Berger Inquiry. It saw northerners take the Inquiry seriously, and present compelling testimony in extensive community hearings. Testimony was accepted in northern languages, and extensive use made of local community media to report on hearings. Expert witnesses faced cross-examination not only by other technical experts, but also by local people whose livelihoods and identities were at stake. Berger's approach took his Inquiry to the people most concerned and made it accessible to them – and was prepared to take the evidence seriously. Having redressed the characteristic imbalance between northern communities and southern entrepreneurs and governments, the Inquiry established a foundation for participation in which northerners' traditional ecological knowledge was recognised and became influential. Berger's recommendation for a ten-year moratorium on pipeline development to allow for settlement of native claims, his inquiry process, and his support for struggling representative organisations of northern First Nations all had lasting implications for the Canadian Arctic:

> Such participation not only helped to provide otherwise unavailable data, it also served as the basis for ongoing institutional changes in the relationship between northern natives and southern populations, and between supporters of exploitative development and balanced development in all of Canada. This change occurred at a consciousness level as well as through organizational forms that developed out of that consciousness.
>
> (Funk 1985: 132–3)

The recognition accorded to the NWT Indian Brotherhood, the Metis Association of the NWT, the Committee for Original Peoples Entitlement, Inuit Tapirisat of Canada and the Council for Yukon Indians by Berger's funding programme sits awkwardly in Fraser's (1997) characterisation of the pre-eminent political dilemma of the era. Although such recognition is affirmative in the sense that she uses that term, its implication is also transformative. Although the assertion of a distinct identity is central to the process of seeking and obtaining this recognition, the intent was simultaneously about cultural assertiveness, accessing economic resources and securing environmental, economic and cultural security. The inability of the dominant developmentalist paradigm to deliver employment and other benefits capable of offsetting the 'unavoidable social consequences' of the pipeline project (Funk 1985: 126) were recognised by Berger, as was the importance of the subsistence economy. Ten years later, in his introduction to a revised edition of *Northern Frontier Northern Homeland*, Berger returned to precisely these themes, linking together cultural, economic and environmental justice as central elements of socially just and sustainable northern futures in the Western Arctic (1988: 4–10).

The James Bay experience in northern Québec

In another example, discussed at greater depth later in this book, the struggle for recognition of the Inuit and James Bay Cree in the wake of Hydro-Québec's proposals to regulate the wild rivers of northern Québec produced resources that supported a dramatic economic, political, cultural and environmental transformation in Canada since the 1970s. Again in this case it is difficult (and in many ways misleading) to try and disentangle the identity politics, the environmental politics and the redistributive politics. Under the terms of the Canadian federal government's transfer of what is now northern Québec to the province of Québec in 1912, the province was required to make treaties with the region. As part of Québec's Quiet Revolution, the provincial government launched a number of state-owned enterprises, including Hydro-Québec.

During the 1960s, Hydro-Québec engineers formulated a proposal to regulate every one of northern Québec's rivers draining to James Bay (Salisbury 1977; McCutcheon 1991). By 1971, when Québec's Premier Bourassa announced the James Bay project as the 'project of the century' as part of a 'fascinating challenge … the conquest of northern Québec' (McCutcheon 1991: 33–4), native people in northern Québec were wondering when they might be consulted about the project of the century (Salisbury 1977; Scott 1995). The Cree began preparing legal action against the project in 1972 in response to proposals to begin work on the La Grande River. In 1973 they succeeded in securing an injunction requiring Hydro-Québec to cease work because the project would 'have devastating and far-reaching effects on the Cree Indians and the Inuit' (Justice Malouf in McCutcheon 1991: 55). Within a week Malouf's ruling was overturned by the Québec Appeals Court, but the seriousness of indigenous claims had been established and the provincial government agreed to negotiate. Canada's first modern treaty, the James Bay and Northern Québec Agreement 1975 (JBNQA) (Québec Provincial Government 1997) was negotiated.

Like many treaties, the JBNQA is interpreted differently by the different signatories. The Cree argue that it was negotiated 'under circumstances that were clearly inequitable, highly pressured and, in a number of key respects, unconscionable' (Grand Council of the Crees 1995: 252), while they see the province as increasingly using it 'to diminish or deny Cree fundamental rights' (ibid.:250). Once again, it is possible to read both contest and ambiguity in these circumstances, as well as simultaneous dimensions of cultural, economic and environmental issues in the Cree struggle for justice. As the Grand Council of the Crees' 1995 document on sovereignty exemplifies, the terms of the treaty remain strongly contested, but it is through the institutions developed through and as a result of the negotiation of the treaty that Cree political and economic autonomy is pursued.

Australian examples

In Australia, too, there is increasing recognition of the traditional indigenous knowledge of Aboriginal and Torres Strait Islander peoples, and a now familiar ambiguity about and contestation over its recognition and incorporation into resource and environmental management systems. Deborah Bird Rose identifies Aboriginal traditional ecological knowledge as the basis for developing an 'indigenous western land ethic' (1988: 386), and suggests that 'Dreaming ecology' provides a basis for a shift in thinking that would demonstrate how 'human and ecological rights are most properly embedded each within the other' (1996a: 49, 86).

The most celebrated Australian examples of the recognition and application of traditional ecological knowledge are the jointly managed national parks in the Northern Territory, where compulsory lease-back was a condition of the granting of land claims under the Aboriginal Land Rights (NT) Act. Uluru-Kata Tjuta, Kakadu and Nitmiluk National Parks are all managed by boards of management in which Aboriginal traditional owners are direct participants. Management plans for these parks increasingly incorporate traditional ecological knowledge as part of their basic orientation.

Toyne (1994: 49), for example, talks of 'the empowering aspects of traditional knowledge being valued as a major contribution to conservation management'. In an important policy options paper for the Commonwealth, Robertson *et al.* (1992: 89) unequivocally assert that 'effective joint management [of national parks] is based on respect for indigenous law'.

In the case of the Uluru-Kata Tjuta National Park joint management arrangements, one of the traditional owners and members of the Board of Management explained that the Tjukurpa is seen as a foundation for the national park and management of environmental, economic and social processes in the park (Tjamiwa 1988; see Figure 1.5 on page 46). While such joint management programmes are strongly endorsed as a model for 'negotiated solutions to some of Australia's deepest conflicts' (Robertson *et al.* 1992: 89), and provide many benefits to the Aboriginal people involved (see for example Young 1995: ch. 6), they were not universally welcomed. Toyne (1994: 48–58) provides a brief account of the Northern Territory government's opposition to the hand-back and joint management arrangements between the Commonwealth and Aboriginal traditional owners.

Similar negotiated settlements of land claims in high conservation areas of the Northern Territory have been reached under the Aboriginal Land Rights (NT) Act for Kakadu and Nitmiluk National Parks and the Coburg Peninsula Marine Park. The Native Title Act 1993 provided a basis for Aboriginal and Torres Strait Islander people to claim ownership of national parks where native title survives, although current amendments proposed by Australia's conservative government would greatly restrict these claims. Litigation of common law rights might further extend native title domains, and is likely to produce more negotiated accommodation of indigenous interests into management of conservation and other areas.

In local areas throughout Australia, indigenous communities are asserting traditional ecological knowledge as a foundation for local environmental management programmes. Outside the framework of major national parks, world heritage sites and conservation reserves, using funding from Commonwealth Landcare programmes, various state and territory initiatives, their own resources and voluntary agreements with other land-users, these communities have established diverse programmes to protect, rehabilitate and manage landscapes, resources and cultural knowledge.

In Central Australia, caring for country programmes of the Tangentyere and Pitjantjatjara Councils and the Central Land Council are integrated with environmental health programmes. At Kowanyama on Queensland's Cape York Peninsula, coastal management programmes base fishery, marine mammal and coastal area management on traditional values and knowledge supported by scientific research. Aboriginal Community Councils throughout Cape York Peninsula have participated in a successful Community Ranger Training programme which assists in training young indigenous people as professional rangers applying both traditional and scientific insights to managing lands under council control. This programme has been used as a model and was applied widely in other areas of Australia throughout the 1990s.

At Yirrkala, local Landcare groups such as Dhimirru are documenting and supporting application of traditional ecological knowledge at the same time as seeking to improve environmental practices of the nearby Nabalco mining operation. In Western Australia, Aboriginal groups have struggled for recognition of their rights and interests, against considerable state government hostility. Gulingi Nangga Aboriginal Corporation in the West Kimberley has argued strongly for application of joint management principles established in the Northern Territory and traditional ecological knowledge to management of marine and coastal areas of the West Kimberleys. Such principles have been applied in the Purnululu and Karijini areas of the East Kimberley and Pilbara regions.

In the Torres Strait, not only have Murray Islanders successfully asserted their continuing native title rights, but fisheries management, environmental planning and regional development planning are all increasingly controlled by Torres Strait Islander organisations, and increasingly rooted in traditional ecological knowledge and cultural values. In NSW and Victoria, where historical dispossession, genocide, industrialisation and settlement have restricted Aboriginal control of their traditional territories and eroded the knowledge base of many indigenous communities, tourism, land management, educational and resource management programmes are all drawing on archaeological, historical and oral tradition materials in developing new insights and approaches.

Resource management, institutional development and justice

Recognition and application of traditional ecological knowledge is an increasingly important element in many resource management systems. The challenge that traditional ecological knowledge represents to dominant resource management paradigms and praxis is, as we have seen, contested and ambiguous. Despite calls from some scientists for the scientific credibility of traditional ecological knowledge to be recognised, there remain many examples where scientific research and management continue without any concessions to or acknowledgment of traditional ecological knowledge. In many cases it is possible to find indigenous peoples themselves appropriating scientific tools such as Geographical Information Systems (Jacobs and Mulvihill 1995; Jawoyn Association 1997; Denniston 1994; Alexander and von Dijk 1996; Duerden and Kuhn 1996; Robinson et al 1994), environmental planning and land capability studies (Young 1995). Environmental philosophers also acknowledge the need for paradigmatic and practical shifts (for example Rose 1996b, 1999), and governments are incorporating indigenous values into planning legislation (for example New Zealand's Resource Management Act 1991; see Chapter 13).

Arguments over intellectual property rights, research ethics and the relationship between ecological, cultural and economic issues in indigenous

Figure 1.5   Incorporating indigenous law into conservation management

Source: Adapted from Tjamiwa 1988.

politics have also emerged as critical issues in the 1990s. The rapid growth of biological prospecting in indigenous domains and the emergence of a powerful new commercial interest in indigenous knowledge and conclusion of agreements such as Merck–INBio Agreement in Costa Rica, have raised serious questions about an era of biopiracy and primitive accumulation based on traditional ecological knowledge. This has led to the development of professional ethics protocols that empower indigenous interests to exercise greater influence over academic and commercial research (see for example McNabb 1993). It has also encouraged calls to implement formal processes that ensure indigenous communities are entitled to receive benefits of research and to hold researchers accountable for delivering such benefits (Fundacion Sabidurua Indigena and Kothari 1997).

In the shifts and struggles over traditional ecological knowledge, it is possible to recognise precisely the challenges identified below (Table 1.3) as ideological, socio-cultural and politico-economic challenges. It is also possible to see in these struggles all of the dimensions of the struggles for justice discussed by Fraser (1995, 1997a, b), I. M. Young (1990) and Harvey (1992). We can see how the indigenous imperative to bring together notions of economic, cultural and environmental justice, to integrate the biophysical, socio-cultural and political-economic domains, places intellectual, political and practical demands on professional resource managers. The values clarity that derives from engaging with indigenous struggles makes fragmented approaches to justice – or a strategy in which one form of struggle or one form of justice is privileged and given some sort of hierarchical priority or sequential preference over another – no longer easily justified. Reducing indigenous rights to anachronistic relics of pre-modern times, or reconfiguring them as a new sort of natural resource available for neo-colonial exploitation, is similarly unacceptable. Yet, the resource management systems that are embedded in these complex and ever-changing relationships with indigenous peoples, indigenous knowledge and indigenous rights are not easily transformed into more just, equitable and sustainable systems. If nothing else (and of course there is plenty else!) the institutions of resource management, environmental planning and regulation are almost all legacies of previous eras of injustice and denial. To take just one example: following Australia's acknowledgment of the persistence of native title, the key institutions for land management, resource management and indigenous administration that had developed on the presumption of *terra nullius* were immediately faced with the need to reinvent themselves as post-*terra nullius* institutions. Few have made this transition to date. Many have fought rearguard actions to reimpose *terra nullius* through various back-door means with the tacit, and sometimes explicit, support of national, state and territory governments which are themselves historically predicated upon the assumption of *terra nullius*.

Jacobs and Mulvihill suggest that institutional change is central to achieving reorientation of resource management systems towards justice and sustainability:

> The institutional infrastructures inherited from the past can and most often do present major barriers to progress. Breakthroughs … depend partly on institutional change. But there is little that is obvious about the design of new institutions.
>
> (Jacobs and Mulvihill 1995: 13)

In the Canadian context, they identify comprehensive land claim settlements, impact assessment arrangements and other processes that are creating opportunities for new institutional outcomes. They advocate what they term 'adaptive' institutions, in which:

- Flexibility, breadth and discretion are valued
- Integration rather than fragmentation is prioritised
- There is only minimal *a priori* specification of operational parameters
- Institutional design is largely in the hands of institutional users
- Arrangements are non-hierarchical, region-centred and stakeholder controlled
- Accountability to more than one authority or constituency is mandated
- Priority is given to anticipatory change.

> (ibid.: 14)

They suggest that groups such as the Inuit Circumpolar Conference and Yukon Wildlife Management Advisory Council provide pointers to the sort of adaptive institutions they have in mind. In Australia, many small-scale Aboriginal and Torres Strait Islander organisations are oriented towards caring for country, and the Indigenous Land Corporation, which is Australia's only major institution that is genuinely post-*terra nullius*, provides similar pointers.

In such institutions, it is possible to see a reorientation of both thinking and practice – paradigm and praxis – towards the core values prioritised in this book. In the simultaneous pursuit of environmental, economic and cultural justice, indigenous peoples' efforts to secure recognition, understanding, respect and application of their traditional ecological knowledge unsettle the binarised conceptions of justice that dominate contemporary discussions in theoretical discourses. While the dominant paradigms and practices of resource management have the power to turn indigenous peoples' worlds upside down – a power that has been demonstrated many times over in recent decades – that power is, as we've seen, both contested and ambiguous. In challenging it, indigenous people have demonstrated the need for multicultural literacy as well as technical competence in resource managers. And they have made it imperative for resource managers to 'see' more clearly the social, political, cultural and ecological contexts in which they operate.

Industrial resource managers and indigenous peoples

Competing views of 'resources', and conflict over how best to husband, manage, conserve and exploit resources on indigenous territories has been a fundamental source of tension between indigenous peoples and other populations. Such tension is not restricted to indigenous/non-indigenous relations, but often underpins relations between local communities and the wider societies of which they are part. In contemporary industrial societies, however, resource conflicts between indigenous peoples and resource developers and nation states are an important arena of social conflict. Energy resources and mineral deposits contained in indigenous lands, for example, have been targeted as one of a restricted number of unexploited sources of high grade available to industry in the late twentieth century. Pollin (1981) suggested the quality of resources still accessible in settled areas of major industrialised nation states by 1980 was generally poor, and that the reservations, treaty lands and other indigenous landholdings in North America, Australia and elsewhere was a key exploration target for North American resource corporations. Gedicks has characterised the struggle for control of the vast energy resources and power generation sites on Indian lands in the mid-West USA as a 'New Resources War' (Gedicks 1982, 1993; Johansen and Maestas 1979). Cramér suggests continuity between centuries of colonialism and contemporary 'cleptocracy – extractive exploitation' (1994: 55). In locations as diverse as Bougainville's tropical island copper-gold mine and Norway's Arctic Alta Dam site (see Chapter 9), indigenous resistance to particular projects, the

Table 1.3 Indigenous challenges to the dominant culture

Ideological challenges	The popular appeal and pervasive influence of indigenous epistemologies emphasising holism, humanity–environment links and spiritual rather than economic matters;
Socio-cultural challenges	The legal and social dimensions of indigenous resistance to efforts to disperse, assimilate or destroy indigenous cultures and related claims to self-determination; and
Politico-economic challenges	The political and economic consequences of indigenous claims of ownership and control of resources and the territories which contain them.

development trajectories planned for these regions, and the general consequences of industrial resource management have become genuine showstoppers for industrial resource management systems. The strategic importance of indigenous resources and continuing primitive accumulation to industrial production, and the resistance of many indigenous groups to continuing dispossession, displacement and destruction, together with the popular appeal of some indigenous values, singles out relations between indigenous peoples and resources as an area of profound importance. Relations between resource managers and indigenous peoples can be seen to face three important challenges – ideological, cultural and economic (Table 1.3).

These challenges may seem far removed from the imperatives of operational management or project planning. Within large resource corporations, such challenges are rarely anybody's specifically allocated responsibility (Howitt 1997b). Yet, as the experience of Bougainville Copper demonstrates, they can become showstoppers (see Box). In the Realpolitik of adversarial legal, political, economic and social relations within enterprises, the profound challenges from disempowered and marginalised groups outside the enterprise are sometimes the most difficult to see, understand and address. Let us consider each of these challenges in turn.

Ground Zero: Bougainville Copper, PNG

In November 1988, landowners around the massive Panguna copper mine on the island of Bougainville in Papua New Guinea commenced a campaign of sabotage against the mine's facilities. By May 1989, despite the intervention of national troops, and the efforts of church leaders, community groups, international mediators and various national and provincial officials, the copper mine was forced to close. The protest against the mining company quickly developed into a war of secession, with demands that the company pay landowners compensation of 10 billion kina for environmental damage and a unilateral declaration of independence for the Republic of Bougainville by the Bougainville Revolutionary Army. After ten years, the situation remains tense, the mine is still closed, and Bougainville's economy and society have been scarred deeply.

The Panguna mine was operated by Bougainville Copper Ltd, a subsidiary of Conzinc Riotinto of Australia, the Australian arm of British-based resources transnational Rio Tinto Zinc Corporation. Development of the mine commenced in 1967, and production in 1972. This mine quickly became enormously important in generating revenues for the new nation of Papua New Guinea after independence from Australia's United Nations sanctioned colonial role was granted in 1975. Connell (1991: 55) reported that the mine had contributed 16 per cent of the nation's internally generated funds and 44 per cent of its exports since 1972. Distribution of the revenues generated by the mine, governed by a mining agreement which was renegotiated in 1974 and again in 1981 laid the foundations of conflict, however, with 60 per cent of total revenues going to the national government, 35 per cent to foreign shareholders in the operating company, 5 per cent to the North Solomons Provincial Government and 0.2 per cent to local landowners (ibid.: 55). Already in 1975, at the time of PNG's independence, dissatisfaction with the balance of power between national and provincial levels had produced secessionist tendencies. The grievances of the Bougainvilleans in general and the local landowners affected by the mine in particular were largely left unaddressed in subsequent years. For a long time, the affected people 'resorted to apathy and stubbornness' in the face of the national administration's 'contemptuous behaviour' (Dove *et al.* 1974: 183).

Filer has argued that the Bougainville crisis reflects social processes that have been entrenched in the process of development itself in Papua New Guinea:

> mines in almost any part of Papua New Guinea will generate the same volatile mixture of grievances and frustrations within the landowning community [as generated at Bougainville], and, all other things being equal, blow-outs will occur with steadily increasing frequency and intensity until there is a major detonation of the time bomb after mining operations have continued for approximately fifteen years.
>
> (Filer 1990: 3)

Gerritsen and MacIntyre (1991) have refined this argument by identifying tendencies within the 'capital logic' of large mining projects – the balance between financial requirements and revenue generation, and the project's implementation schedule. In discussing the Misima gold project in PNG, they point out that the period of greatest impact in the proving, establishment and construction phases, is also the period of greatest financial demand and least revenue. In other words, large-scale projects are least equipped to meet the concerns of affected local people at precisely the time when the effects are likely to be greatest.

Strictly legalistic interpretations of mining agreements, which are often negotiated under circumstances of extremely unequal power between resource companies, national governments and affected communities, have often given resource managers a false sense of security. At Panguna, Bougainville Copper had more than fulfilled its 'legal' obligations under the 1981 agreement. From the company's understanding, they had not only paid the required compensation payments to landowners and others, but had also established the Bougainville Copper Foundation in 1971 as a charitable body to undertake business develop-ment activities, agricultural extension work, and provision of training and service delivery in health and hospital services (Connell 1991: 69–70). Relatively good relations with older members of the Panguna Landowners Association into the 1980s made it easy for many within Bougainville Copper to believe that all was well. In addition, the national government had taken a 20 per cent equity interest in the project in the early 1980s, further entrenching the company's view that nothing further could possibly be required of it because it was so clearly a 'good neighbour', in partnership with both the nation and the local community in the process of development. In 1987, however, a generational change began to occur, with the emergence to maturity of a group of younger local people who had had the benefit of education and, in the case of Francis Ona, military training in Australia. These articulate and critical young people viewed the 1981 agree-ment as completely inadequate and formed a new Panguna Landowners Associa-tion to challenge the arrangements that had developed.

They were better organised, understood the operations of the company more clearly than their predecessors, shared all the concerns of the villagers, and were more militant (Connell 1991: 71).

The new PLA demanded changes. They wanted to see improvements in basic services, increased employment of local people in the mine, greater control over pollution, and a new survey of the area. These demands escalated to include transfer of the national government's equity in the project to the landowners, and payment of massive compensation for environmental damage.

For company managers, the unravelling of relations, and rapid escalation of the dispute to one which intermittently closed the mine until its final closure in 1989 came as an almost incomprehensible shock. When company staff were shot at by the rebels, and heavy-handed military intervention by the national army pro-voked wider support for them among other Bougainvilleans, the company with-drew all its staff, and the national government imposed a complete blockade on the island. Subsequent military action, widespread human rights abuses by both BRA and PNG troops and the lack of provision of even the most basic medical and humanitarian supplies to the people on Bougainville created a great deal of human misery that remains unresolved. The Bougainville crisis has had wide repercussions within the government of Papua New Guinea, in both financial and administrative terms (Saffu 1992), and on the actions of other groups of land-owners affected by large-scale mines. At Ok Tedi, for example, landowners used Australian courts to prosecute BHP Ltd for enormous environmental damage caused by operating a gold and copper mine in the Star Mountains of Western Province near the West Papuan border without a tailings dam (see for example Filer 1993; Hyndman 1991; Radio National 1995a–d; Lafitte 1995). Out of court settlement of this case led to multimillion dollar compensation payments, a massive fund for environmental rehabilitation work and immediate work to develop a tailings dam – a cost which has shocked many resource managers used to using the courts to enforce their own privilege, rather than finding themselves held to account in them (Banks and Ballard 1997).

In the Bougainville case, the process of change together with the impact of poorly managed industrialisation and development, has left a tragic legacy which many of those who facilitated the mine's development have come to regret. Once again, the indigenous people brought into the orbit of industrialisation and development were left at Ground Zero of a process of social disintegration, environmental damage and uncertain futures.

> At the time I signed the agreement allowing BCL to commence mining operations here on Bougainville, you didn't tell me what would happen to my environment. You capitalised on my ignorance and after 18 years here much of my land has been depleted. What happens when the gold and copper finishes? You will leave with your money and I will be left with a barren wasteland. The government stays in Port Moresby and says BCL knows what it is doing and yet we see our environment dying daily. When I was young they fooled me and now I am old and still alive to see the result of my decision I weep. Who cares about a copper mine if it kills us?
> (Former member of Parliament for South Bougainville and Minister for Mines, Paul Lapun, quoted in Connell 1991: 74)

Ideological challenges: indigenous peoples' ecological knowledge

With the emergence of a societal concern with shaping ecologically sustainable human systems, many people have found valuable ideas, guidance and wisdom in the traditional values, knowledge and epistemologies of indigenous peoples. As Suzuki puts it, indigenous knowledge – the 'wisdom of the elders' – seems to offer: 'Something powerful, very relevant and profound for members of the dominant society' (Suzuki, in Knudtson and Suzuki, 1992: xxi).

The combination of increased popular understanding of ecological processes and acceptance of a spiritual dimension to environmental and economic relations creates a potent political space for indigenous groups. While recognising this, it is also important to avoid naïve romanticism that masks the diversity and specificity of indigenous cultures and knowledge with a homogenising sameness that reduces ecologically specific insights to mindless truisms and vague idealisms. Resource managers seeking to develop resources in indigenous territories ignore the potency of the ideological challenge from the indigenous movement at their peril.

The effect of exclusion of 'indigenous knowledge' from resource management systems is not solely a question of spiritual or political crisis. It also includes a wide variety of deeply practical concerns. In many places, scientific knowledge is fragmented and few things of value to industrial societies have been identified. In many Arctic regions, for example, few scientists and fewer science-funding organisations have sustained observations through northern winters. In contrast, indigenous peoples' observations and understanding has spanned seasons and generations (Hobson 1992). As Berger (1977, 1988) acknowledged, the intimate ecological knowledge of indigenous peoples carries its own weight, authority and significance and provides a basis for understanding and relating to these places. Despite the efforts of many advocates of industrialisation and development to marginalise indigenous ecological knowledge, it is increasingly recognised as co-equal with traditional science: in some cases as superior. As one commentator puts it 'traditional knowledge *is* science' (Hobson 1992: 2, emphasis in original):

> Western scientists have a tendency to reject the traditional knowledge of native peoples as anecdotal, non-quantitative, without method, and unscientific. From our scientific ivory towers we tend to ignore basic knowledge that is available to us … . Often overlooked is the fact that the survival of northern aboriginal peoples depended on *their* knowledge, *their* special relationship with the environment, and *their* ways of organizing themselves and their values. Traditional knowledge was passed on from one generation to the next. Today, aboriginal peoples are aware that they must integrate traditional knowledge into the institutions that serve them; it is essential to their survival as a distinct people, and it is the key to reversing the cycle of dependency which has come to distinguish aboriginal communities.
>
> (ibid., emphasis in original)

Recognition of the strengths of traditional ecological knowledge, and the limitations of scientists' ecological knowledge has increasingly led to efforts to incorporate traditional knowledge into scientific research and environmental management and regulation systems (Mailhot 1994; see also Dwyer 1994; Payne and Graham 1984; Gamble 1986; Jacobs and Mulvihill 1995). Sallenave goes as far as suggesting that most environmental impact assessment studies undertaken in northern Canada are 'ineffective' because they fail to address the implications of traditional knowledge:

> At present, most environmental assessments and most monitoring systems for northern development projects neither involve aboriginal communities significantly, nor include northern peoples' vast knowledge of the natural environment.
>
> (Sallenave 1994: 16)

Yet, as the discussion elsewhere in this book demonstrates, the issue is not simply one of harnessing the traditional knowledge of indigenous peoples to the needs of science, development and industrialisation. Traditional ecological knowledge may well improve the efficacy of scientific knowledge in some co-management arrangements. The specific knowledge of particular ecological circumstances is of enormous value in such circumstances. Systems of traditional ecological knowledge, however, are significant in their own right. In asserting the value of traditional ecological knowledge for developmentalism, there is a risk of reducing traditional value systems and cultures to fragmented 'facts' of some utilitarian value, for appropriation and exploitation by developmentalist interests (Agrawal 1995). Indigenous cultures are not something to be subsumed to the service of scientific knowledge. They are a source of specific rights for indigenous peoples. The ethics, values and philosophies underlying traditional ecological knowledge also have lessons whose relevance extends beyond the specifics of particular ecological systems (Rose 1988) and provide a framework for thinking about ecological rights, human rights and alternative trajectories for the planet to those shaped by the imperatives of industrialisation and developmentalism (Christie 1990):

> The relationships between people and their country are intense, intimate, full of responsibilities, and, when all is well, friendly. It is a kinship relationship, and like relations among kin, there are obligations of nurturance. People and country take care of each other. I occasionally succumb to the temptation to sort these relationships into categories – there are ecological relationships of care, social relationships of care, and spiritual relationships of care. But Aboriginal people are talking about a holistic system, and the people with whom I have discussed these matters say that if you are doing the right spiritual things, there will be social and ecological results. The unified field of Dreaming ecology is demonstrated very clearly in the intersection of sacred sites with ecological sanctuaries.
>
> (Rose 1996a: 49)

For Aboriginal people in Australia, the term 'country' encompasses all these complexities – the depth of intimacy with a particular place, the intimate dynamics between people, other species and environmental processes, the rights and responsibilities that inhere in such intimacies, and an holistic perspective on relations between these elements:

> Country in Aboriginal English is not only a common noun but also a proper noun. People talk about country in the same way they would talk about a person: they speak to country, sing to country, visit country, worry about country, feel sorry for country, and long for country. People say the country knows, hears, smells, takes notice, takes care, is sorry or happy. Country is not a generalised or undifferentiated type of place, such as one might indicate with terms like 'spending a day in the country' or 'going up the country'. Rather, country is a living entity with a yesterday, today and tomorrow, with a consciousness, and a will toward life. Because of this richness, country is home, and peace; nourishment for body, mind, and spirit; heart's ease.
>
> (Rose 1996a: 7)

The lament of Australian Aboriginal people separated from their traditional country[6] has been immortalised in the title of the novel *Poor Fellow My Country* (Herbert 1975) and the song 'Gurindji Blues', written by Galarrwuy Yunupingu and Vincent Liangiari (Builders' Labourers' Federation 1975: 12–13). The cry of 'poor fellow my country' or 'poor bugger me' is simultaneously a lament for the state of the environment and for the state of the webs of social relations embedded within the landscapes and the state of the individual alienated from the previously seamless social fabric (see Box). It is, therefore, much more than a lament for the ecological disruption engendered by land degradation, although the specific problems of environmental degradation are obviously an important element of this.

Poor fellow my country: dispossession and degradation

Where are we going to go? Where are we going to get a place to live, to stay? Where? In the air or where, because farmers coming on the land; fisheries are coming on the sea? Where can we find a place now for Aboriginal people? Where? We can't live on the air. Where are you pushing us? ... This is our land. This is our homeland.

(John Baya, in a submission to the Aboriginal Land Commissioner regarding control of entry onto seas adjoining Aboriginal Land in the Milingimbi, Crocodile Islands and Clyde River Area, May 1980, quoted by Rose 1996a: 78)

The white people came into the area many years ago, our ancestors called them spirits because they were white in colour. The white people came mustering cattle ... . But the cattle they can make the country bad, they muddy the water in the lagoons so that we cannot drink it because it stinks, and we do not hunt because we can no longer find goannas and long necked turtles. This country of importance has a new name now and white people have changed it so it is now called 'Manangoora Pastoral Lease'. But we Yanyuwa people, we cannot forget about this country, we are continually thinking about it, this country that was truly for our ancestors, we are thinking about them all the time.

(Yanyuwa people quoted in Bradley 1988: 47)

In several areas, including fire management, ethnobotanical work, species conservation and reintroduction, and joint management of key conservation areas in both terrestrial and marine environments, recent scientific work has confirmed what many indigenous people already understood. Aboriginal peoples' cultural activities and values often have compelling environmental or ecological logic underpinning them (Taylor 1995; Rose 1996a; Flannery 1994; Kohen 1995; Langton 1998). More broadly, these values, and the specific knowledge they encompass, have much to contribute to the task of developing sustainable human systems in vulnerable ecological niches, which requires holistic rather than fragmented approaches to resource use. Aboriginal organisations themselves have taken many initiatives to re-establish their ability to care for country, and have integrated caring for country into their wider struggle for justice and recognition.

Recognising the value of both locally specific environmental knowledge and indigenous cultural and environmental values shifts the focus of relations between indigenous peoples and resource management systems away from either naïve and idealised notions of 'the wisdom of the elders', or a patronising concern with disabled indigenous minorities. Conventionally, the concerns of indigenous people would be dealt with as outside the operational concerns of resource management systems. This new perspective, however, challenges the ideology that would render indigenous groups as external or marginal to resource management systems. It prompts a more pragmatic consideration of just how current and future management practices might build on the strengths of 'best practice' traditions from several relevant approaches, including indigenous traditions, scientific traditions, community participation principles and industrial thinking. Pursuing this line of thinking will equip all resource managers to move a long way from the dominant conventions of industrial resource decision making which construct decision making about resources as an exclusive prerogative of specialised experts, even when they are unequivocally part of the public domain. It also challenges the sort of racist and discriminatory social thinking about 'public domain' resources that is so widespread in industrial societies.

Ground Zero: Ranger Uranium, Australia

In the 1970s, a proposal to develop large-scale uranium mines was debated in Australia. World class uranium deposits were identified on the western boundaries of the Arnhem Land Aboriginal Reserve in the Northern Territory, in areas of enormous value to local Aboriginal people, and of high conservation and heritage value. At the national level, all the major political parties were committed to non-proliferation of nuclear weapons and to a non-nuclear power policy. The question of whether or not to enter the nuclear fuel cycle through mining split the Australian community and contributed to development of a popular environmental consciousness that continues to echo in today's green political movements. For the local Aboriginal people, the issues were clear. This was dangerous and sacred country that should not be disturbed. Their opposition to development of the mines was virtually unanimous and clearly acknowledged in the report of the Commission of Inquiry established by the national government to provide scientific advice on the proposal (Fox *et al.* 1977a, b).

At a time when recognition of Aboriginal land rights in the Northern Territory was being legislated, the uranium province was given special treatment. At a time when new laws were giving other Northern Territory Aborigines a limited veto over mining on their traditional lands, these people's opposition was overruled in 'the national interest'. At a time when an inquiry of similar scope and importance for indigenous rights in Canada was being conducted by Mr Justice Berger to provide advice on a massive oil and gas pipeline proposal in the Mackenzie Valley, setting new standards for indigenous and general community participation (Berger 1977, 1988), the Ranger Uranium Inquiry was buckling to conservative government pressure to remove decision making about uranium mining from the political and into the scientific arena (Howitt 1989a provides a comparison of the two inquiries).

Despite Aboriginal opposition, two mines at Ranger and Nabarlek were approved and developed. A third mine at Jabiluka was approved, but was undeveloped due to a change of government policy. It emerged as a renewed proposal in the mid-1990s. Mining leases were excised from Kakadu National Park, which was recommended for World Heritage Area listing, and traditional ownership of the area was determined, giving Aboriginal people a flow of royalty-type payments from the projects.

Despite rigorous safeguards, the development of 'joint-management' in the national park and a flow of benefits, lingering concerns about the social, environmental and health effects of the mining in the area persisted.

In the early months of 1995, during a record Wet Season, fears of Aboriginal people downstream from the Ranger mine were raised over proposals to release contaminated water from a restricted release zone into the river system that provided most of their bush food resources. Prior to this, the mine had operated for eighteen years without a proper water management plan. Contaminated water was stored on site in areas required later for other uses. Despite earlier undertakings about a zero water release policy, release of contaminated water into the river system was presented as the only option available in 1995. In the process, despite nearly two decades of successful intercultural communication between the mine and the traditional owners, and highly prized 'joint-management' of the world heritage areas between government scientists and traditional owners, old-style 'divide-and-rule' tactics quickly emerged as traditional owners downstream from the mine took legal action to stop the water releases:

> It was made clear to us by the company representative that if the release did not go ahead, the mine might have to shut. Therefore, there would be no more royalty payments to the traditional owners. This created tension between the traditional owners upstream who believe they are not directly affected by the release, and those of us downstream who believe that the environmental risks are totally unacceptable.
>
> (Christine Christophersen, traditional owner of the affected area
> and plaintiff in the legal action to stop the water release,
> in Christophersen and Langton 1995b)

In seeking to justify the release of the contaminated water into the world heritage area, a powerful combination of racism, paternalism, science and law was marshalled against Aboriginal criticism. Bill Neidjie, senior traditional owner and an acknowledged 'expert' in the joint-management system, was ridiculed in the press by the chief executive of the mining company:

> It isn't easy to explain the scientific facts to Big Bill ... . We have been unable to convince him this will not affect his water, his country and his food chain ... we have consulted with Big Bill many times, had experts speak with him, we haven't got anywhere.
>
> (Phillip Shervington, CEO of Energy Resources Australia Ltd,
> quoted in the *Weekend Australian*, 10–11 March 1995)

Scientists sought to explain the 'facts' to the traditional owners again in a meeting in mid-February:

> They explained it to us this way: 'It is not advisable to drink a glass of this water, but it is okay to release it into the river'. Someone else from [the Northern Territory Department of Mines and Energy] explained it to us in his version of New Guinea pidgin, a quaint story about a spoon and a pot of stew: 'If you spit on a spoon, it is not okay to eat off that spoon, but if you mix the spoon in a pot of stew, it is all right to eat that stew'.
>
> (Christophersen and Langton 1995b)

This science, based on the careful monitoring over the mine's life which allowed a situation to develop in which all the contaminated water on the site could not be contained within the site and 'had' to be released into the environment, simply failed to listen to the local people:

> Bill Neidji ... carries the authority formed by a culture indigenous to the very land your feet walk on today. His authority has been shaped by decades of obligation, of ceremony and learning from the Old People of many, many generations. His authority has been created by personal actions and deeds and by strength and by teaching, kindness and most of all, by immense dignity ... He knows of no scientist or federal government representative who is dependent on land as a food source downstream of a mine release site. He knows of no scientist or government representative who want to live downstream of the proposed release site.
>
> Yet he is being asked to place his trust in a science that he knows very little of. He does not. A science that relies on Aboriginal knowledge to show them how and why things happen in relation to land management and resources. A science who with all its cleverness, cannot communicate in either his father's or mother's language. A science that is almost always unintelligible when spoke(n) in English to him. A science that has only monitored information relevant to the area from 20 years ago.
>
> He is also being asked to place his trust in a government decision. He does not. He knows of our history and Governments when dealing with Aborigines. He knows what is going on today, he believes it is not a good story. His personal experience of dealing with government has seen inevitably, a NO turn into a yes, right before his eyes. The use of the English language and sometimes the use of government still confound him, when it becomes clear that it was never the intent, even though it was said.
>
> Big Bill Neidji, Senior Traditional Owner of Gagadju Country, Bunitj Man is asking that you listen to him. To trust him and his authority and the science that he knows.
>
> (Christine Christophersen speaking on behalf of her uncle Bill Neidji in Darwin in March 1995, tabled in the Australian Senate on 2 March, Hansard: 1355)

Despite undertakings that all releases of water from the site would be tightly controlled, and undertaken in consultation with the traditional owners, there have been leaks and *ad hoc* releases, with *ad hoc* consultation. For people downstream from the mine this has created:

> ... fear and a feeling of disempowerment ... People on hunting and fishing trips are fearful of the waterways where the releases are made – areas which were once loved food resource areas. These social impacts have gone unnoticed by the relevant authorities. And it is clear from the tenor of the current debate, that they are regarded with some contempt.
> (Christophersen and Langton 1995b)

So once again, more than forty years after Ground Zero at Emu (see page 6), Australian Aboriginal people were exposed to radiation hazards in the interests of the wider nation. Once again, scientific evidence was used to disguise the political and economic dimensions of the juxtaposition of risk and indigenous people:

> They say the situation has two easily identifiable opposing standpoints: on one side there are the entirely rational, infallible scientists and their mates; and on the other side are the incredibly stupid blackfellas, with their ungrounded fears who are once again holding up development. The company and the NT government have not progressed out of the 1970s in understanding what their responsibilities are to the public and that public includes the traditional owners as a special category of people to whom commitments were made by the Federal government in 1977. This argument is not about science versus non-science. What it is really about is 'profits before people' dressed up to look as if it is about science.
> (Christophersen and Langton 1995a)

Socio-cultural challenges: the right to a place in the landscape

Competing land uses and conflicting resource management systems are not simply reflections of competing vested interests, nor competing views of the utility of 'country' for society. In many cases these conflicts reflect much deeper ontological schisms between worldviews – between ways of seeing the world and ways of thinking about peoples' places within the world. The dominance of industrial values in shaping resource management systems denies the cultural integrity and fundamental rights of indigenous peoples (and, of course, many others) to identity, self-determination and legal protection.

Indigenous peoples' resistance to the terms of their incorporation into colonial and post-colonial nation states has produced a struggle for legal recognition. In various jurisdictions, this has produced significant enforceable rights. In the case of the Anglo-Commonwealth countries, McHugh argues that the emerging jurisprudence on Aboriginal title, for example, actually constrains the power of the Crown in quite new and significant ways (McHugh

1996). International human rights standards, and the political effectiveness of the international indigenous rights movement have internationalised the arena in which indigenous peoples' socio-cultural challenges to resource managers are played out (see for example Jhappan 1992).

For many resource managers, indigenous peoples' property, even where title is recognised under non-indigenous law, continues to be treated as if it were some sort of public asset. Indigenous territories have been incorporated into national spaces so that they can be developed and set to work in advancing the 'national interest' (Howitt 1991a). Of course, examples can be found in all nations with indigenous people whose sovereignty and autonomy are restricted.

One area where the tension between indigenous and non-indigenous worldviews is particularly strong is in defining and managing 'wilderness' areas. The long history of separation between 'man' and 'nature' in Western philosophical traditions has produced a categorical distinction between the 'natural' and the 'social' (Fitzsimmons 1989). For people immersed in Western-style thinking, the very idea of 'wilderness' involved places 'where the hand of man has never set foot' (Brower 1978). In developing the idea of wilderness as something to be valued and protected, American conservationists involved in areas such as Yosemite National Park were idealising a landscape from which Native Americans had been forcibly removed (Wilkinson 1993:162–86). Even for many 'progressive' Americans, Indians have been so effectively cleared from the landscape that they are left out of important historical critiques of people-land relations in the US. Even in the work of influential left-wing geographers such as Neil Smith (1984a) and Ed Soja (1989) Native Americans are absent from the US landscape. Indeed, Smith goes so far as to unproblematically refer to the whole Lower East Side of New York as 'Indian Country' (1994: 93), not because of a Native American presence, but because of an imagined frontier between some vaguely implied white, middle-class, middle-American mainstream normalcy and homeless people.

Similarly, in Australia the areas targeted as having high wilderness value are often a result of generations of human intervention to maintain a particular ecological regime (see Box). Initial efforts to develop an Australian approach to wilderness protection and management were dogged by widespread ignorance of the impacts of Aboriginal management practices on Australian ecosystems (Kohen 1995). As Langton (1995) points out, Australian concepts of wilderness were inherently linked to the now repudiated and always false legal notion of *terra nullius,* which reinforced racist notions of justified expropriation of indigenous lands (see also Robertson *et al.* 1991). The principle of *terra nullius* was spelt out in the Gove land rights case, where Yolngu Aboriginal people in northeast Arnhem Land were denied the right to stop the Federal government approving bauxite mining on their land. The Yolngu were deemed to have a system of law whose property rights were unrecognisable by English common law (Williams 1986a; Blackburn 1970). In this view, Australia was an empty land belonging to nobody prior to British settlement. This idea, which has so long been at the root of Australian property systems (Reynolds 1987, 1996), was overturned in the 1992 *Mabo* decision. Yet it continues to be influential in conservative politics, and to influence popular understanding of many intercultural issues. For example, it combines with racial stereotypes linking 'primitive' Aborigines to 'primitive' ancient landscapes (Head 2000). Such 'primitive' technologies were clearly, it is argued, incapable of affecting the sort of changes and environmental controls involved in 'wilderness' management. The clearances of Aboriginal people from many of Australia's best known 'wilderness' areas in the southeast of the continent simply reinforced the notion that wilderness involved an absence of human influence in the minds of even many progressive environmentalists. The picture is further complicated when indigenous people insist on exercising their rights to use wildlife and other resources that environmentalists want to 'protect' (Langton 1995; see also IWGIA 1991 for a relevant account of similar issues in the Arctic).

Wilderness or 'Wild' Country?

As part of the documentation of [environmental] degradation [in the northwestern part of the Northern Territory], I made a short video of some of the most badly affected areas. I asked [Daly Pukara]one of the senior custodians of this country what he called the degraded area. He looked at it for a while and said, 'It's the wild, just the wild.' He then went on to speak eloquently of the lack of care in this area and to contrast this wild country with another area he termed 'quiet' ... . [He] is telling us that his country is becoming a 'wilderness' – a man-made and cattle-made wilderness where nothing grows, where life is absent, where all the care, intelligence and respect that generations of Aboriginal people have put into the country have been eradicated in a matter of a few short years. In contrast, he tells us that country that is cared for, that is unspoilt by the encroaching wilderness, is 'quiet'.

(Rose 1988: 386)

The response of Aboriginal people, and of indigenous peoples' organisations around the world, to environmentalists' efforts to further displace or restrict indigenous sovereignty, has produced a redefinition of 'wilderness':

wilderness has come to mean a landscape that is valued because it is undeveloped by colonial and modern technological society. In this sense, 'wilderness' does not represent a perpetuation of notions of 'wasteland' and *Terra Nullius* used against indigenous people. Rather it encompasses a view that all that the land contains – including indigenous culture – is to be respected, appreciated and sustained. Given the correlation between remaining wilderness areas and land which retains cultural importance for indigenous people, wilderness protection in Australia may not be properly achievable unless prior ownership and current Aboriginal and Islander aspirations are comprehensively addressed. These remain national concerns, urgently requiring resolution at a national level.

(Robertson *et al.* 1991: 18)

In such examples, it is possible to see indigenous peoples asserting their right to exist within the geographic (and by implication in their holistic terms, the political, economic, cultural and social) landscapes of contemporary life. At the same time, such examples emphasise the political nature of these geographies.

As Berger (1977, 1988) emphasised in the title of his influential report, *Northern Frontier, Northern Homeland*, however, cultural relationships to places inevitably have political implications. Indigenous territories are not just indigenous homelands. For members of the dominant society, these places (often perceived by them as 'spaces') are 'wildernesses' and 'frontiers' (see Box). In nations such as Australia, there simply is no wilderness in the sense that many people understand that term.

Behind the story of the mines and the oil rigs lies the question: are the native peoples merely a curious cultural backdrop to the activities of Western man, or are they the peoples for whom the North was made? A lack of understanding and of sensitivity to native peoples and native values is endemic in European-derived political systems. What is remarkable is that despite the attempts to separate native people from their language, history and culture, they have retained their distinctive identity.

The Dene, the Inuit and the Metis are advancing land claims proposals and proposals for new political arrangements in the Northwest Territories. Whatever the outcome of these proposals, they are evidence of a renewed determination – and a new capacity – on the part of native peoples in the North to defend what they believe is their right to a future of their own. They are engaged in a search for self-determination and in the development of new political institutions. As well, they have undertaken the defence of the northern environment.

(Berger 1988: 10)

For many of the masculinist myths of conquest, such indigenous domains are 'virgin' territories waiting to be 'taken' by those with sufficient strength (or money, or power) to secure proprietary rights to them. In resource terms, such gendered images take on further overtones as oil, mineral and biodiversity 'explorers' constitute indigenous territories as 'virginal', and target them for priority action (see for example Trigger 1996; Willems-Braun

1997). Thus we find national identities constructed around images of imagined frontiers, where indigenous people are present only as an object of conquest, or a barrier to national destiny. Such images abound in the USA, Canada, Australia and Latin America. They underpin much of the orientalist literature of conquest and exotica, and they drive the ideologies and political programmes of nationalist, racist and supremacist movements in many places. These images are not, of course, uncontested within the dominant society – class struggle, for example, teaches that similar experiences can be interpreted in opposing ways within the same cultural group. Frontier settlers in agricultural, forestry or mining settlements are likely to interpret the 'frontier' quite differently to urban-based 'New Agers' seeking reconciliation with nature. In the work of Harman (1981), Salisbury (1977) and Berger (1977) we can find valuable analyses of the juxtaposition of frontier images and indigenous territorial interests in areas of Australia (Western Australia's northern regions) and Canada (James Bay Cree homelands, and the Mackenzie Valley of western Canada).

Indigenous peoples' right to a place in the landscapes of industrial society's resource frontiers has, of course, been strongly challenged by the beneficiaries of the primitive accumulation that occurs in these locations, and their ideological supporters. Harman records how developmentalist ideologies about Western Australia's resources frontier 'direct and legitimate state intervention' (1981: 167) based on the creation of jobs and wealth that benefit the state as a whole. Harman identifies several key elements of the ideological justification for state action on the frontier – often at great cost to Aboriginal people whose rights and concerns are rendered 'invisible and irrelevant' (ibid.: 180). She suggests (Figure 1.6), for example, the following issues:

- Expanding the number of jobs and the level of income for people within the state;
- The expansion of settlement and extension of civilised social control (of both uncivilised Aborigines and undisciplined workers) to facilitate development;
- The settler population's inescapable destiny in building a new state and the closely-related protection and extension of states' rights;
- Contributions to 'nation building';
- Contributing to development in the underdeveloped world through the provision of resource commodities;
- Protecting capitalism and democracy;
- Advancing the cause of civilisation.

Despite such ideologies, indigenous peoples have persisted in their efforts to maintain and expand acceptance of their identity, rights and responsibilities. In many places, indigenous peoples are formally involved in the management and use of a range of resource industries including conservation areas, forestry and mining areas, urban areas and various multiuse zones. If indigenous ownership

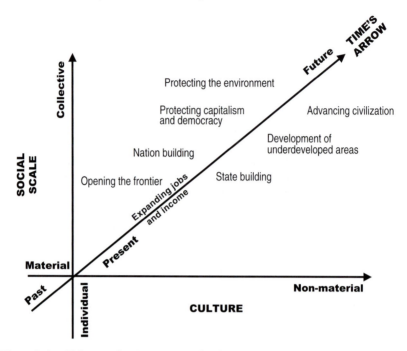

Figure 1.6 Values underpinning state developmentalism

Source: Based on Harman 1981: 178.

is recognised, and the value of indigenous knowledge is acknowledged, the aspirations of people to return to their country and to care for it inevitably intrudes upon non-indigenous notions of land and resource management. The right to a place in the landscape inevitably implies a range of other sorts of rights to manage, influence and benefit from the use of that landscape.

Most importantly, indigenous peoples throughout the world argue that these rights involve a right to collective self-determination – a right to decide what happens to them and their property. Nation states whose existence is predicated upon colonial dispossession of indigenous peoples often dispute these rights, alleging that such claims constitute mischievous threats to national unity and territorial integrity. It is in the intellectual and geographical spaces created by this tension, however, that competing claims to resources and cultural identities, and the need for a new professional literacy, and to rethink resource management systems is to be found and dealt with.

Politico-economic challenges: the right to resources

Resource co-management, and even autonomy in some areas and issues, leads us to some of the most difficult challenges facing relations between resource managers and indigenous peoples. Indigenous claims that their indigenous identity gives them some specific, even decisive right to own, control and benefit from various resources in their traditional territories represent a major challenge. In some jurisdictions, the residual rights derived from prior sovereignty are well entrenched (if persistently challenged) elements of public decision making. While the practical benefits for indigenous groups of such rights might continue to be challenged by structural racism, it is clear that such rights must be addressed by resource managers seeking to utilise water, minerals, timber, fish and wildlife, and energy resources from many tribal territories. This is certainly the case in the United States (Deloria 1988; Jaimes 1992; West 1992; Churchill 1988, 1995). Similarly, in New Zealand, parts of Canada, Papua New Guinea and elsewhere (Fleras and Elliot 1992; Churchill 1995; Little Bear *et al.* 1984; Notzke 1994; Cant *et al.* 1993; Renwick 1991), the ownership of resources in indigenous estates is entrenched in statute, treaty and common law. In other areas, such as Australia and the non-Treaty lands of North America, recognition of indigenous peoples' 'rights' continues to be strongly contested by resource developers.

Yet, despite the rhetoric of opposition to indigenous rights (Heilbuth and Raffaele 1993; Howard and Widderson 1996), many resource companies which are such vigorous opponents of indigenous self-determination in Australia have been able to accommodate tribal governments in North America. BHP, for example, which opposed the rights of tribal people affected by pollution from their Ok Tedi mine in Papua New Guinea to protect their rights with legal action in Australian courts, complies with the environmental regulations of the Navajo EPA in its coal operations in the western United States.

As McHugh (1996) observes, despite the naïve notion that the nation state, or in former British colonies, the crown, is unencumbered by restraints resulting from indigenous rights, there is an emerging international jurisprudence to the contrary:

> For generations public authorities in the Anglo-Commonwealth countries of Australia, New Zealand and Canada assumed that resources of a public character (minerals, fisheries, waterways and the like) were vested in the Crown without any tribally related legal qualification ... . Yet it has become clear since the mid 1980s that aboriginal claims in the Anglo-Commonwealth countries ... raise legal issues that not only have a direct bearing upon resource development but which are also of fundamental constitutional importance.
>
> (McHugh 1996: 300)

The scope, structure and vision of this book

This book seeks to rethink resource management systems. It aims both to deconstruct and reconstruct the ways in which resource managers understand both the focus and the context of their work. Its principal audience is young, university-based trainee professionals in the general field of resource management. It also seeks to offer something to those who are already working in the field as practitioners, beneficiaries, administrators, regulators, opponents or victims of its currently dominant paradigms.

In pursuing these tasks, the book straddles a discursive and political space between critique and alternative. It aims to provide both a critique of the 'old' and critical advocacy of an alternative. These two things are developed side by side as the 'narrative' of the book unfolds. In this sense, the book is trying to do something a little different from a more conventional text. To some extent, the text itself is intended to introduce its readers to the polyphony and uncertainty, the complexity and dynamism that is seen as underpinning better resource management. It is intended that the text will, to some extent, introduce the diverse sections of its audience to each other. I hope that this might allow each to gain some clearer insight into and understanding of the rationalities that underlie the actions, concerns and priorities of the others. This is not to disclaim an authorial position. There should be no doubt that I am seeking to put forward a very strong, highly political and subjective position in this text, and that I believe passionately that the argument developed here needs to be taken seriously by others. The difference from many other university textbooks is that I do not think it is possible to put forward simple ('textbook') answers to the extraordinarily complex problems that are within the compass of this book and the field to which it refers.

In the field of resource management there are multiple voices, each of which needs to be understood by others in the field. The problems of 'representing' such polyphony within a written text have been hotly debated in academic circles (Crang 1992; McDowell 1994; Marcus and Fischer 1986). Of course, it is not possible to simply represent here even a small portion of the enormous diversity of indigenous thinking; some effort is made in the pages that follow to let some of the participants in the stories being told speak for themselves to some extent. Drawing on a range of published and unpublished sources, including works of fiction, polemic, biography, poetry, and analysis, as well as my own field notes and interview records, and the work of various colleagues, I have tried to set up a dialogue of a different sort within the text. In these excerpts, often contained in boxes at the margins of my own text, readers will be guided towards sources for ideas; positions and ideas put forward in the main text will be clarified, reinforced and challenged; alternative readings of information provided in the main text will also be outlined; and questions for group discussion and further investigation will also be raised. In this way, it is hoped that the convenient fiction of an authoritative 'textbook' narrative might be a little unsettled in this book, and readers led towards their own engagements with the issues involved.

The idea of the world being turned upside down by the decisions of those involved in resource management is a central one in this book. In some ways, the book itself aims to turn upside down (or at least slightly disorient) the taken-for-granted worldviews of many of its readers. As an educator, I would prefer for the text to engage readers in pedagogical dialogue rather than as a didactic, monotonic and authoritative authorial address. Much of the book originated as lectures and discussions in which students were active co-explorers of these themes and issues rather than passive recipients of a singular authoritative, refined and pre-ordained wisdom. Some sections derive from fieldwork with Aboriginal groups seeking to transform specific aspects of local relations with development narratives. It has been written in same the spirit, even though genuine dialogue between author and reader is not possible in any direct sense. The pedagogical problems of accepting polyphony and displacing the 'expert' from the centre of our textual and educational narratives are relatively new issues in university teaching (McDowell 1994, Howitt 2000). My intention is not to undermine the value of expertise *per se*, nor the credibility of particular experts (particularly not myself as author). Rather I aim to open the foundations of this credibility to a critical gaze that is constructed in processes that extend beyond a narrowly defined academic or professional peer group and to encompass a much wider audience of human peers.

This book has both empirical and theoretical (practical and conceptual) objectives. Specifically, it aims to:

- Provide a sound and practical conceptual framework for understanding complex issues in contemporary resource management from several vantage points;
- Discuss the particular experience of indigenous peoples in the rapidly changing world of resource geopolitics;
- Demonstrate the relevance of critical human geographical perspectives to the process of rethinking (and reshaping) industrial resource management systems in ways which are consistent with the core values of social justice, environmental sustainability, economic equity and cultural diversity;
- Rethink the assumptions and implications of the currently dominant developmentalist paradigms in industrial resource management.

Key competencies

In the real-world employment markets in the field of resource management, many employers want applicants for jobs to 'demonstrate competencies' in specific areas. This book seeks to facilitate development of important competencies common to many areas of professional resource management. Most generally, it seeks to contribute to development of general critical skills – skills in reading, observing, analysing – and skills in synthesising diverse and complex materials in coherent arguments. Readers who use this volume as part of their formal studies will hopefully also be encouraged to hone their writing, listening and speaking skills more directly than is possible in this format. More specifically, the book will facilitate some competence in several key areas (Table 1.4), which should appeal to those who need to put forward a strong resumè to prospective employers. The knowledge, skills, understandings and values developed here will also equip readers with some of the

Table 1.4 Key competencies targeted in this book

Social impact assessment in relation to both the assessment of impact of resource projects on indigenous peoples, and in the wider arena of impacts of resource projects on affected social groups and localities.

Social theory relevant to effectively understanding social conflict over the use and management of natural resources.

Human geography as a disciplinary foundation for participation in multidisciplinary teams in practical areas of resource management, including a critical understanding of both its strengths and limitations.

Skills in reviewing, researching and responding to relationships between place-based conflicts over resource management and wider social processes and wider scale issues of environmental and social change.

Specific knowledge and understanding of the relevance of the experiences of indigenous peoples to the work of resource managers.

The importance of ethics and values issues in practical resource management.

---

competencies needed to allow you to provide effective support in various possible advisory roles.

Basic structure of the book

The purpose of rethinking the processes and procedures of resource management in this book is defined here in terms of achieving outcomes from resource management that contribute to improvements in the four core value fields identified at the outset (social justice, ecological sustainability, economic equity and cultural diversity). This focuses analytical attention explicitly on the nexus between biophysical, socio-cultural and politico-economic domains (Figure 1.7). It brings issues in the social domain from the background of general context of resource management to the foreground. Recognising the importance of purpose in orienting one's overall approach to the tasks of resource management emphasises the naïvety of claims to 'objectivity'. In such politically contentious and potentially divisive arenas, claims to objectivity are not sustainable. A broad framework is needed to situate various efforts to analyse, explain and participate in the activities of resource management. It is this framework that is central to the idea of a resource manager's 'toolkit'.

The book is structured around the three basic components of a conceptual 'toolkit' for professional resource managers (Figure 1.1, page 9):

- New ways of seeing; which produce a need for
- New ways of thinking; which lead us towards
- New ways of doing.

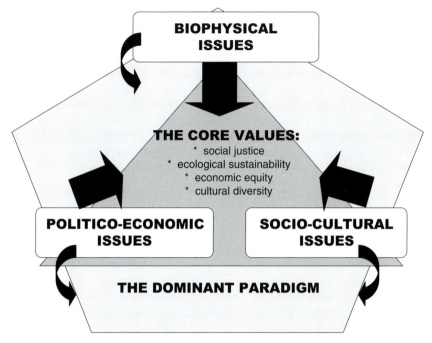

Figure 1.7　Tension between the 'core values' advocated in this book and the dominant paradigm of resource management in dealing with biophysical, politico-economic and socio-cultural issues

In addition, the book includes a section in which case studies addressing aspects of the argument are explored, and a discussion of how the issues raised may lead to an integrated praxis which encompasses social justice, ecological sustainability, economic equity and cultural diversity within the professional arena of resource management.

Ways of seeing

Visualising complexity is an essential skill for resource managers. This is one of the skills that needs to be in your toolkit. Many professionals' cultural and educational training, however, is rooted in scientific, religious and political systems in which complexity, uncertainty and change have been interpreted as threatening stability and order. Consequently, this skill is neither highly esteemed nor easily developed.

Part of the problem is that in pursuing simplicity and clarity, Western-style scientific thinking typically minimises the extension of interaction to a relatively narrow range of direct causes and effects. In many fields, stability and confinement are assumed as a 'natural' state of affairs. For example, in neo-classical economics, general equilibrium theory assumes that economic systems move towards a state of general stasis and balance. In such theories, the challenge in scientific terms is to explain change, and in management terms to avoid it! Change is treated as a result of external interference, an anomaly. Even in approaches using ideas such as 'dynamic equilibrium', the tendency towards stasis or balance rather than flux is emphasised.

What is needed, then, are ways of envisioning complexity, interaction and change as normal parts of our experience, rather than as uncomfortable inter-regnums between periods of 'normal' stability and isolation. We need to have ways of transcending singular, even insular views of human experience in order to encompass the breadth and depth of human experience in dealing adequately with the operational demands on real-world resource managers. We need, in short, new ways of seeing.

Ways of thinking

Having 'seen' the world differently – having 'seen' the things conventionally placed in the category 'externalities' as integral; having 'seen' the things conventionally rendered invisible – it becomes imperative to develop a conceptual framework which allows us to think about resource management differently. Exploration of the epistemological and ontological implications of what has been 'seen' is undertaken in Part III of the book.

The realm of social theory is often far removed from the conventional cur-riculum of many resource management programmes at universities and col-leges. The certainties and stability of 'data' are more familiar to students of resource management than the uncertainties and open-endedness of 'theory'. The often obfuscatory discourses of social theory are as alien to many resource management students as the discourses of indigenous cultures. It is argued here, however, that it is in precisely such unfamiliar terrain that we must seek the conceptual tools with which to rethink resource management, and to reorient the practice of the field towards the core values highlighted here.

The dominant debates and challenges in social theory in recent years – debates between modernism and post-modernism, between various sorts of determinists and anti-essentialists, between competing sorts of dualism; and challenges from feminism, realism, and so on – all reflect the need to carefully consider not only what is the content of social theories, but also how they are constructed. Again, the juxtaposition of indigenous experience and the domi-nant 'scientific' paradigm in resource management is instructive. Many critical ontological issues are clearest in that juxtaposition, and the challenge of bridg-ing the gaps in understanding that result from different ways of seeing things is starkest.

In more general terms, the need for clarity and critical reflexivity about ontological and epistemological positions, including one's own, is crucial to the project advocated here. In many dogmatic approaches to social theory, theory is seen as a container within which whole totalities fit, and from which 'correct' interpretation – the 'Truth' – can be divined. In contrast, the approach adopted here is eclectic and pragmatic. It envisages a theoretical framework as scaffolding for elevating us to see more of the world more clearly. Using the work of political philosopher Bertell Ollman (1976, 1990, 1993) and recent debates about non- and anti-essentialist epistemologies in geography (Graham 1988, 1990, 1991, 1992; Gibson-Graham 1996, 2000), foundations are laid for a practical accommodation of epistemological diversity in resource management systems.

Case studies

Having provided a basis for 'new ways of seeing' and 'new ways of thinking' about resource management, the book moves on to a series of practical case studies. Clearly, it is beyond the scope of a single work to deal with anything but a small selection of examples of the nature of contemporary indigenous experience of industrial resource management systems and the need to 'rethink' the whole practice of resource management. The cases examined here have been selected to illustrate key concerns and demonstrate alternative trajectories for more preferable futures.

The case studies include material from my own field-based research, and secondary studies drawing on diverse material from other scholars and activists. They include examples from several industries and from many parts of the world where indigenous and industrial resource management systems are juxtaposed. Each study puts forward an argument about indigenous experience of resource management practices – the ways in which states' rights and indigenous rights are juxtaposed; the foundations and importance of indigenous claims to resource rights; the role of transnational resource corporations and government legislation; the implications of colonial relationships, including treaties, in contemporary resource management systems; and the prospects for new ways of 'doing' resource management.

Ways of doing

The underlying reason for exploring vision and theory in this book is to contribute to rethinking and reshaping of professional practice – to reorient what resource managers actually do. The position developed here is that new ways of seeing and new ways of thinking lead to new, and in the terms defined here, better ways of doing the everyday work of resource management. The implications of this notion are explored in Part V. The weakness of many resource management systems is their failure to address the social, cultural and political complexity as competently and comprehensively as they tackle ecological and engineering complexities. This book attempts to demonstrate that they can be incorporated into 'real-world' resource management systems. It is necessary, in other words, to demonstrate practical new ways of doing resource management that reflect new ways of seeing and thinking.

This is done in a series of discussions of professional practice and methods.

The field of Social Impact Assessment (SIA) is considered in detail, and the implications of adequate consultation and participation in SIA for indigenous groups are considered in terms of a framework for negotiating outcomes of resource management decisions. This approach is also developed in other fields of 'applied' resource management such as the development of public and corporate policies, legislation and non-government organisation agendas.

From theory to praxis

Finally, the book considers the implications of such rethinking of resource management systems for professional practice. Issues of professional ethics and accountability are explored. The need for resource managers to develop thoughtful dialectical relationships between theory and practice is advocated. In particular, the sort of literacy required for resource management practitioners (whether professionally employed, engaged in community-based activism, or in other ways) who might contribute to more just, sustainable, equitable and diverse futures is considered. The need for literacy in the complex constructions of 'landscapes' as well as the technical complexities of sub-fields and specialist disciplines is demonstrated. The book concludes with a discussion of optimism. It is argued that optimism is the most important element in a resource manager's toolkit. Specifically, a critical and engaged optimism is discussed as central to praxis in which futures are not simply extrapolated from past patterns of injustice, inequity, ecological damage and cultural genocide into bleak and inhumane futures, but are built through responsible action that reflects commitment to justice, sustainability, equity and diversity.

Part II

Ways of seeing

The Second Coming

Turning and turning in the widening gyre
The falcon cannot hear the falconer;
Things fall apart; the centre cannot hold;
Mere anarchy is loosed upon the world,
The blood-dimmed tide is loosed, and everywhere
The ceremony of innocence is drowned;
The best lack all conviction, while the worst
Are full of passionate intensity.

Surely some revelation is at hand;
Surely the Second Coming is at hand.
The Second Coming! Hardly are those words out
When a vast image out of *Spiritus Mundi*
Troubles my sight: somewhere in the sands of the desert
A shape with lion body and the head of a man,
A gaze blank and pitiless as the sun,
Is moving its slow thighs, while all about it
Reel shadows of the indignant desert birds.
The darkness drops again; but now I know
That twenty centuries of stony sleep
Were vexed to nightmare by a rocking cradle,
And what rough beast, its hour come round at last,
Slouches towards Bethlehem to be born?

William Butler Yeats
(from *Michael Robartes and the Dancer*, 1921)

2 The problem of 'seeing'

Ideas about environment, population and resources are not neutral. They are political in origin and have political effects.

(Harvey 1974: 273)

Understanding the nature of environmental problems and how they might be solved requires more than a scientific appreciation of environmental processes. It demands an understanding of how societies work, and how collective action within those societies is both organised and constrained.

(Johnston 1989: 199)

Is seeing really believing?

The proverb 'seeing is believing' has the power of truism in the tyrannical world of Western common sense. The visual arts constantly remind us that 'seeing' involves perception. Seeing is never unmediated ingestion of 'objective' reality. There is no simple nor automatic relationship between 'what you see' and 'what you get'. What one sees is always mediated by how one thinks. Interpretation of what one sees depends on a wide range of environmental, individual and social factors. Visualisation is always contextual. In Fred Williams' etchings of Australian trees, for example (Figure 2.1), one faces a genuine difficulty in separating the forest and the trees. This ambiguity is neither mere illusion nor artistic manipulation of perspective to draw us to new insights. Rather it is a window on the co-existence of simultaneous realities – simultaneous meanings and competing perceptions. The alternative visions co-exist. In one well-known illustration (Figure 2.2), most observers are initially confronted by either the old hag or the young woman in a hat. Most people can visualise the alternate image when it is pointed out to them; but which one is the 'correct' image? If 'seeing is believing', which image is the 'right' one; which reality are we to believe in? In these illustrations, the images are mutually constitutive. One does not exist without the other. They cannot be disentwined. As Escher's 'Day and night' images (Figure 2.3) demonstrate so clearly, this is not just a matter of illustration, but bears some relationship to material realities. Many aspects of material reality interpenetrate and mutually constitute each other in a similar fashion, with the one being inseparable from the others.

Figure 2.1 Seeing the wood and the trees. The work of Fred Williams (1927–82) offers a view of Australian forests in which the trees and the spaces in between them are visually entwined: (left) *Forest* (1958; etching, aquatint, engraving, pencil; 20 × 14.5cm); (right) *Red Trees* (1958; etching, aquatint, engraving, drypoint; 20 × 14.6cm)

Sources: *Forest*: National Gallery of Australia, Canberra; *Red Trees*: Gift of James Mollison 1987, National Gallery of Australia, Canberra.

Figure 2.2 Ways of seeing: simultaneous realities. This perceptual illusion based on an illustration by Toulouse-Lautrec simultaneously presents a young woman and an old woman

Figure 2.3 Day and night. Escher's memorable image provides a powerful visual metaphor for a relational view of the nature–culture divide that so powerfully divides environmental and cultural politics

Source: M.C. Escher, © Cordon Art B.V.

John Berger's influential review of artistic expression in terms of 'ways of seeing' (1973) moves beyond the issues of perception of art and leads us into the realms of the cultural, social and political constructions of human experience. When we view great art, we are not simply engaging with our perception of reality, but also engaging with an artist's representation of reality – their way of seeing things. As part of a socially and culturally (and often economically) constructed audience, we become part of a dialogue or discourse which constructs the nature and meaning of the artwork for our society. We confirm, critique, alter perceptions and social opinions of the quality, value and meaning of a painting, a piece of music, a novel or a poem. Our initial response is, perhaps, shaped by existing critical opinion. We read a review of a film that puts us in a hostile or a receptive frame of mind; we are familiar with a particular writer's or artist's earlier work and expect to like or dislike the latest addition. We've been told a particular composer or performer is inaccessible and we're surprised by the unexpected emotional impact an unknown piece of their work has on us. But in these cultural matters, value judgements are well accepted; 'seeing is believing' is easily replaced with 'I know what I like'. In the world of art and literary criticism, we recognise that tastes and perceptions change. We recognise that culture shapes responses to art, and that art and culture mutually constitute each other in complex ways. And we accept that there are many ways of seeing the same object, many ways of responding to expressive, abstract or representative realities. In dealing with artistic representations, then, it is not distressing to abandon the common sense truism that 'seeing is believing'.

Are the complex material landscapes of resource management really so different to this? Is there really a single, objectively 'correct' way of seeing a resource management system? Is seeing really believing when it comes to professional resource management? For many professional resource managers, the answer to such questions continues to be an unequivocal 'Yes'. And they can point to the unambiguous indicators they use as criteria against which objectivity is measured – for example the market, scientific instruments and experimental success all provide such criteria. Yet the reduction of complex realities to such measures misses something important. Even the most complex measures require dissolution of the relationship between the observer and the object of their observation. They involve reduction of what is being observed to an object, disconnected from its ever-changing temporal, spatial and cultural context in order to avoid subjective or extraneous interference with their 'objective' examination. This approach assumes complexity away and makes it easy to mistake the re-presented simplicity as reality, and to assume (because such measures are 'objective') that seeing is believing. If they were not so tragic, the results could be hilarious (see Box 'Seeing is believing').

The practical challenge to resource managers is to 'see' the dimensions of resource management all together – to visualise the simultaneity of cultural, economic and ecological domains; to be critically aware of what various models and approaches render important, and what they render invisible. The visualisation of complexity, dynamism and simultaneity is a skill which runs counter to many approaches to systems management, where the emphasis is on relatively simple models as the key to grasping complex relationships and processes.

Seeing is believing

We've all heard them. Traveller's tales that regale us with the 'real' story. They're incredibly hard to dissuade from their opinion, their conclusions, because 'We've been there. We've seen it. We know what it's like. After all, you know, seeing is believing', they say. And then they trot out some worn anecdote that confirms a well-known 'fact'. I faced this as a young researcher returning from the field and trying to open people's eyes to the harshness of Australia's frontier towns. The tourists who had been there were riled. 'No', they said. 'We've seen it! Those Aborigines in Kununurra sitting in the street drinking. In front of their kids. Just sitting around like they're waiting for something. Sitting doing nothing and drinking. No wonder ... ' Well, I'm sure you know how it goes from then. And they always ended up saying 'Well. What are you trying to say? Seeing is believing, you know!' And I would try to point out the flaws in their vision of Kununurra. Aboriginal station workers are often brought to town in the back of a truck by the station owner or manager and told 'Oh, I don't know when I'll be going back, but I'll pick you up on the corner'. And, of course, if they're not on the corner when the truck goes past, they simply don't get home. They can't wait in the hotel, they can't wait at someone's house (there's housing shortages anyway), they can't wait at a bus stop. So the whole family waits where they've been told for who knows how long. Their drinks are kept in paper bags and in the shade. The kids play while they're waiting. And if the station owner remembers, after he leaves the hotel, or dinner at his friends' house, they pile into the truck for the dusty trip back to the station. Oh yes, the tourist glimpsing the scene from the tinted coach window has 'seen' precisely what they expected to see. Aborigines doing nothing and drinking in public places. Wasting taxpayers' money. For them, the power relationships, the poverty, the history of dispossession, repression, violence and theft in this place, the history of resistance and persistence, were all invisible. Well, of course 'seeing is believing'. But just what do we 'see', and what do we miss?

Towards 'peripheral' vision?

In broad terms, then, the task of achieving 'better' resource management outcomes requires resource managers to have

- A clear idea of what 'better' means; and
- A 'better' toolkit – a range of both conceptual and practical 'tools' that facilitate 'better' outcomes.

The contents of this metaphorical toolkit need to include more than scientific ideas and technical tools. Crucially, as suggested in the passages quoted at the beginning of this chapter, it needs to include some of the basic tools of critical social science. It is also necessary to combine these basic social science tools with a mature understanding of the intellectual, political, scientific and geographical contexts in which they might be applied. It has already been argued that part of the problem is dealing with complexity (see Figure 1.2, p. 15).

When this complexity is examined a little more carefully, it is quickly apparent that social complexity is very different to the sorts of systemic complexity familiar in systems engineering and the physical sciences (such as geology, metallurgy or chemistry). The complex dynamics of resource management systems are also quite different to the sorts of complexity addressed in some of the social sciences such as economics and accounting. In these fields 'complexity' is generally addressed by elegant models which simplify complexity by holding certain things 'constant' while others are changed in particular ways (Coveney and Highfield 1995). Even in the ecological sciences, where for example animal behaviour cannot be held 'constant' in quite the same way as the behaviour of chemicals or atoms in more conventional physical science models, the complexity involved is fundamentally different to the complexity of social systems in which little, if anything, can be held 'constant'.

In human systems, complexity is often a product of constant multidirectional and dialectical (that is, interacting) change. In modelling such complexity (as indeed with all 'scientific' models) one needs to acknowledge that the model is no more than a metaphor – an intellectual abstraction from the reality it attempts to represent. That is, even the most powerful scientific models are a way of seeing, a way of representing reality, and not the reality itself. While academic debate may value theoretical neatness and conceptual tidiness (Wallman 1977), the real world – in this case, the Realpolitik of resource management – is rarely neat, tidy or easily modelled. An important task in rethinking resource management, then, involves seeing the elements of resource management systems in a new way.

Osherenko has advocated the need for a new vision, a new way of seeing fundamentals in a different context. In discussing impact assessment in the remote Arctic, she argued:

> A number of explorers envisioned the future potential of the Arctic for resource development and as transport corridors between continents … . Their vision arose in an era of conquest and colonization in which many explorers approached the Arctic with the central paradigm of the day – that humans could dominate over the elements of nature. Some, who subscribed fully to views of European superiority and advancement over the indigenous people, perished. Others … valued and used the knowledge and experience of the Arctic residents. These explorers had what I call *peripheral vision*: they were able to view the world around them with appreciation for different lifeways and adaptations to the environment.
>
> (Osherenko 1993: 115, emphasis added)

For professional resource managers at the turn of the twenty-first century, the need for peripheral vision is urgent. In too many contexts, professional life has been dominated by short-term imperatives constructed in the marketplaces of bottom-line profitability, short-term political outcomes and project advocacy. On the peripheries of the global marketplace, however, we find a range of costs (and potential benefits) lying outside the professional's conventional frame of reference; outside the accounts delivered to annual meetings of shareholders, outside the presentations of politicians, and outside the understanding of many whose decisions, actions and omissions create them. It is these peripheries that need to be included and empowered in visions of resource futures, if the goals of 'better' resource management are to be adequately addressed.

Social science and resource geopolitics

Once the imperative for this sort of peripheral vision is recognised, it is obvious that social processes and relationships, the basic focus of the social sciences, are not of marginal relevance either to the 'scientific' management of resources or the contemporary geopolitics of resources. Rather they are integral to understanding, responding to and participating in resource and environmental management systems. Despite the rhetoric of the interests

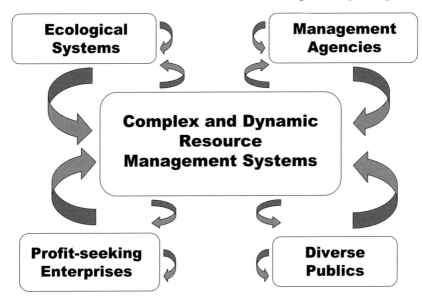

Figure 2.4   A model of resource management systems

Source: Developed from Gale and Miller 1985; Mercer 1991.

privileged and empowered by industrial resource management systems, no single set of criteria can adequately define what is rational or optimal in resource management. As Leftwich's definition of politics (1983: 11) points out, decisions about resource management always involve political relations of co-operation and conflict.

Some models of resource management suggest that a combination of free markets and scientific expertise is sufficient to guarantee rational, even optimal, utilisation of resources. Such models are underpinned by ideas of a value free 'science' (biophysical or social) with neutral methods for correct resource management. A critical approach to the social sciences readily debunks this notion. In rethinking resource management, a critical literacy in the social sciences provides strong conceptual foundations for a more holistic and socially oriented model of resource management systems. It also provides a useful foundation for empirical analysis of resource issues away from a narrow view of both 'resources' and 'management' towards a more inclusive concern with the dynamics and interactions that characterise the relationships and processes which contribute to resource geopolitics.

Resource management systems are not limited to the natural ecological systems within which natural resources exist and from which human societies extract them. Following Gale and Miller (1985), Mercer (1991) suggests that resource management systems should be thought of as also including management agencies, profit-seeking industries and a variety of publics (Figure 2.4). In other words, participants in resource management systems are not simply involved in the management of natural systems for profit, but should be seen as dealing with interaction between complex environmental, economic, political and social processes (see also Figure 1.2, page 15). It is clear that the toolkit needed by resource managers to handle the work generated by this interaction needs to extend beyond a narrowly scientific realm. All participants in these systems need to have some broad social literacy as well as scientific, economic or other technical expertise.

One of the problems we face in tackling this complexity, and responding to the difficult issues involved, is that the currently dominant paradigms of resource management in all three worlds of the old world order – the capitalist First World, the nominally socialist Second World and the 'developing' Third World – emphasise production and trade of resources above all other aspects of these complex systems. Other consequences of their operations are rendered invisible, unimportant or simply unfortunate necessities to achieve a common good defined by the systems' beneficiaries. Many important aspects of the very real complexities faced in everyday management of activities in the field have been literally structured out (or never structured into) the basic resource management models that dominate professional education. Instead, these basic models have entrenched the naturalised, common-sense notions of what resource management is about – maximising the production of raw materials for their use values and, increasingly, for their exchange values. In the process, they render invisible and unimportant for professional practice many of the things that are most significant in the relationships between resource industry activities and their host communities.

In order to improve management outcomes, to achieve better outcomes in terms of social justice, ecological sustainability, economic equity and the protection of cultural diversity, resource managers need to be highly critical of the information they rely on, information sources and the uses of information in their professional activities. It is also reasonable to expect a high level of critical self-awareness in professional approaches to resources and resource management. In other words, we should take nothing for granted. We should interrogate carefully all the data, information and opinion we receive. We should be in the habit of checking it carefully for inaccuracies, over-simplification, myopia, faulty thinking, ideological blindness and so on . The task is not the impossible one of excluding bias, but the important one of detecting it and taking it into account (Williams 1986b).

While all models necessarily simplify reality, it is both essential and reasonable to question what is simplified out, and what is left in or prioritised in any model-building exercise. In most cases, we use existing understanding of what is most significant to simplify complex realities by producing, or abstracting, simple categories that act as convenient labels to reflect entrenched priorities. In the case, for example, of 'natural resources' such as timber, fish or minerals, we simplify the complex totality of matter by emphasising the usefulness or financial value of some of its components. But we rarely question just how it is that we 'see' trees as separable from the forest ecosystems in which they are embedded. Nor do we easily question just how it is that we can separate some fish species from marine ecosystems, or some mineral types from the geological totalities around them. The point is that the dominant models of resources have become 'naturalised' or invisible as models. The categories used to simplify the complex totalities of 'forests', 'marine environments' or 'landscapes', have become invisible as categories, and instead become the things themselves. For most people, this means that the idea of questioning whether or not something is a resource has been rendered as quite simple. 'Natural resources', however, are really only notionally separable from the complex totalities of which they are part. They are modelled as a distinct category or entity for a range of socially constructed purposes – but they are not 'natural' or 'common sense' categories.

It is very easy for this social construction of resources, this abstract intellectual separation, to be mistaken for a real and categorical separation actually present in the world itself. This separation is then quite easily entrenched as an unquestioned (and unquestionable) 'common sense' that effectively defines what can be included in and what is excluded from our models of resource management. Just as 'natural resources' themselves are integral parts of biophysical and ecological systems, so are the other elements in a resource management system only notionally separable from the various politico-economic and socio-cultural systems of which they are part. In separating them from the complex totalities of which they are part, we cannot afford to suspend our critical judgement. We need to be critically aware, at every step, of how, why, and with what consequences, our processes of abstraction, of simplification and categorisation, are proceeding. And we need to be open to criticism from alternative perspectives that might throw a different light on the nature of the task or the consequences of our particular approach to it. It is here that the necessity for a social as well as a biophysical scientific literacy as an essential component in a resource manager's toolkit becomes apparent.

Whether one considers the dynamics of relatively local scale, place-based conflicts over resource management systems and decisions (for example air and water quality disputes in urban areas throughout the world; concerns about the environmental or social consequences of specific mining projects; arguments about the balance between jobs and environment over many sorts of resource-based development projects; the direct environmental consequences of local oil spills and so on), or issues constituted at wider geographical scales (for example the key issues of global climate change – the greenhouse effect and ozone depletion; management of fisheries in international waters; terms of trade in international commodity markets; cross-border pollution from industrial sources, etc), the intertwining of the social and the natural is inescapable and can be ignored only at the risk of substantial 'mis-management' of resources. Dealing with management of a fishery such as the South Pacific tuna fishery only in terms of ecological imperatives would render invisible the complex processes of international relations generated in the negotiation of the United Nations Law of the Sea Treaties (Parry 1994;

Rogers 1995), and the subsequent impacts of bilateral and multilateral fisheries agreements within contested territorial waters on the traditional resource management systems of local cultural groups. Similarly, dealing with forest management issues in Southeast Asia only in terms of the need to produce either firewood or building materials has led aid programmes to introduce plantations of fast-growing eucalypts into areas where local communities have traditionally relied on forests for these things plus a variety of other resources, including forage, animal habitats, medicinal materials, refuge, and spiritual and cultural observances (Hirsch 1993; Chandrakanth and Romm 1991; Shiva 1992). Likewise, the reduction of the complex geographies of the real world to the abstract 'level playing field' of the General Agreement on Trade and Tariffs and the World Trade Organisation does substantial damage to the human and non-human activities that occupy and rely upon the real-world landscapes smoothed out in the process of levelling the field.

The image of the level playing field is one that dominates conventional professional education in this field. It is an image that denies geography and the geopolitical domains that affect resource management so profoundly. In the wake of the Cold War, the limited notion of geopolitics as an issue of great power diplomacy over territorial issues has quickly expanded to acknowledge the 'place-based politics of identity and the new cultural politics of difference and diversity' (Howitt 1996: 4). In post-Cold War resource landscapes, industrial production systems have pushed the planetary system as a whole, and some local environmental systems close to or beyond the limits of survival. In these landscapes, the fundamental elements of geopolitical analysis – territory, identity and power – are relevant to a wider range of issues than simply international relations. Within and between localities, within and between communities, indeed, within and between all geographical scales, these fundamental elements shape the everyday dynamics of resource management systems. At one level, one can see in global models such as Ekins' global problematic (1992: 4–13) an abstract simplification of the interaction between planetary scale processes and social and economic processes operating at much more local scales, including the nation state. Such abstractions are most powerful, however, when they are not simplistically global. It is all too easy to see the global arena as simply dominant. This is certainly a common failing of much of the currently popular globalisation literature. Local, sometimes very local, cases of resistance or responses to 'global' crises have much wider repercussions, and themselves shape and change wider scale relations. Even cursory consideration of many new social movements that have shaped the agenda of the United Nations, the World Bank and many transnational resource companies in recent years provide a glimpse of this new domain of geopolitics. Once this new geopolitics is acknowledged, it is obvious that both education and analysis must proceed in a multiscale rather than simply global (or local) way. We need to 'see' different scales of analysis and operation simultaneously rather than in a fragmented way.

The post-Cold War transformation of global geopolitics brought about by (and reflected in) the collapse of the former Soviet empire, has shaken many of the economic, political and epistemological foundations of the system of world order that was negotiated at the end of the Second World War (Ward 1992; Taylor 1993; Chomsky 1994). It has also given rise to new geopolitics that are constructed at narrower scales and often focused on the powerful combination of cultural identity, territoriality and repression, and their interaction with issues of resources and economic independence (see also Jonas 1994; Jhappan 1992; Kelly 1997).

This transformation came in a period of accelerating global integration in some spheres. Deterritorialised transnational corporations and institutional structures focused on the few privileged nation states that increasingly dominated international trade: a few global media and information technology companies dominate international information systems. Similarly, increasing integration of Europe, development of the North American Free Trade Agreement, the Asia-Pacific Economic Forum and the persistence of the North Atlantic Treaty Organisation beyond the end of the Cold War, provide opportunities for greater political integration through trade and cultural processes. Even in the NGO sector, globalisation is occurring in many fields including human rights, environmental protection, indigenous rights and cultural action. These globalisation processes are not, however, monolithic, homogenous or uncontested (see for example Barnet and Müller 1974; Dicken 1998; Bryan 1987; Fagan and Bryan 1991; Fagan and Webber 1994). In many areas of the world and domains of social affairs, however, globalisation is resisted by national level policies, or undermined by disintegration, regionalism and fragmentation in others.

The contradictory tendencies to be found in the complex processes of globalisation, national development and localisation affect the day-to-day operations of resource management systems. It may be tempting to abandon the notion of complexity in favour of one or other of the all-encompassing versions of 'truth' marketed by competing ideologues in the bazaar of ideas. We have already acknowledged, however, that the world around us actually is complex, and complexity can be abandoned only at great cost to our real-world effectiveness as resource managers.

Conventional analytical approaches to resource industries can capture some complexities. Political or economic approaches, for example, can direct attention to the roles of corporate players, particularly the global resource corporations, nation states and interstate institutions, both as producers and consumers of resources, as well as the roles of trade unions in resource industries. All of these undoubtedly play influential political and economic roles in resource management systems. Yet conventional approaches also tend to shift attention away from those groups and issues that are marginalised from the core institutional and geographical framework of industrial production – including indigenous peoples. In dealing with what Michael Peter Smith (1994) referred to as the 'globalization of grass roots politics', it emerges that interconnections between places and scales – a core concern of geography – urgently need attention. We need to understand better just how resource management systems bring people and places in all parts of the globe into new relationships with each other.

In advocating this approach, let me argue two key points. First, we need to recognise that the interaction of environmental, social, cultural, economic and political factors across geographical scales already has considerable practical significance. Second, we need to reorientate the way we view the relationship between the *context* and *focus* of resource management. In visualising a more complexly interacting set of biophysical, politico-economic and socio-cultural systems, we open opportunities for exploring new ways of responding to the circumstances in which resource managers find themselves. By moving beyond the narrow conventional focus of resource analyses, things conventionally dismissed to the peripheral role of 'context' take on new importance and meaning. We might even 'see' the whole domain of resource management differently, with a 'peripheral vision' that foregrounds things so often treated as outside the system of resource management and outside the scope of resource management professionals. In the process, we find that not only are the skills of the social scientist of direct relevance even to technicians in resource management, but that the knowledge of those who are marginalised by the normal operation of the systems might also help us intervene to produce 'better' outcomes.

Geography matters in resource management

The contested landscapes of resource management sometimes seem to be occupied by a very strange array of visionaries, vandals and technical wizards. They certainly contain a complex array of vested interests, conflicting agendas that are often hidden or camouflaged, contradictory intentions and priorities, and disempowered, marginalised and oppressed 'victims of development' (Seabrook 1993) and 'victims of progress (Bodley 1990), both human and non-human. Resource management systems simultaneously produce both commodities and power, and have been linked to the construction of political, economic and social power, wealth and privilege throughout human history. Current resource landscapes, however, do not just reflect (and experience constraint from) contemporary priorities and imperatives. They also reflect (and are constrained by) the consequences of actions and omissions in many previous periods of development, investment and struggle. They are part of complex geographical and historical processes. Actually, in resource geopolitics, geography, both as a reality and as an intellectual discipline, actually matters!

Geographies of resources are clearly undergoing rapid transformation. Transformation of geography is envisaged in at least three senses:

1 Some degree of regional restructuring, a change in internal and external boundaries between nations and peoples;
2 Some degree of a change in power relations (economic, political, military), with much attention given to the changing balance between the 'West'

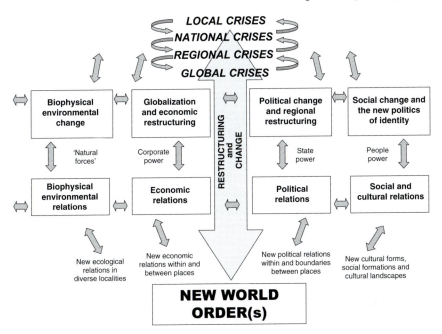

Figure 2.5 New world order, power and local restructuring

and the former 'Communist' bloc, but with recognition that these changes will also be reflected in a variety of ways at a variety of scales;

3 Some degree of new relationships between economic and ecological processes, although this is one of the key points of difference between 'top-down' and 'bottom-up' visions of the new world order.

From a geographer's viewpoint, several important geographical elements can be identified in conceptions and visions of a new world order – spatial dimensions (regional restructuring), social dimensions (power relations), and ecological dimensions (relations between society and nature) (Figure 2.5). Each of these elements has been an important theme in the discourses of social theory in recent years.

In her early 1980s critique of positivist quantitative geography's reduction of complex geography to a notion of 'space', itself conceived as a single measurable dimension of 'distance', Massey (1984a: 4) suggested that separating the social and the spatial dimensions of human experience was a misconception. Geographical models which proposed 'spatial' causes for 'spatial' processes missed something of great importance: '"The spatial" is not just an outcome; it is also part of the explanation', Massey argued. This is an insight with consequences not only for geographers, who need to become more widely literate in 'the social', but also for other social scientists, who need to become more literate in 'the spatial'. It is also significant for resource managers, whose concern is bound up in complex geographies of resource management systems. It is interesting to reflect on the extent to which space has become an important concern of social theorists within and outside geography since she wrote the following:

> While geographers struggled to learn other disciplines and apply their knowledge to the understandings of spatial distributions, the other disciplines continued to function, by and large, as if society existed on the head of a pin, in a spaceless, geographically undifferentiated world.
>
> (Massey 1984a: 4)

The spatial domain deeply penetrates and co-constructs the social domain, and vice versa. Similarly, the domain of nature, often treated as external to society also needs to be seen as interpenetrating and co-constructing the spatial and the social dimensions of experience. This leads Massey to argue that we need to reconceptualise geography not just as space, but as a genuinely complex phenomenon which genuinely matters in social life (see also Leftwich 1983: 12–13). In developing her notion of why this complex geography matters, Massey (1984a: 5) highlighted a range of attributes of geography which influence wider social relations:

- distance;
- differences in the measurement, connotations and appreciation of distance;
- movement;
- geographical differentiation;
- notions of place and the differences between places;
- symbolism and meaning which different societies, and different parts of the same society, attach to all these things.

> The uniqueness of a place, or a locality, … is constructed out of particular interaction and mutual articulations of social relations, social processes, experiences and understandings, in a situation of co-presence, but where a large proportion of those relations, experiences and understandings are actually constructed on a far larger scale than we happen to define for that moment as the place itself …
>
> (Massey 1993b: 66)

Berdoulay (1989) takes this further, challenging what he calls 'our customary epistemological approach' landscapes:

> The landscape is coded by society. Usually several codes coexist, as they are linked to different spheres of life, be they social, political, cultural, or economic … . Meaning can then be read in the landscape … . In such an approach landscape is viewed as an autonomous level of creation of meaning. While ultimately social processes are responsible for its production,

nevertheless, meaning in the landscape (and thus in place) comes from its own organization. In this perspective, we must take into account physical and organic processes as well. What counts, in fact, is the spatial level of interaction and concatenation [interconnection or linking] of all these various processes, which produce the landscape.

[This] view of place ... calls for important changes in our customary epistemological approach to landscapes and regions ... [and] opens the way for disturbing our persistent conceptual categories in order to fully consider meaning as a geographic process.

(Berdoulay, 1989: 131, 136)

If we take Chomsky's basic point about the continuities in the world order symbolically established by Colombus' accidental invasion of the so-called New World (1993: 3), our customary epistemological approach has buried the concerns of the conquered. The enormous cultural diversity of the worlds conquered by Europe (and its post-colonial offspring) has been reduced to a singular exotic and homogenised 'Other'. Yet 'a few of the conquered have somehow survived'[1] and their resistance to and contestation of the 'top-down' imposition of the geography of the new world order continues to shape outcomes at all scales. In order for us to understand the implications of this for our new way of seeing resource management systems, we need to deconstruct another fundamental, and often taken-for-granted category – the nation state. We also need to reconsider its role in resource management systems.

Anderson (1992) suggests we need to discard four important misconceptions about nationalism and nation states to engage critically with notions of a new world order. He suggests that, from an historian's perspective, it is more appropriate to think of the processes producing integrated states as aberrant. The violence out of which modern nation states were forged is a reminder of the importance of frontier violence in the dispossession of many previously sovereign and autonomous peoples around the world. He also argues that there is a frequent assumption 'that in some way "small" countries with limited resources in raw materials and labour are somehow not real countries in the face of the industrial giants and the exigencies of the world capitalist economy' (ibid. 41). that 'transnational corporations have somehow made nationalism obsolete' (ibid. 42).

As Dicken's Global Shift (1998) demonstrates, transnational corporations have changed the nature of economic and geographic relationships at an international scale, but they still perform on stages embedded in national jurisdictions. Anderson also notes that the cultures and practices of TNCs, even where they have internationalised all three spheres of production, exchange and consumption (see also Bryan 1987; Fagan and Bryan 1991; Fagan and Webber 1994) continue to reflect the power and continuity of nationalist ideologies. Anderson further suggests that the final point in this argument is that there is 'some inscrutable connection between capitalism and "peace" such that the "free market" is instinctively juxtaposed not merely to the command economy, but to war' (1992: 42).

> Despite the end of the Cold War, dangerous convergences already born in the last century show every sign of continuing to develop: market led proliferation of weapons systems; mythologisation of militaries as ... symbols and guarantors of national sovereignty and ethnicisation of officer corps.
>
> (ibid.: 46; see also Anderson 1983)

Like Anderson, Chaliand reminds us that the nation state is a very recent construct, dating not from the mists of antiquity as many nationalists would have us believe, but from the late eighteenth century. Chaliand also reminds us that the struggles of minorities and indigenous peoples for recognition and rights is a central element of the overlapping and interacting local, regional and global crises:

> Minorities fight for ever smaller and smaller sized nation-states of their own to protect their human rights from ravagers, as they see it, of the wider nation-state or states in which they exist. The Kurds, the Protestant Irish, Tamils and Eritreans, to mention just a few examples, illustrate this simple point; everywhere minority peoples are fighting with their lives against great military odds.
>
> (1989: 1)

In the light of the terrible human toll accompanying the regional restructuring involved in the disintegration of the former Yugoslavia, the reconstruction of Cambodia and Palestine, the civil wars in, for example, Lebanon, Eritrea, Somalia, Mozambique, Rwanda, Sri Lanka, Georgia, Azerbijan, Aceh, Ambon and Timor, the need to challenge nationalism is urgent. The extent to which many of these tragic circumstances reflect very specific local, regional and international disputes about resources, reinforces the relevance of the work of 'political imagination' in constructing alternative ways of seeing the place of resources in contemporary geopolitical relations.

Difference, diversity and struggles for justice: the case of indigenous knowledge

> We come in peace, they said, to dig and sow.
> We come to work the land in common and to make the wasteland grow
> This earth divided, we will make whole
> So it can be a common treasury for all.
>
> Leon Rosselson[2]

Given the centrality of resources in the construction of power and politics, there can be no doubt that resource management systems are deeply implicated in diverse struggles for justice. Fraser (1995, 1997a) suggests that politics in the late twentieth century could be characterised as a dilemma between two

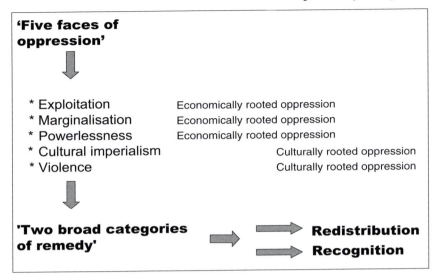

Figure 2.6   Redistribution/recognition: a bifocal approach to justice and oppression

Source: Based on Fraser 1997a: 198–9.

sorts of struggle for justice. She examines two interpenetrating political imaginaries, one rooted in materialist socialism, the other in a new cultural politics she labels 'post-socialist'. In these two political imaginaries, 'justice' has quite different orientations. In materialist socialist positions, Fraser suggests, struggles against material inequality and liberation from economic exploitation are prioritised. She summarises the strategic focus of these struggles as 'redistribution'. In 'post-socialist' cultural politics, struggles for recognition and liberation from cultural domination, struggles around racial or gender inequality and the politics of difference have been given greater prominence. Fraser summarises the strategic focus of these struggles as 'recognition'. While noting that such distinctions are an analytical convenience rather than categorical reality, Fraser identifies the 'redistribution-recognition dilemma' as one of the 'central political dilemmas of our age' (1997a: 13). She also identifies 'broad approaches to remedying injustice that cut across the redistribution-recognition divide' (ibid.: 23), which she labels 'affirmation' and 'transformation'. Fraser defines 'affirmative remedies' as those that are 'aimed at correcting inequitable outcomes of social arrangements without disturbing the underlying framework that generates them', and contrasts them with 'transformative remedies' that are 'aimed at correcting inequitable outcomes precisely by restructuring the underlying generative framework' (ibid.: 23).

Fraser argues that to shift from a single focus redistributive politics to a 'bifocal' concern with culture *and* political economy represents 'an important step forward in political theory' (1997a: 190). She suggests that Young's influential characterisation of 'five faces of oppression' (I. M.Young 1990; see also Harvey 1992) (Figure 2.6) does not escape an implicit endorsement of this recognition-redistribution binary. In response, Young asserts the need to 'pluralize categories and understand them as differently related to particular social groups and issues' and accuses Fraser of 'adopting a polarizing strategy' (1997: 149). Likewise, feminist theorist Judith Butler treats Fraser's discussion as reducing identity politics to a 'merely cultural' domain, which she equates with a neo-conservatism that is unresponsive to the unsettling of conventional political, economic and cultural readings of society implicit in queer theory (Butler 1998: 44). While debate over Fraser's original paper has been heated, she suggests readers such as Young and Butler misrepresented her bifocal framework as a binarising approach. Rather than seeking to prioritise either the economic or the cultural, Fraser seeks to provide a matrix that provides for a way of reformulating the redistribution-recognition dilemma (1997a: 27; also 1997b, 1998).

This debate, in which Fraser emphasises the need to 'conceptualize two equally primary, serious, and real kinds of harm that any morally defensible social order must eradicate' (1998: 141), reflects a wide emergent concern in social theory with identity politics. Within post-modern discourses, there has been much emphasis on what West labelled the 'new cultural politics of difference' (West 1990). Bhaba, for example, mounts a strong defence of 'difference' against a radically depoliticised notion of 'diversity' (1994: 31–9). Can the diversity of indigenous experience (and identities) be adequately encompassed in such debates? Using the example of indigenous peoples efforts to assert the contemporary relevance, value and integrity of traditional ecological knowledge, I want to show that these binary distinctions, however elegant, neglect key elements of real-world geopolitics of resources. Rather than bivalent or bipolar models, the core values of justice, equity, sustainability and diversity are employed in the discussion below to envision a more ambiguous, polymorphous, complex and demanding political space in which resource geopolitics are played out.

The absence of any conception of environmental justice in Fraser's discussion of post-socialist dilemmas of justice is a significant shortcoming. Its absence from the critiques of Young and Butler is also disturbing. Despite their assertions of differences between their positions, there is much common ground, and no fundamental disagreement on the need to integrate the socio-cultural and politico-economic domains. What remains absent, however, is the realm of environmental dimensions of social justice (and the implication of multiscale, and intergenerational dimensions of justice). Fraser correctly advocates the need to overcome the false antitheses between the binaries implicit in the tension between the socialist and post-socialist political imaginaries, but limits her attention to this bilateral split and identifies the following as the 'crucial "post-socialist" tasks':

> First, interrogating the distinction between culture and economy; second, understanding how both work together to produce injustices; and third, figuring out how, as a prerequisite for remedying injustices, claims

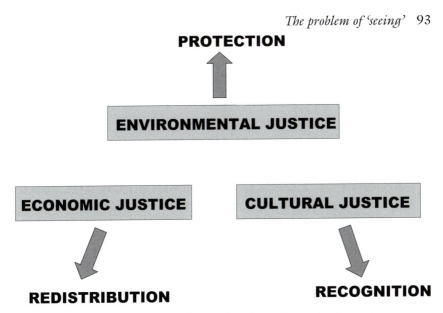

Figure 2.7   Visualising the contradictory directions of struggles for justice

> for recognition can be integrated with claims for redistribution in a com-
> prehensive political project.
>
> (Fraser 1997a: 3)

Harvey (1992: 600) adds 'ecological unsustainability' as a sixth face of oppression in his discussion of Young's work. His more recent emphasis on 'difference' and continuing preference for a rather economistic reading of notions of
environmental justice (1996) suggests this is an indicative rather than substan-
tive critique of the bivalence implicit in Young's approach. The value of Fraser's typology is that it indicates the extent to which single-minded strategies
for recognition and redistribution may be in conflict with each other. In the
realm of environmental politics, single-minded strategies of preservation or
protection create similar contradictions, tensions and conflicts (Figure 2.7).

For Aboriginal groups during the 1990s, the need for strategies targeting
recognition (particularly recognition of land and other indigenous rights) and
redistribution (delivering economic justice to indigenous Australians) has
rarely been understandable in isolation from each other or from Aboriginal
groups' own assertion of the need for strategies that target what they call 'caring for country' (Young *et al.* 1991). The approach to questions of justice and
equity that comes from this experience suggests a need to visualise a political
landscape in which diversity rather than difference is prioritised. Similarly, it
pushes us to encompass geographical as well as social diversity in our political
vision. We must grapple with outcomes that do not predicate 'justice' in one
place on entrenching injustice to another. We must grapple with cross-scalar
issues, so that just outcomes at one end of a scale (national employment or

Figure 2.8   An integrated vision of struggles for justice based on indigenous experience
in Australia

revenue benefits from a resource project) are not predicated on the creation
of unjust outcomes at another (e.g. local environmental health, cultural
marginalisation or other negative effects). In contrast to representations in
which strategies targeting different aspects of dimensions of justice are visualised as contradictory, many of the Aboriginal groups I work with target integrated visions of justice in which it is quite concrete environmental, economic
and social relations that need to be engaged with (Figure 2.8).

For Aboriginal groups, the social, environmental and economic relations
involved are not abstract theoretical concepts. For them, they are very concrete notions. They include:

- The specific people-to-people relationships that constitute their particular
  culture, law and tradition and are closely interwoven with individual and
  collective identities;
- The specific people-to-country relationships that constitute the rights and
  responsibilities that provide foundations for social and economic activity;
- The relationships between people (individuals and groups) and specific
  elements of the landscape (wildlife, sites, biophysical forces and processes)
  and their mythic representations;
- The contemporary relationships with non-indigenous interests and institutions, including governments, industries and individuals.

In this context, strategies to secure recognition, to secure improved material
conditions and to secure sustainable environmental relations must be conceived

(and pursued) as integrally related to each other. It is also possible that within the overall struggle for recognition of indigenous rights (including not only cultural rights but also economic and environmental rights), different historical circumstances will require different priorities to be set or different issues to be targeted for different groups. In this sense, the notion of a master narrative of social change – a blueprint designed as a 'top-down' guide for specific actions and interventions in resource management systems – must be rejected. So, in indigenous politics, the situation that Fraser sees as characterising the 'post-socialist' condition, and requiring in her terms a 'critical approach [that] must be "bi-valent", ... integrating the social and the cultural, the economic and the discursive' (1997a: 5), is an everyday reality:

> An absence of any credible overarching emancipatory project despite the proliferation of fronts of struggle; a general decoupling of the cultural politics of recognition from the social politics of redistribution; and a decentering of claims for equality in the face of aggressive marketization and sharply rising material inequality.
>
> (Fraser 1997a: 3)

This rejection of a singular political project not only leads directly to the complex material conditions of indigenous Australians' struggles for justice, but also to discursive emphasis on 'difference' and 'the Other'. In other words, the material circumstances make it imperative for us to deal not only with historically specific social, economic and environmental relations, but also with the discursive construction of social reality in critical social theory. Bhaba suggests the need to conceptualise an 'articulation of forms of difference' and recognition of multiple 'modes of differentiation' (1994: 67). In the discourse of colonialism, he argues:

> [such] modes of differentiation, realized as multiple, cross-cutting determinations, polymorphous and perverse, always demanding a specific and strategic calculation of their effects ... [create] a form of discourse crucial to the binding of a range of differences and discriminations that inform the discursive and political practices of racial and cultural hierarchization.
>
> (Bhaba 1994: 67)

In other words, colonial discourses conflate differences of many sorts (class, race, gender, language and so on) in order to differentiate it from an imagined and privileged colonising subject. In reversing the conventional idea of the 'Other' in anthropology, Kaliss (1997) asks just what sort of 'Other' it was that arrived in Europe's 'New World' in Columbus' ships in 1492. How is it that the diverse indigenous cultures of North America (and indeed the entire constellation of European colonialism) can be encompassed as the singular 'Other' of a discursively unified 'Europe'?

Despite the appropriation and discursive construction of ideas of a binarised difference by colonial structures of material and discursive power, Bhaba rejects the term diversity in favour of difference as a means of avoiding the pit-falls of relativism. He sees cultural diversity as a term torn between its use as a liberal descriptive term for 'pre-given cultural contents and customs' and a 'radical rhetoric of the separation of totalized cultures that live ... safe in the Utopianism of a mythic memory of a unique collective identity' (1994: 34). In trying to understand indigenous peoples' experience, neither of these dis-cursive forms is adequate. But neither is the bipolar simplicity of a self–other difference. The complex and dynamic processes of articulation of multiple modes of differentiation – the simultaneous differentiation along axes of gender, culture, language, age, history, sexuality, skin colour, class, economic circumstances, political orientation and so on – means that dealing with the real geopolitics of resources requires the material and discursive construction of economic, social and environmental relations to be engaged with rather than assumed. In contrast to Bhaba's rejection of diversity as an inadequate term, I would suggest that neither diversity nor difference can be understood in terms of pre-given content. Rather, both need to be addressed in terms of the political, material and discursive possibilities they open up. In terms of resource managers, this orients us towards exploring the historical and geo-graphical context of real-world social, economic and environmental relations within resource management systems as the basis for achieving 'better' resource management outcomes. In practical terms, the simultaneous opera-tion of overlapping modes of differentiation in any particular resource man-agement system embeds resource managers in complex contexts that are better dealt with in terms of diversity rather than difference.

Seeing power in resource management systems

> October 11, 1992 brings to an end the 500th year of the Old World Or-der, sometimes called the Colombian era of world history ... . The major theme of this Old World Order was a confrontation between the con-querors and the conquered on a global scale. It has taken various forms, and been given different names: imperialism, neocolonialism, the North–South conflict, core versus periphery ... . Or more simply, Europe's con-quest of the world.
>
> (Chomsky 1993: 3)

The exhaustion of key natural resource supplies in industrial societies, and the loss of access to others through political upheaval, local revolts and market forces, has created a renewed interest in the reserves and badlands left to indigenous peoples (Pollin 1981). Gedicks (1993: 5) argues that this consti-tutes a 'new resources war' comparable to the frontier land wars of the nine-teenth century. The juxtaposition of indigenous and industrial interests in resource management systems also juxtaposes top-down and bottom-up approaches to resource management. It reveals an overlapping and interpenetration of important themes from the theoretical discourses of geography and social science and the wider political discourse about interaction and change in the contemporary world – an interaction between space (in the form, for example, of territory), society and resources.

Shiva (1992) suggests that the dominant solutions to resource management problems imposed from the top-down increase regulation of the Third World's (and indigenous peoples') resources by the global system's dominant powers. They are, Shiva argues (1992: 35):

> feeling the threat of erosion of this control unless they become even more controlling and even more militarised in guaranteeing security to themselves – at the cost of other people's security.

Drawing on the implied metaphor of the earth as a common home for all humanity, and using the example of the International Tropical Timber Forestry Action Plan, she goes on to say:

> After all, planet Earth does demand that we live as world citizens and we do need political formations that will allow this. But negotiations and discussion of the new environmental order and international control is unfortunately biased in two ways. The first bias is that they are choosing resources that lie in the Third World [and indigenous peoples' homelands] and are controlled by local communities and taking them into international control ... . The second aspect is that the real issues of regulation needed at the global level, which are issues of regulating global enterprise ... is really needed internationally and it is not what is being talked about.
>
> (1992:35)

Shiva suggests documents such as the Tropical Timber Forestry Action Plan (1985) need to be turned on their heads in order to arrive at solutions that go to the heart of the overlapping world-scale crises that produce the argument for a new world order. Because they invert the conceptualisation of the problem, documents such as the Forestry Action Plan have 'become the problem rather than the solution; the real problems are not identified' (Shiva 1992: 35). These global plans are generally put into action by international agencies to control Third World resources and:

> identify local people as the biggest threat to the resource, even though commonsense would say that if the people and the resource have been there together for millennia, the relationship is one of balance, not one of destruction ... . (And) since local people are treated as the biggest threat, global agencies are treated as the biggest solution. And then they come up

with solutions that actually undermine the capacity of local communities to conserve.

(Shiva 1992: 35)

These imposed, top-down global solutions, while ostensibly protecting biodiversity, directly undermine the conditions for social, economic and cultural diversity (Gray 1991). They directly threaten the viability of local and regional economies that operate consistently with the principles of ecological sustainability. In these models, value is produced from the death of living systems (conversion of forests into timber; conversion of earth into metals) rather than on nurturing them. And even the superficial benefit of preserving biodiversity is turned to the service of those who are already enriched and empowered within industrial production systems through the emerging industry of bioprospecting and genetic engineering (Parry 1996).

Top-down solutions from global agencies such as the World Bank, the United Nations and others are not, however, uncontested. Affected communities, who have their own visions for alternative futures unimagined, and often unimaginable, by the global technocrats, inevitably respond to, conflict with, accommodate and circumvent centralised top-down scenarios. Their actions and responses – sometimes chaotic, sometimes co-ordinated, sometimes effective, sometimes defeated; sometimes naïve, sometimes sophisticated – create another sort of pressure for a new world order: a bottom-up plethora of alternative futures. This is what Mercer refers to as the 'diverse publics' (see also Figure 2.6, page 91).

In both sorts of new world order – the imposed centralisation and the plethora of more local scale alternatives – geography, in at least two senses, matters. On the one hand, geography in the form of the unique and varied characteristics of different places, the specificities of social and environmental relations and processes and the interactions between them, constitutes an important part of the setting in which top-down and bottom-up processes of social change are played out. These geographies clearly matter in resource management. On the other hand, the relationships between places also matter. Geography, in the sense of distance, interaction and differentiation, also matters. It is in this arena that global agencies, international relations and the 'free market' are constructed.

Shiva provides us with an important element of a new way of seeing the problems and solutions; a new way of thinking globally and also acting at wider-than-local scales:

Each action that some community takes is global, because it has a global impact. Everything is ecologically linked. It has global economic impact because the destructive forces against which they fight, whether Sarawak tribes or the people fighting the Namarda dam in India, are fighting global interests who have a certain vested interest in destruction. Therefore local communities in action are actually rolling back that global interest and putting it within the ecological, economic and ethical constraints within which it should function.

(Shiva 1992: 39)

The nature of the power that enables nation states, global corporations and international agencies to impose top-down solutions on diverse local communities needs to be understood more clearly in our analysis of resource management. To contribute to the targeted deconstruction and reconstruction, it is necessary to have some way of analysing power relations. Power has been at the centre of debates in social theory for decades.

The displacement and destruction of traditional systems of resource management has historically been an important part of the geographical expansion of industrial production systems. It is also clear that this process continues as an important element in the current world order. Primitive accumulation, dispossession and alienation, plunder of natural wealth, particularly from marginalised and minority groups, the imposition of military force and political domination to guarantee access to resources; all these things characterise the resource management systems with which we are dealing.

Power, of course, is one of the central themes of the social sciences. In debates about social theory over the last twenty years, power has often been a central issue. The work of Foucault, the French social theorist, provides an influential perspective that has challenged many conventional views of social power; he particularly emphasises the all-pervasive nature of power. For Foucault, every social location was a site in which power was at work. In his later work, he began to explore the ways in which location and space – in our terms, geography – might shape the ways in which power is constructed, exercised and resisted. The ubiquity of resistance to power was also important to Foucault's vision of power.

Through the work of Foucault and others there has been an explosion of interest among social philosophers and social theorists in the impact of space on social relations. This has led to an increased dialogue between human geographers involved in theoretical work, and wider social theoretical debates (see for example Harvey 1989, 1993; Soja 1989; Pudup 1988; LeFebvre 1991; Massey 1984a,b, 1993a,b, 1994a; Said 1978; Foucault 1980; Cosgrove 1978, 1992; Graham 1990). For resource management, this active dialogue between human geography and social theory provides a useful perspective on the relationships between the limited focus of professional education and the wider operational context of professional practice.

Michel Foucault (1980a,b; also Fraser 1989) has been perhaps the most influential of recent writers on power. For many readers, Foucault's analysis is dense and difficult to apply to everyday circumstances such as resource management. In contrast Galtung, a Norwegian peace researcher, provides a 'mini-theory of power'(outlined in Galtung 1973: 33–44; see also 1980: 61–72) which is a useful and easily accessible way of seeing power in resource management systems. Galtung's elegant little model of power enables us to look at the relative power and powerlessness of various groups involved in resource management systems. It also challenges dangerously flawed common sense notions of power and conflict.

Galtung distinguishes two essentially different concepts of power – power over others (the common-sense notion of power); and power over oneself (autonomy) (see Box). In the common-sense notion of power-over-others, the more power X has over Y, the less power Y has over X. In this sense, a balance of power can be understood as either a book-keeping or a mechanical balance. In either case, the distribution of power involves the empowerment of some at the expense of others. In contrast, autonomy, the 'ability to set goals that are one's own ... and pursue them' (Galtung 1973: 33) does not require disempowerment of anybody. Of course, those already enriched and empowered by the existing system are able to exercise both kinds of power, and those disempowered and impoverished by the existing system are able to exercise neither, but the distinction allows us to see that questions of power are not just about winners and losers.

Galtung's mini-theory identifies two basic sources of power: what one is or has, and where one is within a structure – resources power and structural power – and three channels through which power is exercised: ideological, remunerative and punitive:

> Ideological power is the power of ideas. Remunerative power is the power of having goods to offer, a 'quid' in return for a 'quo'. Punitive power is the power of having 'bads' to offer; also called force, violence. In the first case, one is powerful because the power sender's ideas penetrate and shape the will of the power-recipient. In the second case, one is powerful because one has a carrot to offer in return for a service; salary for work, beads for a signature on a scrap of paper giving away a country or two, tractors for oil. In the third case, one is powerful because one has a big stick ready if the object does not comply so that one can destroy him or his (sic) property.
>
> (Galtung, 1973: 33–4)

For professional resource managers, it is perhaps in the area of sovereign control of natural resources, often labelled 'national' resources because of their importance in supporting the nation state, that the implication of issues of power and the need for professional literacy and professional ethics are most starkly apparent. These issues cannot be conveniently pushed aside as too political or outside the ambit of professional education: they actually constitute the very systems in which professionals practice. In the model of resource management systems discussed in this chapter (Figure 2.6, page 91), the political, economic and cultural processes that are central to nationalist ideologies, nation states, international agencies and the global setting of the new world order are also implicated in the construction of resource management systems.

Nowhere is this clearer than in questions about the ways in which governments assert a right to grant interests in, and acquire benefits from, natural resources in the traditional estates of indigenous peoples (Connell and Howitt 1991b; Howitt 1996). In the Australian case (see Chapter 7), for example, state and territory governments claim that colonial acquisition of sovereignty produces a contemporary right to grant mining rights and to levy royalties on mineral production from lands in which indigenous people claim prior sovereignty. The recognition of native title as part of Australian common law in 1992 (Bartlett 1993a) left unresolved the question of what residual rights in resources Aboriginal peoples might derive from prior sovereignty (Reynolds 1996). Australian land management systems have historically developed to support and legitimate resource-based capital accumulation and the expansion of settlement. Bartlett (1993b: 118) suggests that resource interests such as mining companies grew used to a system which made industrial interests paramount and rendered Aboriginal interests invisible. Since the confirmation of persisting Aboriginal native title rights in the Wik decision in late 1995, the Australian Commonwealth, State and Territory governments have confirmed their willingness to prioritise mining and pastoral interests over indigenous rights. Their difficulties in legislating to extinguish the residual rights and interests of Aboriginal landowners have slowly seen a shift towards negotiated settlements (see Chapter 8).

For indigenous peoples whose estates and resources were alienated to provide the foundations for massive private and state wealth, the legitimacy of pastoral and mining leases and other business interests in land and resources that are granted without negotiations with indigenous peoples is tenuous at best. In many cases, these operations represent an unwelcome occupying presence on indigenous territories. In the 'new resources war', reassertion of resource claims, land claims, sea claims and the right to self-determination focus political and legal attention on the basis for nation states to impose calamitous conditions on indigenous communities and populations. It has often been acknowledged that the level of civilisation of a nation can be judged from its treatment of minority groups. In challenging the legitimacy of state claims for the power to create interests in publicly owned resources, the debate over indigenous rights has wide-reaching implications for all resource managers, and indeed for the constitution of national identities and nations.

Part III

Ways of thinking

3 Complexity in resource management systems

Conceptualising abstractions and internal relations

The conceptual problem and the realm of theory

It is common that modern resource management professionals focus on just a small fragment of the processes that accompany production of a particular commodity. The dominant scientific-technocentric paradigm simplifies complex realities in specific ways: it fragments, subdivides, specifies, objectifies and atomises. It conceives the task of managing resources as technical – technical experts are required to make judgements in order for 'good management' to happen. Key participants in industrial resource management systems rarely have a sense of the whole production process, let alone how production is embedded in wider social processes or the implications of various aspects of social, political, ecological and cultural context. For many professionals the observation that resource management systems simultaneously produce both commodities and power carries little significance. The fragmentary nature of their work renders the nature and exercise of power invisible and apparently irrelevant to their immediate professional concerns. It also makes many of the ethical, social and environmental consequences of the processes involved (including the consequences of their own actions and omissions) invisible for them.

The new way of seeing the field of resource management advocated in the previous chapter makes industrial resource management systems, resource localities and resource landscapes less clear-cut and less manageable than they once seemed. The task of managing resources should also seem more difficult, disorienting and uncomfortable in comparison with the neat and orderly systems and models of the dominant paradigms. This 'new world' should no longer be totally invisible. This way of seeing constructs a vision of a world in which interaction and change are constant, multidirectional, interdependent, complex and continuing. Having 'seen' this, however – once we can envision this complexity and dynamism – we face the challenge of thinking about it without becoming paralysed by overwhelming complexity and detail. Theories, models and frameworks valued for neatness, efficiency and simplicity are unlikely to prove adequate to the demands of this new way of seeing (see also Wallman 1977).

The conceptual challenge, then, is to develop new ways of thinking. We need to build a framework to think about, examine, analyse and act upon a much wider set of issues and relationships which are not conventionally seen as directly relevant to the work of professional resource managers. We need to do this because the actions (and omissions) of professional resource managers are embedded in wider social processes; because resource management decisions are affected by both their material and discursive contexts; and because both context and focus matter in shaping better resource management outcomes.

A series of theoretical questions are, therefore, central to this book:

- How might we think rigorously, coherently, openly and constructively about the complexity within which real-world resource management is undertaken?
- How might we usefully identify and think about the relevant processes of interaction and change?
- How might we maintain a practical focus on operational management issues in resource management systems while simultaneously taking into account the wide range of issues impinging on us?
- How might we realistically move our criteria of accountability away from dehumanised, reductionist, quantitative measures towards more qualitative concerns for the core values at the heart of this book – social justice, ecological sustainability, economic equity, and cultural diversity?

Responding to these questions leads to some of the central debates and concerns of contemporary social theory in general, and requires re-evaluation of the relationship between resources, society and philosophy. The path taken here develops a 'relational' model of industrial resource management systems, and then uses it to reconsider polyphony in resource regions, and particularly the place (and dis-placement) of indigenous voices in the narratives of resource localities. In taking this path, my position is clearly founded in the work of 'process' philosophers such as Whitehead ([1925] 1997, 1985) and Ollman (1976, 1990, 1993, see also Harvey 1996 ch. 2), but the practical orientation to indigenous experience leads me in a different direction from that expounded by Harvey.

The practical challenge centres on the need to move from a way of simply 'seeing' the interactions – being able to recognise, acknowledge, identify, categorise and describe them – to formulating a coherent and rigorous way of thinking about and analysing them, a way of practically engaging with and responding to them. In the most practical terms possible, many of the elements that are excised from conventional models of resource management are potential 'showstoppers' – issues capable of producing catastrophic disruption to even the most 'well-managed' and orderly commodity production system.[1] We need ways of intervening in geopolitical realities that facilitate constructive transformations of resource management systems, rather than either their catastrophic disruption or their catastrophic continuation.

Put simply then, the argument is this. Once the linkages between the decision-making processes involved in industrial resource production and their complex and dynamic social, environmental and economic contexts are 'seen', new 'ways of thinking' about resource management are needed. There are practical, ethical and intellectual imperatives demanding systematic, rigorous, coherent and constructive approaches to analysing and responding to the diverse interactions, linkages and complexities. This is the task of developing 'theory' in resource management. We need conceptual frameworks in which to situate the information we have, with which to make sense of it and through which to apply it to material and discursive realities.

Theory in resource management

For many students and practitioners of resource management, the world of theory is limited to much narrower issues than those tackled here. A resource economist might consider aspects of theoretical economics to model commodity markets, price movements and cost structures. A project engineer might use theories of materials science to calculate load stresses and minimum strength requirements for a processing plant. A fisheries scientist might rely on theories of marine ecological processes to set seasonal catch quotas. But this is not the sort of theory that is needed to deal with the wider context being addressed here. We must deal with the broader issues of social, environmental, cultural, political and economic interaction – the core concerns of social theory. For many people with a general background in environmental sciences, an interest in resource management based on a general concern with environmental issues, or an operational interest in particular resource industries, social theory can be difficult, confusing and alienating. In entering the discursive spaces of social theory, it is important to keep our purpose clearly in mind. It is all too easy to be sidetracked into specious debates about terminology, nuances and dogma. So let me state my purpose very clearly: the aim is to build a coherent theoretical framework that will allow us to:

- Think simultaneously at multiple scales (world markets; national policies; local communities; micro-environmental niches);
- Rigorously analyse linkages between systems that are conventionally kept separate (corporate boardrooms and local community forums; governmental policy processes and biophysical environmental processes);
- Respond practically to complex processes of interaction and change within holistically defined resource management systems.

The unambiguous purpose, then, is to improve practical management by providing a framework in which actions, decisions and their complex consequences can be more fully debated and carefully considered. Like the goal of 'management' itself, the purpose here is not and cannot be narrowly 'scientific' (description/explanation/prediction) but is more broadly applied. The target is pragmatic and effective intervention in situations in ways that enhance outcomes in terms of the core values identified previously. This is further developed later, with discussion of 'geography', polyphony and the place of localities and communities in resource management systems. Places are conceptualised as complex sites of interaction which are constructed and reconstructed at multiple scales, and where links between predominantly 'local' and predominantly 'global' imperatives shape lives and opportunities.

Within resource localities, there are multiple voices, oriented towards multiple goals, imperatives and concerns, each exercising some influence on the local trajectory of resource management, and each raising theoretical issues. Resource localities are conceptualised as places where, because of the presence of resource industries, either through direct employment or in myriad other ways, 'men and women struggle through everyday life producing and consuming products for and from world markets' (Hadjimichalis 1994: 239), while simultaneously struggling with issues of local, regional, national and personal importance in the domains of politics, culture and identity (*inter alia*).

The conceptual problem is how to envision both the focus and context of resource management as part of a more holistic, complex, dynamic and human totality. Specifically, we need a way of integrating into resource management practice many issues that are rendered invisible by the dominant paradigm. By classifying these issues as unimportant parts of the 'context' of professional practice (externalities), the dominant paradigm creates time bombs for resource managers and the host communities. The problem can be illustrated by considering a hypothetical mining operation. A conventional approach to management issues might conceptualise key issues as encompassed within the boundaries of a mining lease. But of course the space defined by the lease and its mineral resources is actually an intersection of overlapping, interacting, sometimes reinforcing and sometimes contradictory relationships which impinge on the 'practical' management tasks. Where these interactions are recognised in the dominant paradigm, it is links to markets, technology and expertise that typically expand horizons beyond the lease boundary. Such linkages typically scale-jump over local and provincial linkages and emphasise connections to national and international domains. Nevertheless, the hypothetical mining lease is easily conceptualised as a discrete, separate space, disarticulated for practical purposes from the wider world except through the links from its corporate owners to world markets. Using the way of seeing constructed above, however, this hypothetical lease can be seen as a nexus of many things other than local resources, corporate strategies and world markets. For example, operational management on this hypothetical lease will be constructed by and will contribute to many things. Consider for example:

- Interactions among competing industrial resource systems;
- Interactions between industry and government(s);

- International trading relations, both bilateral and multilateral;
- Competition between companies in the same industry sector;
- Competition between industry sectors and competing technologies and applications;
- Interactions between industrial production and non-industrial systems;
- Relations between production and consumption cycles;
- Development of corporate strategies, including takeover and merger processes, industrial concentration, vertical integration and so on;
- Changing patterns of end-uses for specific commodities (militarisation, power consumption, greenhouse gas emission policies etc.);
- Changing regulatory requirements and performance standards (environmental, human rights, consumer protection, workplace health and safety etc.);
- Relations between non-industrial resource management systems (for example subsistence and recreational) and the people involved in them and the impacts of industrial production;
- Ecological relations between the mine-site and surrounding ecological systems;
- Pre-development and construction impacts of particular projects on particular people and places within and adjacent to the lease area;
- Operational impacts of particular facilities, processes and industries, including 'downstream' impacts beyond the resource locality;
- Complexly interacting effects of boom–bust cycles, technological, market and corporate changes, and of closures of resource projects;
- Links between industrial resource production, regional development, cultural dynamics and government programmes;
- Unanticipated biophysical or ecological interactions between the industrial production system and its host environments, including political responses to these interactions;
- Environmental, cultural and social concerns and responses of affected local communities or community sectors.

Clearly, this list could be expanded. The point is that there are diverse, often contradictory and certainly interacting linkages within and beyond such a mining lease, which affect management options. These linkages occur within the biophysical, politico-economic *and* socio-cultural domains.

The image of the discrete mining lease offers a metaphor for the issues facing wider resource management systems. In this hypothetical example, we glimpse how diverse ecological, social, cultural, political and economic issues affect operational management. Competition from other industrial sectors, such as forestry, tourism or downstream acquaculture may all require mine management to justify priority being given to mineral exploitation. Regulatory authorities may require multiple-use management of a lease area consistent with maintaining not only maximum mineral production, but simultaneously optimising forest products and tourism exploitation of the same space. New legislation aimed at restricting unacceptable behaviour by mining operations in other sectors (such as legislation to reduce 'high grading' of metallic ore deposits to maximise short-term profits at the expense of long-term optimisation of resource use), may affect management options in mines where high grading is used occasionally to meet specific customer requirements. Discovery of new high grade, low cost deposits of the same mineral may change market conditions, requiring mine management to find ways of reducing overheads, cut corners, increase production, or even close existing operations.

The hypothetical mine might also interfere with other commercial or subsistence hunting, agricultural or food-collecting systems. Recreational fishers may compete with the mine for priority on use of local watercourses; local subsistence farmers might compete for land, water or access; local hunters might require reservation of key lease areas; discharge of tailings might interfere with ecological systems used for food gathering; tourists may visit the local area for its scenic qualities. In some cases, non-industrial user groups might respond in ways that directly impinge on management options for the mine – sabotage, mine occupation, legal action, media campaigns and so on. Even if local environmental damage is demonstrably unrelated to the mine's operations, mine management may need to respond to such campaigns. Social impacts of particular facilities, for example the unanticipated consequences of building new houses for workers in an existing community, or of building a new community for workers, and limiting access to community facilities (health, education, water, shops etc.) to employees and their families, may all become issues with direct implications for mine management as social issues become industrialised through trade union action or socialised through community political action. Operational managers might inherit the consequences of poorly planned construction processes generating ongoing problems, such as poorly constructed tailings dams leaking or even failing under operational conditions; poor security during construction leading to influx of illegal small-scale mining operations outside the control of the mine managers; poor behaviour of construction contractors affecting the credibility of operational managers. Predicted environmental impacts may extend further downstream from a mine than anticipated, leaving negotiated compensation packages open to legal challenge. There may also be unpredicted geological problems such as harder-than-predicted ores requiring replacement of mining equipment more often than existing cost structures can withstand. An exemplary mine management system put in place for a new 'best practice' mine may face a sudden collapse of its anticipated markets as a result of new technologies, new materials or general market conditions. Or perhaps management of an existing mine will be required to perform to new standards consistent with a new mine within the same corporate structure, or constructed by new authorities in new territorial administration arrangements. It might also be found that global environmental changes (such as the greenhouse effect) lead to entirely new environmental standards that affect markets for commodities such as coal. A mining lease with technically exemplary management might suddenly find itself subject to a hostile corporate takeover, or a speculative raid on sharemarkets. Such actions can suddenly change an operation from the centrepiece in a small company to a marginal operating unit in a larger company responsible only for a bottom-line production or profit figure. If a mine is acquired by a competitor, it may be closed in order to control market competition. As a result of elections, a lease operated as a joint venture with a national government can suddenly move from being a stable to a volatile partnership. Nationalisation of competing mines in another country can lead to instabilities and uncertainties outside electoral cycles. The activities and policies of international producer organisations can also lead to dramatic changes in policy and economic settings. Strategies to protect habitat and conditions for an endangered species of bird may suddenly be undermined when a mine faces operational difficulties that see sterilisation of the resources in a section of the lease for habitat protection render the whole operation uneconomic. At the same time, the cost of not contributing to habitat protection, in terms of credibility and reputation, make accessing those resources within the lease problematic. A catastrophic environmental accident in another mine run by the same company or producing the same commodity, or even in a downstream user industry, can dramatically affect the credibility of even exemplary environmental managers.

This hypothetical shows that all sorts of issues conventionally treated as externalities in terms of operational management can become 'showstoppers'. All the instances referred to above are drawn from real experiences in the international mining industry, yet many mining industry representatives support continued adherence to narrow models and narrow visions of the scope of 'good' management. For them the problem is how to maintain a sense of 'good' while protecting the prerogatives of 'management'. Politicising management has long been seen as a dangerous path, and even senior corporate figures have faced censure when they have overstepped the boundary between management and politics. What this shows is that management, even at the scale of a single mining lease, is intensely 'political' (in the sense that Leftwich defines political), and intimately connected with the wider worlds contained within and containing the mining lease itself.

The realities of resource management are not asocial, ahistorical or aspatial. In turning upside down the taken-for-granted worldview of privileged, educated, technocratic resource managers, it is apparent that resource management systems themselves are socially constructed. They are embedded within socially constructed realities. The terrain constructed and occupied by industrial resource management systems is, both physically and ideologically, contested terrain, and the task of theorising this terrain is highly political. As foreshadowed, the political orientation of the approach developed here is towards transformation of resource management systems in ways consistent with the core values – social justice, environmental sustainability, economic equity and cultural diversity. Earlier discussion considered how traditional ecological knowledge is implicated in the generation of such transformational politics and the generation of new sorts of political space (Chapter 1, see also Ruiz 1988). The experience of indigenous peoples, however, cannot be isolated from the wider context of conflict within this contested terrain. It is not just at the level of politics that conflict is constructed and played out. Parameters of conflict are also constructed at deeper levels of epistemology and ontology. There *are* different ways of thinking about these issues. To engage with the material and discursive geopolitics of resources, a conceptual framework that acknowledges this is needed.

Industrial resource management systems do not occur in some ethereal *terra nullius*, waiting to be colonised and developed for the unequivocal good and betterment of humanity. Resource management systems and the commodities they produce are integral to the construction, maintenance and reconstruction of global order: they simultaneously produce both commodities and power. They are related to each other *and* to wider social processes, forces and relations. The way we think about the world inevitably shapes the way we think about resource management issues, and humanity has constructed a variety of ways of thinking about the world – diverse ontologies and epistemologies.

Given the current balance of power that exists, industrial resource management systems generally operate in favour of the core institutions and regions of international capitalism: the major global resource corporations, major nation states and powerful groups and core regions within them. The landscapes in which industrial production of resource commodities occurs are enormously diverse, ranging from the high Arctic to remote deserts, urban fringe areas, marine environments and tropical forests. In all locations, in terms of their complex geographies, these territories are, both literally and metaphorically, *terra mater* – the treasured human landscapes of people in their communities, the nurturing Mother Earth of diverse cultures and the source of social, cultural and personal identity. These places and peoples are not disembodied components of an abstract commodity production process. They have pasts, presents and a range of possible futures built on traditions, values and worldviews very different from those of efficient resource technocrats in industrial commodity trading systems. But to the technocratic way of seeing these places seem to be empty spaces. The idea that these 'empty' (but resource rich) landscapes might already be occupied, valued and used – that 'their' *terra nullius* might already be somebody else's *terra mater* – is alien and unimaginable. Yet at times, these alternative realities do intrude into the comfortable technocratic models of neatness and efficiency. At Bougainville, Narmada, Nam Choan, in tropical rainforests of Southeast Asia and Latin America, in the hearings of the Waitangi Tribunal and the Mackenzie Valley Pipeline Inquiry and in diverse settings around the world, grassroots visions have persistently exploded, blockaded, resisted, negotiated, mediated, and participated their way into the geopolitics of resources. They have also struggled their way into the academies where resource managers are educated; they force their way into negotiations and policy debates; they find voices in different media. How can resource managers make 'sense' of this polyphony? How are we to acknowledge and engage constructively with it?

In this contested terrain, virtually every element of both vision and reality is subject to conflicting interpretation and alternative meanings. Material *and* discursive reality is contested. What is a 'resource' to one group of people is a 'wasteland' to another; what is valued by one group is invisible to another; what is rational to one group is unaccountably strange to another; what is criminal behaviour to one group is heroic self-defence to another. Because resource management systems simultaneously produce both commodities and power, the issue of power – social power, political power, economic power, military power and so on – is always integral to these landscapes. Resource landscapes are always landscapes of power. Natural resources are not the only important elements in the power structures of resource management systems, but power is certainly an inescapable element in all resource management systems. To better understand the contests that occur within this terrain, we need to return to our simple model of a resource management system (Figure 2.4, page 81) in which we identified four basic elements in a resource management system: the resources themselves, management agencies, profit-seeking enterprises and diverse publics affected by the system.

The basic model

In our earlier discussion of Figure 2.4, it was established that these elements, and the relations within and between them, constitute 'resource management systems'. The basic model presented as Figure 2.4 could be expanded by specifying other elements such as the media and technology, and by specifying the content of various elements (for instance Figure 3.1). For example, the 'diverse publics' category might be refined by specifying the presence of indigenous groups, trade unions, environmentalist groups, religious groups, local community action groups and so on. For each of the elements identified a considerable body of practical and theoretical literature exists. To understand each of these elements, and their interactions, we need to draw on this literature, and to integrate that knowledge into a more holistic rather than fragmented conceptual framework.

To understand the ecological systems within which natural resources are embedded, we need elements of ecological, geological, chemical and other biophysical scientific theory and knowledge; to understand the political and governmental systems within which management agencies operate and are constrained, we need to draw on elements of political science and other studies of governance in political economy, geography, sociology and so on. To deal with the activities of profit-seeking enterprises, it will be necessary to draw on the work of economists, political economists, industrial relations analysts, economic geographers, organisational sociologists, psychologists and others. To deal with the 'diverse publics', which include a range of both

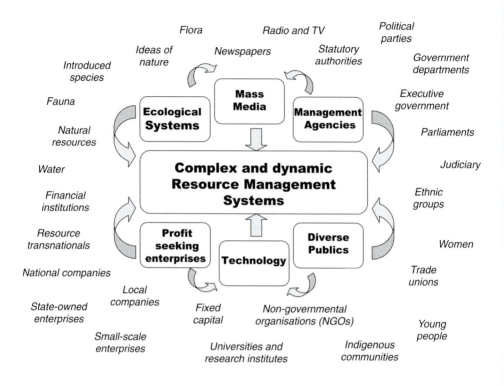

Figure 3.1 A more complex representation of the context of resource management systems

organised and chaotic elements, conventional community organisations such as trade unions and progress associations and 'new' social movements organised around issues of environment, race, gender, culture and so on (the new cultural politics of identity and locality), we will be drawn towards work in cultural and social geography, cultural studies, literary analysis, sociology, anthropology, political economy and other fields. In tackling this broad (and rapidly growing) literature, we will have a purposive focus – the deconstruction, demystification and critical analysis of resource management systems (Figure 3.2). We are not seeking to become technical experts in all these fields, but to draw on them constructively, critically and thoughtfully in the pursuit of a practical goal – better resource management.

In considering each basic element in our model (Figure 2.4), it is clear that simple categorisations of management agencies, profit-seeking enterprises, ecological systems and diverse publics are problematic. For example, in developing regulatory arrangements for resource industry operations, governments ('management agencies') will often involve representatives of resource companies

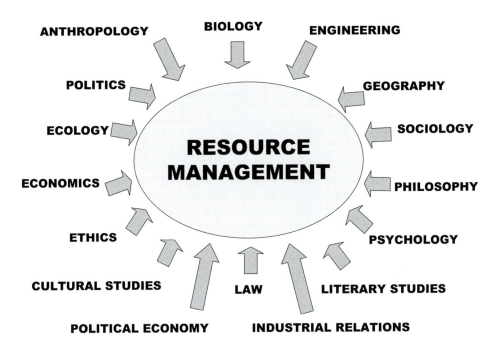

Figure 3.2 Synthesising the insights of diverse disciplines to improve resource
management outcomes

('profit-seeking enterprises'), industry bodies (part of the 'diverse public', but
clearly not easily dissembled from their constituent 'profit-seeking enterprises'). Progressive governments might include consultative procedures to
include trade unions (who as workers are constituents of the 'profit seeking
enterprises' and as organisations part of the 'diverse public' arena), consumer
groups and other public interest advocacy organisations ('diverse publics'). In
the process, the helpful analytical distinctions represented as discrete elements
in the diagram become intertwined. Indeed, in some regulatory agencies
(such as Australia's Joint Coal Boards and Great Barrier Reef Marine Park
Authority; the United States Office of Surface Mining and Rehabilitation;
and a variety of environmental regulatory authorities), blurring of categories
is institutionalised. Does this mean that the original basic model is flawed?
Not necessarily, but we must step further into social theory to tackle the
conceptual issues.

Conceptual tools for resource management

> The ontology to which each of us subscribes is so utterly familiar that few of us ever hear or use the word. Our reality is so certain to admit no alternatives, that we have trouble in accepting it, with any sincerity, as *just one* possible version of reality, believing it in our heart of heart, to be *the only* possible reality.
>
> (Christie 1992: 1)

In trying to develop a conceptual framework for resource management, there is no suggestion that there is just one single way of thinking about issues in resource management. This is not advocacy of a new dogma or 'correct line'. In academic terms, there is no proposition here that students should be required to use a compulsory way of theorising or discussing the topics raised in this book to succeed in their assessment tasks. In other words, the purpose in raising these difficult theoretical, philosophical and epistemological issues is to demonstrate a practical way of thinking which allows individuals not only to 'see' the complex context within which resource professionals operate, but also to begin constructing personal ways of addressing those complexities.

All our experience of the world is mediated by ideas. We can neither experience nor represent reality directly:

> our knowledge of the real world is mediated through the construction of concepts in which to think about it; our contact with reality ... is contact with a conceptualized reality.
>
> (Ollman 1976: 12)

Concepts and the ways societies conceptualise reality directly affect how political relationships evolve and how resources are identified, utilised, distributed and consumed. In different eras, in different places, with different intellectual and cultural tools, very different approaches to mediating our experience of reality and of constructing the concepts we utilise to think about material reality have developed. Not only is there enormous cultural diversity on the planet, but there is also considerable ontological diversity. Different peoples genuinely put the world together differently. This has enormous significance for understanding and managing the contemporary geopolitics of resources.

These diverse ontological frameworks cannot be divorced from their historical and geographical contexts. The intertwined rise of the Enlightenment, Western science, European colonialism and the geographical expansion of capitalism provided a powerful platform for the uneven development of wealth and power on a large scale. It is easily assumed that the efficacy of Western science in producing wealth and power demonstrates an inherent superiority over other approaches to construction of knowledge. The persuasiveness of Western science is strengthened by its claims to objectivism. The idea that science can reveal or discover pre-existing facts or the really real, dangerously veils the extent to which 'all scientific statements are a product of the imagination' (Christie 1992: 17). Claims to objectivity obscure the social negotiation of knowledge and uneven access to the spectacular wealth and power generated by scientific knowledge. These claims tend to 'naturalise' the knowledge produced, and the institutionalised knowledge-power systems built upon it. Rose (1999) provides a powerful view of the way in which Australian Aboriginal peoples' 'cultural construction of subjectivity' – their ontological assumptions about the nature of themselves and their relations with the wider world – and the effect of colonisation on them. In rejecting the term 'post-colonial' in favour of 'deep colonising' Rose highlights the continuing 'ecological and spiritual brutality of the regimes of violence within which we are all entrapped' (1997: 9):

> While it is demonstrably the case that many formal relations between Indigenous people and the colonizing nation have changed in the past three decades, as have many of the institutions which regulate these relations, it is also the case that practices of colonization are very much with us … many of these practices are also embedded in the institutions meant to reverse the processes of colonization. Colonizing practices embedded within decolonizing institutions must not be understood simply as negligible side effects of essentially benign endeavours. This embeddedness may conceal, naturalize, or marginalize continuing colonizing practices.
>
> Furthermore, practices of colonization are so institutionalized in political and bureaucratic structures and policies, that they are almost unnoticed.
>
> (Rose 1999: 182–3)

Rose's paper demonstrates just how intertwined these ontological, political and ecological issues are. It also indicates how easily knowledge–power structures that atomise and objectify complex and dynamic relationships can be naturalised.

In contrast to the obfuscation of much scientific discourse, some knowledge–power systems clearly acknowledged that knowledge is and must be socially negotiated. Christie (1992) provides a comparative account of an only slightly exaggerated Western scientific approach to knowledge making, and the approach of Yolngu people with whom he worked for many years as a teacher-linguist in Northeast Arnhem Land. Yolngu science, he notes, makes neither objectivist nor atomistic assumptions about the world, and is not anthropocentric in orientation. Rather than accepting the knowledge–power claims of any speaker, Yolngu listeners seek to avoid the pitfalls of naturalisation:

> This … is precisely what Aboriginal science is constantly vigilant to maintain. Everyone is agreed to have an ex-centric [rather than

anthropocentric] view of reality, so every time a Yolngu speaks the
community will ask: Whose interests are being served by positing this
shape for reality at this particular time? and What other possible claims are
rendered absent or forced to the margin by this claim?

   When we remember that every scientific statement carries with it a field
of privilege and power for those who prosecute it, and at the same time
renders all other possibilities absent, we perceive a need of great urgency
to reshape the knowledge making business in western society so that everybody and every thing is given a voice. We need to be asking the same
questions as Aboriginal knowledge makers. Whose interests does this particular way of constituting reality serve? And what other possibilities may
we be forgetting about.

(Christie 1992: 17–18)

But, how does one go about the task of building the concepts with which to
represent material reality? Ollman asserts that abstraction is central:

> Everyone ... begins the task of trying to make sense of his or her sur-
> roundings by distinguishing certain features and focusing on and organiz-
> ing them in ways deemed appropriate. 'Abstraction' comes from the Latin
> abstrahere, which means 'to pull from'. In effect, a piece has been pulled
> from or taken out of the whole and is temporarily perceived as standing
> apart.

(Ollman 1993: 24)

In the discussion that follows, I consider the process of abstraction and how it
is implicated in the task of rethinking resource management in the contemporary world. Like many others who challenge Western science's virtual monopoly on knowledge making, and its socio-cultural, politico-economic and
biophysical consequences, I adopt a dialectical approach which emphasises
representation of reality through processes, flows and relationships, rather
than focusing on elements, things, structures and systems. Harvey adopts a
similar position (1996: 49), but often overemphasises process to the virtual
exclusion of 'things' – despite his own reference to the lesson of quantum
theory in physics (ibid.: 50), which suggests that material reality might be
better understood as simultaneously both 'thing' and flow.

   This discussion leads to consideration of the difficulties faced in
operationalising a dialectical approach to knowledge–power issues in resource
management. It is argued that geography – as both material and discursive
reality, both a 'thing' and a discipline – actually matters in the task of rethinking resource management. To this end, the discipline's foundational concepts
are reviewed and debated before moving on to more familiar applied resource
management examples.

Categorical and relational models

In the first instance, it is helpful to distinguish between two different approaches to thinking about social experience – categorical and relational. Categorical approaches dominate everyday experience and common sense thinking in the West. In these cultures, children are encouraged to sort and classify things from a very early age. In the process, they acquire a way of thinking in which increasingly precise definitions of things are available to create mutually exclusive categories for everything. A taxonomic classification of species is a good example, with increasingly precise definitions producing a discrete, separate and mutually exclusive category for each species, down to each individual in each species.

In categorical models, the categories used to describe and explain 'reality' are seen as neutral vehicles to carry specific parts of a larger story. The content of individual categories is understood to be a form of its real subject matter. In most categorical models this relationship between a category and its subject matter is assumed to be self-evident and limited to a specific part of reality. For example, most economists treat capital as a category that describes a specific form of money or investment. This approach not only creates discrete, mutually exclusive categorical definitions for things, it also deals with interaction and change as a product of external relationships between discrete and separate things. It is generally assumed that both the nature of the real thing and the content of the concept will remain unchanged and stable unless and until some external influence causes it to change.

In a categorical model, a thing's relationships with other things cannot be part of its ultimate definition. In terms of things like the model of resource management systems, a categorical model, in its basic form, emphasises the 'boxes' (the categories) rather than the 'arrows' (the relationships – Figures 2.4, 3.2).

An alternative approach *is* available. Relational models, that don't require relationships between things to be distinct from how we define them, open new opportunities for building ways of thinking about resource management systems which address the lack of holism, cultural openness and political inertia of conventional frameworks. The term 'dialectics' is another way of talking about this approach. This term might be more familiar to some readers, but for others it may come with a considerable baggage of confusion and ideology. Nevertheless, it is also possible to distinguish between dialectical (relational) and non-dialectical (categorical) thinking (see Ollman 1993). It is this approach that I want to spend some time considering now.

Rather than emphasising the identification of 'causes' and 'effects', relational models take dialectics as a foundation for thinking about social experience. Dialectics is a method for dealing with and thinking about interaction and change. Ollman identifies four basic features of dialectical thinking (1993: 13; Table 3.1).[2]

Table 3.1 Basic features of dialectical thinking

Identity/difference	Unlike non-dialectical notions, where things can be either identical to or different from other things but not both simultaneously, in dialectical thinking it is recognised that elements of both identity and difference co-exist in most things, depending on how you look at them, or why you are looking at them.
Transformation of quantity to quality	At some point, continuing quantitative change will produce a qualitatively new entity (and the bundle of relations tied up with it).
Interpenetration of opposites	The intimate relation that exists between opposites such as 'positive and negative', 'cause and effect' and so on, and the point that the truth of any contrasting observation depends on the point of view of the observer.
Development through contradiction	A view of history which sees 'a present ... (as) part of a continuum stretching from a definable past to a knowable (if not always predictable) future, in which alternative futures are in tension and contradiction with each other, and the process of working out these contrary movements in bundles of relations is what shapes the context in which theorising occurs.'

Source: Adapted from Ollman 1976: 54–6.

The philosophical roots of dialectics can be traced through Marx to Hegel and further to Spinoza, but for our purposes I want to build on two basic sources – the work of Ollman (1976, 1990) and its use in recent work by political economists (Resnick and Wolff 1987, 1992) and human geographers (Harvey 1996; Graham 1988, 1990, 1992; Graham and St Martin 1990; Gibson-Graham 1996; Gibson-Graham *et al* 2000). This detour into philosophy, particularly Marxist philosophy, may seem a little unexpected in a treatise on resource management. It is worth, perhaps, restating that my purpose is to develop a conceptual framework for thinking about something complex and dynamic, and handled badly by conventional philosophical frameworks.

The philosophy of internal relations: challenging ideas of 'cause' and 'effect'

Everyone recognizes that everything in the world changes, somehow and to some degree, and that the same holds true for interaction. The problem is how to think adequately about them, how to capture them in thought. How, in other words, can we think about change and interaction so as not to miss or distort the real changes and interactions that we know, in a general way at least, are there? ... This is the key problem addressed by dialectics; this is what dialectics is all about.

(Ollman 1990: 27)

Ollman, through an examination of Marx's approach to conceptualising social relations, provides a coherent philosophy of internal relations. Although Ollman's purpose is to interpret the writings of Marx, his work provides some useful insights into the underlying problem confronting our own concerns. Ollman aims to construct a framework for understanding a complex and dynamic totality (in his case capitalism, in ours contemporary resource management). In learning from his approach to his task, there are valuable lessons for our approach to the task of rethinking resource management systems.

For Marx, at least in Ollman's reading:

> epistemological priority [was given] to movement over stability, so that stability, whenever it is found, is viewed as temporary and/or only apparent, or ... as a 'paralysis' of movement. With stability used to qualify change rather than the reverse, Marx – unlike most modern social scientists – did not and could not study why things change (with the implication that change is something external to what they are, something that happens to them). Given that change is always a part of what things are, his research problem could only be *how, when* and *into what* they change and why they sometimes appear not to change.
>
> (Ollman 1993: 31, emphasis and parenthesis in original).

So Ollman is suggesting that Marx's conceptual framework starts from an acknowledgment of recognition of interaction and change as a characteristic of complex systems, rather than as something unusual requiring explanation. This sounds like what we are looking for – but how is it actually done? Can Ollman's approach to what he calls the 'philosophy of internal relations', be put into practice in ways which might improve professional practice in resource management? Ollman's approach has faced criticism, which he has addressed at length (1976: 256–76). Through both empirical and conceptual discussion, I will argue that this powerful and revealing approach not only can be operationalised but is perhaps one of the most valuable tools to be added to the contemporary resource manager's toolkit.

If one considers the idea of capital, which is one of the concepts our model of resource management systems has in common with the work of both Ollman and Marx, some of the implications and strengths of a relational approach to defining 'things' can be seen. In a dialectical or relational approach, capital is not treated as a neutral, categorically distinct concept that describes an obvious and pre-existing objective reality. Rather, capital is defined not only by how it appears and functions, but also by how it develops and how it interacts with and relates to other parts of the social totality. So what it *does* (in various social, political and cultural terms as well as its formal economic functions), how it develops, and its links with other elements of society including labour and the state, must also be considered as part of what it *is* – part of the definition of capital. In other words, capital's relations with nature, labour and the state become part of the operational definition of capital. This means that it also becomes necessary for us to actively consider just what these relations involve rather than assuming their nature and implications.

The inclusion of historical and geographical dynamics in this way of defining things further strengthens its way of entrenching interaction and change in our analysis of and responses to resource systems. In contrast to a conventional view in which history is something that happens *to* things rather than part of their nature, Ollman suggests that history (and I would suggest also geography) is part of what things actually are:

> History for Marx refers not only to time past but to future time. Whatever something is becoming – whether we know what that will be or not – is in some respects part of what it is along with what it was. For example, capital … is not simply the material means of production used to produce wealth, which is how it is (used) in the work of most economists. Rather it includes the early stages in the development of these particular means of production, or 'primitive accumulation, indeed whatever has made it possible for it to produce the kind of wealth it produces in just the way it does … .
>
> (Ollman 1990: 32)

Elsewhere, Ollman lists elements and relations that Marx used to build a definition of capital. He included the capitalist, the wage labourer, the products and machinery used to produce products, the commodities which go into the products, value and money and so on (1976: 14). All these social relationships – relationships between people and their lives under a broad social order – are contained in Marx's broad conception or abstract notion of 'capital'. Coupled with a view of history that encompasses past, present and future we begin to see that:

> Each social factor [is] internally related to its own past and future forms, as well as to the past and future forms of surrounding factors. Capital, for Marx, is what capital is, was and will be.
>
> (Ollman 1976: 17–18)

In other words, the totality of the 'system' (whether capitalism or a resource management system) is contained within each of its constituent elements, and each of the constituent elements is present in the totality in all of its permutations; the relationships between the various elements are internalised both within the total system, and in each of its constituent elements. This is the notion underlying William Blake's powerful poetic vision of the world in a grain of sand that is considered as a metaphor for a relational view of scale relations elsewhere in this discussion. For Ollman this approach to the process of abstraction provides a crucial nexus between observation and theorisation, between practice and theory, and between material and discursive realities. As he puts it:

> These abstractions do not substitute for the facts, but give them a form, an order, and a relative value ... frequently changing his abstractions does not [for Marx in Ollman's analysis] take the place of empirical research, but does determine, albeit in a weak sense, what he will look for, even see, and of course emphasize.
>
> (Ollman 1993: 39)

Again, there is a glimpse here of an approach to thinking about things that touches many of the issues that have already been posed. This process of producing, applying and refining categories is precisely the process of observing, conceptualising and theorising reality. Ollman suggests that the categories and concepts produced by this approach contain within themselves, and are themselves part of, the complex and ever changing totalities under examination. One of the implications is that it becomes easier to recognise that any explanation provided by a theory is not independent of the social relations (and history and geography) that produce it.

This way of constructing our knowledge of change and interaction involves ways of thinking about some foundational concepts that are very different to those that are common in scientific thinking such as 'causation' and 'determination'. If change, and the potential to change, is understood and even defined as an *internal* characteristic of things and their relations with other things as part of a holistic structure, then logically prior and independent causes which produce logically independent and subsequent effects cease to be logical. Instead, we have a complex set of causal processes and determining influences. In many cases complex causation will inevitably involve mutually influential relationships, where things effectively cause each other. From a starting point rooted in dialectics, this is hardly surprising – it reflects the widely used double-ended arrow of many flow diagrams. Yet in discourse, where definitive causation is privileged, such representations of reality seem illogical. In systems that depend on decisive decision making linked to definitively defined and systematically predicted relationships between cause and effect, this approach to observing, classifying and responding to interaction and change challenges ontological privileging of certain taken-for-granted structures, relationships and processes such as managerial prerogatives, structures of power and privilege, and the role of the state. While this approach opens a myriad of ways through which to influence the nature or direction of change, it can also demand a more thorough and extensive engagement with both empirical observation and conceptual abstraction.

Any particular example will present specific problems for some readers precisely because of the way in which our ability to understand it will depend on both contextual and specific knowledge. It is clear from Ollman's reading of Marx's critique of capitalism, however, that no single cause (in politico-economic, biophysical or socio-cultural domains) is adequate as a theoretical summary of the complex, multidirectional interaction and change we know occurs in resource management systems. Frameworks rooted in economic determinism, environmental determinism or cultural relativism *must* be inadequate for our purposes. A less restrictive approach to causation is needed in these complex hybrids of natural and social processes.

Within Marxism, including its dissident traditions, debate about dialectics has produced both helpful and dead-end ideas about the difficult issue of causation. Marxism is often characterised as a simplistically deterministic philosophy – Marx's suggestion that the economic base determines the legal and ideological superstructure ([1851] 1975). The work of Resnick and Wolff (1987, 1992) offers a non-classical view of Marxism which proposes a 'multiplicity of different economic essences' (1992: 131). Like Ollman's, this view builds on a careful rereading of Marx that rejects naïve, simplistic and deterministic interpretations of relations between economic dimensions of social experience and other, non-economic dimensions.

One label for this version of Marxism is anti-essentialism, meaning that no singular set of relationships or processes is seen as providing the 'essence' of explanation of a complex social totality; no essentialised summary provides an adequate approximation of reality. Another label is overdetermination. This term was introduced in the 1960s by the French Marxist philosopher Louis Althusser (1969). Althusser found the term awkward, but persisted in its use because it highlighted the complex relations involved in causation in dialectical thinking. Writing of the central contradictions that shape social change, Althusser concluded that the 'general "contradiction" ' that derives from the relations of production which characterise a social formation:

> is inseparable from its formal conditions of existence, and even from the instances it governs; it is radically affected by them, *determining, but also determined in one and the same movement,* and determined by the various levels and instances of the social formation it animates; it might be called overdetermined in its principle.
>
> I am not particularly taken by this term overdetermination … but shall use it in the absence of anything better, *as both an index, and as a problem …*
>
> (Althusser 1969: 101, emphasis added)

This image of an element in a complex totality (whether it be a flow, a relationship, a structure or a thing) simultaneously appearing to function as both cause and effect, containing within it both its own conditions of existence and possible divergent future configurations, opens up possibilities for very different accounts of resource management systems. By opening up a multitude of determining and overdetermining relationships in any concrete set of circumstances, this perspective also offers the possibility of a multitude of avenues for intervention in a system to achieve better outcomes. In this view, then, any single event is overdetermined by *all* other things, and as social participants/ researchers/analysts, our entry point into the social totality is itself overdetermined by our complex (and constantly developing and changing) personal histories and current circumstances. According to the feminist geographer Julie Graham:

> Overdetermination posits the mutual constitution of all social and natural processes. It provides a radical conceptual alternative to forms of determinism, and to all other attempts to reduce complex realities to simpler essences at their core.
>
> In an overdeterminationist theoretical setting, knowledge is a social process which is constituted by all other social and natural processes and which in turn participates in their constitution. Fully embedded in social and natural life, it cannot lose touch with other aspects of the world.
>
> (Graham 1992: 147)

Graham, and Katherine Gibson, advocate an approach to overdetermination labelled anti-essentialism, because it rejects the possibility of essential, ultimate or pre-determined causal influences in any complex system:

> From an anti-essentialist theoretical perspective, no aspect of the social or natural world merits ... special ontological status. Every aspect of reality participates in constituting the world and, more specifically, in constituting every other aspect. This mutual constitutivity is what is meant by the term 'overdetermination', used in the sense put forward by Resnick and Wolff.
>
> The notion of mutual constitution provides an alternative to mechanistic conceptions of cause and effect, in which independent and static conceptions entities are sporadically set into motion. It also provides an alternative to dualistic notions of dialectical interaction and interpenetration. An overdetermined site or process is complexly constituted by an infinite multiplicity of conditions; it changes continually as those conditions change; it is pushed and pulled in contradictory directions as its myriad conditions change at different rates and in different ways. It has no essence, no stable core, no central contradiction. Instead it is decentred, existing in complex contradiction and continual change.
>
> ... 'Nothing less than everything is a sufficient explanation for anything' (Schell 1991: 9). There is no social or natural process that is truly independent of any other. The context of any event constitutes and specifies it, and every aspect of life makes up that context. To understand a process or event involves theorizing the way in which every other process contributes to its contradictory development.
>
> (Graham 1992: 142; see also Gibson-Graham 1996)

This is an important perspective because it alerts us to the possibility that in pursuing and constructing knowledge about resource management, one enters an arena of conflict over pasts, presents and futures as surely as if we were taking up mining leases over sacred religious sites, or felling the trees of the last rainforest hosting particular endangered species. There are important links between these material realities and the discursive tools we use to represent and engage with them. Not only do resource management systems simultaneously produce commodities and power, but the knowledge systems used to understand them are also, of course, simultaneously power structures.

Focus and context in abstraction

Two important terms in discussing our work as resource managers are focus and context. Ollman emphasises the importance of abstraction as a way of getting at the same issue – how does one abstract a particular focus from its complex context? Ollman identifies three specific but interrelated modes of abstraction that shape the way we conceptualise reality. 'The process of abstraction', he says

> which we have been treating as an undifferentiated mental act, has three main aspects or modes, which are also its functions vis à vis the part abstracted on the one hand and the system to which it belongs on the other. That is, the boundary setting and bringing into focus that lies at the core of this process (conceptualising reality through abstraction) occurs simultaneously in three different, though closely related, senses. These senses have to do with extension, level of generality, and vantage point.
>
> (Ollman 1993: 39)

In the *extension mode*, the process of abstraction sets notional boundaries to interaction in terms of space and time, in terms of history and geography. While acknowledging the totality of social relations, we abstract an historical and geographical focus, and limits to the extent of our investigations.

The *level of abstraction* brings 'into focus a particular level of generality for treating not only the part but the whole system to which it belongs' (ibid. 1993: 40). Ollman uses the metaphor of a microscope to clarify this, suggesting that this mode of abstraction provides different powers of magnification to view and analyse the particular qualities of a part of a system (in time and space) and its function in the system more generally.

As well as establishing the extent and level of generality at which one thinks, abstraction also 'sets up a *vantage point* or place within the relationships under examination from which to 'view, think about, and piece together the other components in the relationship' (Ollman 1993: 40 emphasis added). In other words, we abstract a conceptual position or vantage point which views the interrelated elements that compose the totality under examination, whether it is capitalism or a resource management system. In doing so, we establish (consciously or unconsciously) a perspective on both the focus and context of our endeavours. In this book, for example, the vantage point of indigenous experience is abstracted as the basis from which to investigate the dynamics of resource management systems, even though it is acknowledged that this vantage point and its abstraction from the whole, is neither impartial nor objective.

Because his focus is on Marx's use of abstraction, Ollman does not specifically address the question of why particular abstractions might be chosen in preference to others. He takes Marx's purpose as given and works from there. The *purpose* of any particular analysis, however, provides the basis from which specific abstractions are pursued – the *focus* for constructing the conceptual framework appropriate to a specific situation. Marx's purpose was to provide a politically incisive critique of mid-nineteenth century capitalism; while that particular purpose continues to provide some insights in terms of historical, philosophical or methodological interest, it is hardly the most directly relevant exemplar for resource managers entering the twenty-first century. In my own research, my purpose has generally involved provision of a politically relevant critique of the social impacts of mining in northern Australia on indigenous communities. My work over recent decades has focused on consequences for indigenous peoples, whose incorporation into industrialised economies reflects the appropriation of their geography (dispossession and resource colonisation) more than appropriation of their labour. My investigations of consequences for indigenous groups of state and corporate exploitation of resources in Australia has applied dialectical methods developed from analysis of specifically capitalist class relations (that is, capital–labour), in the wider context of capitalist social relations. In Ollman's terms (1990: 23–5) the 'vantage point' of my work – its purposive focus on the mechanisms of Aboriginal marginalisation in the mining sector of contemporary Australian capitalism – provides a particular (and partial) view of the totality of contemporary capitalism. It offers 'a vantage point or place within the relationship from which to view, think about and piece together the other components' (Ollman 1990: 42). It also orients my work to a relatively concrete level of analysis rather than more abstract analysis of 'capitalism in general', or 'class history in general' (see also Gibson and Horvath 1983: 126; 1984).

This vantage point frames the abstractions used and developed. As capital is the central vantage point from which Marx seeks to view the internally related parts of the capitalist mode of production, Aboriginal marginalisation has been central to my view of resource management systems in the Australian mining industry. This purposive focus has involved:

1 *Certain abstractions of extension.* Historically, these extend back from the present to the time of local dispossession and further to the period of colonial occupation of Australia, and forward to the decolonisation of Australian

indigenous territories; geographically, they extend to incorporate the international trading systems in which the mineral commodities produced in the specific localities under review are sold, processed or consumed, the localities in which relevant decisions are made, and even to the localities with which they 'compete' in international markets.

2   *Certain abstractions of generality.* Constructing a framework which is specifically relevant to indigenous Australians affected by resource-related development processes, and also to indigenous people's movements around the world, taking into account a range of regional, industrial and jurisdictional variations.

3   *Certain abstractions of vantage point.* Giving marginalisation a central role in the configuration of the relationship between indigenous groups (rather than workers or women or consumers, for example) and the resource management systems in the Australian mining industry.

As a geographer, I have approached the issue of localities, place and environment with a concern for the interplay of society and space. My applied research has given rise to a range of theoretical issues of wider relevance. Consideration of one of these areas – the question of geographical scale – provides a window on the abstraction process and its implications for applied research and management.

Geographical scale and resource management systems

Following Horvath's lead (1991), I would suggest that five co-equal concepts provide the foundations of geography's disciplinary project. They are (Figure 3.3):

- space–time;
- place;
- environment;
- culture; and
- scale.

Detailed accounts of space–time, place and environment can be found in many geographers' work (see particularly the work of Harvey, Massey and Soja for example). Despite its disciplinary importance, geographical scale remains a remarkably chaotic concept and has been subjected to renewed and vigorous debate only recently (see particularly the work of Neil Smith, also Jonas 1994; Howitt 1991c, 1998a; Swyngedouw 1997). A brief examination of the relevance of geographical scale to the task of rethinking resource management systems will provide an enhanced conceptual toolkit for resource managers and demonstrate the relevance of social theoretical work to the operational demands of resource management.

Many commentators have identified important issues of geographical scale

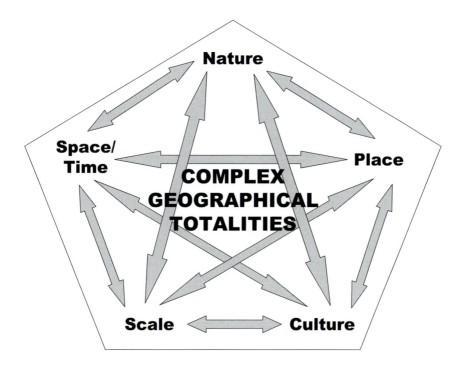

Figure 3.3   The five co-equal foundational concepts of contemporary human geography

Source: Based on Horvath 1991.

in debating how to link the unique features and characteristics of different places to processes and structures which operate, or are constructed at, geographical scales other than the local. In the domains of political activism, this is the issue that underlies the dilemma of how to 'think globally and act locally', how to simultaneously see both forests and trees, or how to frame local action to have positive global consequences. Scale is also one of the key issues that underlies the tension between holism as a philosophical principle and globalism as a politico-economic orientation and globalisation as a concrete process in contemporary resource management systems. These matters have been widely discussed in geography and beyond in recent years. Jonas, for example, has suggested that 'the language of scale is too powerful to be treated simply as a "dimension" of spatiality' (1994: 257). Yet scale often remains a 'contentless abstraction' (Jonas 1994 quoting Sayer 1984: 89–90), and it often remains quite unclear just what sort of 'thing' scale is understood to be in these discussions. In the mid-1950s, McCarty *et al.* put the problem this way:

Every change in scale will bring about the statement of a new problem, and there is no basis for presuming that associations existing at one scale will also exist at another.

(Quoted in Haggett 1965: 263)

In the early 1980s, Taylor suggested that a three-scale typology (global, national, urban) was 'natural' (1982: 23) and could be used to build a 'political economy of scale'. More recently, Neil Smith has spoken of an 'extensive silence on the question of scale (1992: 72), and has argued that 'one of the most pressing theoretical tasks [for geographers] today is to make explicit the relationship between scale and process' (1988b: 321). Neil Smith's emphasis on the 'social construction of scale' (1984b, 1992; Smith and Ward 1987) has produced considerable debate, but it seems to me that there is a risk that while we can grasp the process of social construction, we remain rather unclear about just what sort of thing the scales being constructed might be. Debate within geography has seen a diverse new language of scale emerge. Indeed, a recent paper by Swyngedouw (1997) introduces many new terms into what we might think of as the 'scale vocabulary',[3] but remains unclear about just what this thing called scale actually might be. In contrast, Agnew seems quite clear. He suggests that scale is simply a matter of 'the spatial *level*, local, national, global, at which the presumed effect of location is operative' (1993: 251, emphasis in original). Agnew's definition, however, raises as many questions as it answers – what is meant by 'level'? what is the content of terms like 'local', 'national' and 'global'? Similarly, Taylor's proposition that there are just three 'natural' scales raises questions of in what ways these particular scales are 'natural'. What happens to the non-urban local in this framework? What else is made 'invisible' in this schema? How were these scales 'naturalised'? Like many others, Harvey also emphasises the importance of scale, and quotes Smith's concern about 'grossly underdeveloped' theory in relation to geographical scale (Harvey 1996: 41, 203). He concludes that the theory available 'seems to imply the production of a nested hierarchy of scales (from global to local) leaving us always with the political-ecological question of how to 'arbitrate and translate between them'' (1996: 203–4). Harvey concludes that there is considerable confusion at precisely this point.

It is not only within human geography that the difficulty of conceptualising geographical scale has been a concern. In the mid-1980s, the anthropologists Marcus and Fischer identified the task of 'taking account of world historical political economy' as an important challenge for contemporary ethnographic research. Their articulation of the problem echoes our concern with change and interaction:

how to represent the embedding of richly described local cultural worlds in larger impersonal systems of political economy … . What makes representation challenging … is the perception that the 'outside forces' in fact are an integral part of the construction and constitution of the 'inside',

the cultural unit itself, and must be so registered, even at the most intimate levels of cultural process.

> (Marcus and Fischer 1986: 77; see also Marcus 1986;
> for a contrasting view see de Walt and Pelto 1985)

Scale has also become a topic of debate within resource and environmental
management, not only in the work of bottom-up activist-theorists such as
Shiva (eg 1988, 1992) and Ekins (1992), but also in the more formal literature. Wood (1992: 27), for example, in dealing with the tensions between
regional development and global environmental concerns about tropical
deforestation, develops the concept of 'ecopolitical scale' in which 'four
expanding layers of ecopolitical interaction: local, national, multilateral
and global' have highly spatial characteristics. Fox goes even further, suggesting that 'scale is fundamental, albeit often unrecognized, in most
resource management problems' (Fox 1992: 289). His contribution adds
usefully to the 'scale vocabulary', but the dilemma highlighted by Haggett
(1965) remains.

A lot of confusion results from trying to deal with the notion of 'scale' as a
categorical theoretical issue, divorced from the political rigours of practical
relational engagement in material realities. Like each of the other foundational concepts of human geography, scale cannot be conceived in isolation
from the other elements. A notion of scale that is interwoven with equally
robust notions of space–time, place and environment is needed. We need to
recognise that there is a range of interrelated, though different meanings of
scale (Smith 1992; Jonas 1994) that need to be addressed in clarifying the
concept. As with any relational concept, precisely what it means in any specific
circumstance will depend on the context in which it is used. It is, however,
possible to identify three interacting aspects of geographical scale: size, level
and relation (Howitt 1998a). Most debate focuses on only one of these
dimensions, asserting that scale is exclusively about either size or level.
Agnew's definition, for example, foregrounds just level. It reduces scale to a
limited, unidimensional abstraction. There is relatively little work published
on scale as a relation (although see Howitt 1998a; Kelly 1997). Common misconceptions about geographical scale persist in much of the literature.

Hierarchy, nesting and constructionism

The idea that scale involves a hierarchical nesting of places from the global to
the body (with a range of variations) is common in many discussions. Starting
with categorical notions of separate and discrete spaces (or individuals), it is
assumed that each succeeding scale label subsumes, both in terms of causal
power and territory, those below it. That is, that 'global' is assumed to consist
of the sum of all 'locals'. This approach can be seen in Haggett's attempt to
provide a series of map scales to apply to specific sized areas for analysis (1965:
264) (Table 3.2).

Table 3.2 Comparative scale terminology: early examples

Approx size (sq. miles)	Fennemann 1916	Unstead 1933	Linton 1949	Whittelsey 1954	Map scale for analysis
10^{-1}			Site		
10		Stow	Stow	Locality	1:10 000
10^2	District	Tract	Tract	District	1:50 000
10^3	Section	Sub-region	Section	Province	1:1 000 000
10^4	Province	Minor region	Province		
10^5	Major division		Major division	Realm	1:5 000 000
10^6	Major region	Continent			

Source: Haggett 1965: 264, from the following original sources: Fenneman, N.M. (1916) 'Physiographic divisions of the United States', *Annal of the Association of American Geographers*, 6: 19–98; Linton, D.L. (1949) 'The delimitation of morphological regions', *Institute of British Geographers Publications* 14: 86–7; Unstead, J.F. (1933) 'A system of regional geography', *Geography* 18: 175–187; Whittelsey, D. (1954) 'The regional concept and the regional method', in James, Jones and Wright (eds.) *American Geography: Inventory and Prospect*, Association of American Geographers and Syracuse University Press: 19–69.

This notion of scale as a nested hierarchy treats scale as if it is a bucket (of a specific size) to contain spaces and places. Descriptive regional geography spent considerable effort searching for 'natural regions' and defining boundaries, and the nested hierarchy notion tends to adopt a similarly categorical approach. It risks assuming that simply accumulating small-scale parts and adding them together produces a bigger part and a larger scale. This assumption fails empirical testing, whether one is dealing with biophysical, socio-cultural or politico-economic systems or complex geographical totalities. Jonas (1994: 261) has suggested that it would be 'constructive to view the relationship between the different scales as nested rather than hierarchical'. Neither Jonas nor Swyngedouw, who quotes him supportively (1997: 142), clarify just what this 'nesting' involves. Both reject conflation of scale and any simple hierarchy of causation or influence. Neither gets beyond the term nesting. Both, however, recognise that scales operate simultaneously rather than hierarchically. In other words, the notion of geographical scale should not be conflated with either chronological sequence or a chorological hierarchy.

Several writers have emphasised the importance of the social or political construction of scale (Smith and Dennis 1987; Delaney and Leitner 1997; Kelly 1997; Silvern 1999). While each throws light on processes involved in this construction, just what sort of thing it is that gets constructed remains unclear. It is clearly not just a matter of things 'global' being greater than the sum of their 'local' parts. Nor is it just a matter of relative autonomy of processes operating at a given scale. The point is that the flows (ecological,

economic, cultural, information, ideological and material), processes and relationships that characterise a particular scale are not restricted to that scale, nor
do they simply stop at any notional scale boundary.

So, scale is not the sort of 'thing' that fits into categorical, hierarchical clas-
sifications, although it is often used that way. Some aspects of scale might be
appropriately addressed as hierarchical (more or less complex, larger or smaller
sizes, etc.). Some aspects might also be appropriately conceptualised as nesting one inside another (for example, administrative functions, legal systems,
corporate structures). But neither of these descriptions by themselves adequately encompass the nature of scale. Oversimplifying or overextending
these aspects of scale simply renders invisible the ways in which the complex
flows, processes and relationships at various scales interpenetrate each other in
myriad ways and help to construct, constrain and affect each other. Some of
these elements may have hierarchical relationships with others, but that
property should be a matter for empirical investigation rather than theoretical
assumption.

Micro-scale as a microcosm

Closely related to the idea of scales as a categorical, nested hierarchy is the idea
that something 'local' in scale is worthy of attention because it provides a
microcosm of the 'global'. It has been assumed, for example, that 'small
events' provide a microcosmic window on 'big structures' in globalisation
processes in the economy (Storper 1988). In applications of spatial analysis in
other disciplines, this is a common problem. In a collection of local-scale
anthropological studies (De Walt and Pelto 1985), this conflation of local and
microcosm is particularly apparent. It is also apparent in many aspects of policy-
making in indigenous affairs in nation states where hegemonic racism reduces
indigenous diversity to an homogenised indigenous other. In Australia, for
example, one is able to trace a series of situations where a particular 'local' incident becomes an exemplar of a 'national' problem, in need of a 'national'
solution. The cycle of identifying a problem, establishing an inquiry,
proposing a solution and implementing a new programme is familiar (see
Figure 3.4). Downscaling from these national perspectives to the specificities
and complexities of diverse local circumstances – usually in a context where
there are inadequate resources for the task at hand – often fails to produce
desired results. Similar problems arise when one tries to construct a simple
statistically representative sample for survey purposes in populations that are
not stratified in 'conventional' ways. The notion of representativeness disconnected from specific context is a nonsense for many people (Christie 1992).
Applying individuated health care services to indigenous bodies in populations
where the boundary between body, self, community and country is conceptualised in entirely different ways is similarly problematic (Rose 1999). So, the local
is not significant only because of what generalisations it might permit about
circumstances at a wider scale. Indeed, lessons for resource managers may lie not

Figure 3.4   The cycle of reproducing 'local' marginalisation through 'national' policy
processes

in generalities but in specificities. What makes one situation different (better
or worse) than another? Why do similar policy frameworks, operational guidelines or legislative requirements produce different outcomes in different
circumstances?

Similarly, wider scales cannot be understood as simple accretions of more
specific or localised scales. One does not study the 'global' in order to read off
what will be happening in any particular setting at a narrower scale – any particular nation, locality or body. As one moves upscale or downscale in terms of
analysis or relations, entirely new issues come into play. For example, one may
move from a perspective on specific relations between a mining company and
indigenous communities (or others) affected by its operations in a setting governed by domestic policy and legal frameworks to a transformed geopolitics in
the international arena. Indeed, the process Smith (1993), Kelly (1997) and
others refer to as jumping scales would make no strategic sense at all if the relationship between scales was simply one of macro- and microcosm.

Scale labels as categorically distinct

Most people are familiar with the labels commonly used to signify geographical scale, for example local, urban, regional, national, international and
global. Their widespread use makes it easy to think that these things exist nat-
urally as objective categories or conceptual givens, rather than as socially constructed concepts (Haggett 1965; Smith 1984b: 122). The use of these terms

as scale labels has naturalised them, rendering their metaphorical aspects all but invisible. It is rare, for example, to see a scale label produced as a deliberately constructed term on the basis of empirical investigation, or for specific political purposes (although see Kelly 1997). In this situation, the language used to deal with scale and the process of abstraction that produces and refines them becomes no more than a pre-ordained label for a thing that is assumed to exist. The inadequacy of this situation can be illustrated easily at the level of the national scale. The relationship between the nation state and the national scale is hardly ever placed in an historical context which problematises the extension given to the national as a scale label. Yet clearly it applies to quite different geographical areas in the case of Luxemburg and the Russian Federation. Similarly, did the change of sovereignty in Hong Kong in 1997 stop the label 'national' applying to relationships and processes underway in that territorial entity overnight? What actually changes in that situation? What is the appropriate scale language to use there? In the case of terms etymologically related to 'nation', Anderson (1983, 1992) advocates an historical perspective which problematises the nature of the nation state itself, as well as the scale label applied to it. The need to take an historical perspective on the categories that are adopted as scale labels is equally important in other cases. Precisely who and what is encompassed by various groups' sense of local (or community, or nation) at various times? What do the scale labels international, global and universal signify? Clearly, there are important relational aspects to be accounted for in considering how particular scale labels are constructed and used in any particular study.

It is also only a short step from failing to consider the abstraction process which renders particular scale labels more or less appropriate in different circumstances, and naturalising specific scale categories as universally relevant, to rendering different scales as categorically distinct from each other. For example, the notion of 'thinking' at one scale and 'acting' at another suggests that the two scales are categorically distinct. As Jonas (1994) and Swyngedouw (1997) note, there is a need to deal with things that operate simultaneously at different scales rather than conceptualising the link as hierarchical.

Non-dialectical representations of scale

It is often assumed that the basis for adopting a preferred scale of analysis is that it provides access to causal relationships. For example, those who believe that the internationalisation of capital is the ultimate determinant of social change will assume that global-scale analysis can examine causal elements in a system, while local-scale analysis is limited to contingencies. In contrast, a dialectical view of the notion of geographical scale, in which processes and relationships at one scale influence and are simultaneously influenced by processes and relationships at all other scales in a system, emphasises relationships rather than categorical distinctions between scales. The four elements of dialectics (identity/difference; quantity/quality; interpenetration of opposites; and development through contradiction), along with the overdeterminationists' addition of multidirectional rather than just bipolar relations, are all relevant here.

Conceiving something as only local or regional or global is to misunderstand how the same thing (climatic phenomena, general historical forces such as colonialism, or specific economic elements such as interest rates), can simultaneously play different (reinforcing, contradictory, tangential) roles at different scales. Incremental quantitative changes in scale (upscaling or downscaling) produce a qualitative change in that different processes and relationships come into focus (or move into context) with changes of scale. Mutual penetration of relationships and processes at various scales is clearly important in understanding the construction of (and points for intervention in) resource management systems. This is the point of Shiva's comment on each action within the scale of a single community having global relevance (1992: 39). The dialectician's notion of development through contradiction is also relevant because it is clear that significant concrete change can result from the tension in processes and relations across different geographical scales. For example, tensions between the World Bank as a global scale agency, responding to and shaping global-scale markets for various resource commodities, and local-scale indigenous communities and their traditional resource management systems have been a source of considerable change at several scales (IWGIA 1991).

Theorising geographical scale

Some commentators suggest that human geography's task is to produce a theory of geographical scale. From a relational viewpoint, this is a misguided idea. From the vantage point of a conceptual framework centred on marginalisation and aimed at empowerment – an applied peoples' geography – it is ill-conceived. The notion of a separate theory of scale continues to assume epistemological separation of the social and the spatial, reinstating categorical notions of distinction over relational notions of interaction.

Resource managers operate at a wide variety of scales. Global climate systems and community forestry programmes are all relevant concerns to readers of this book. We are often faced with proposals to transfer lessons from one scale (such as community resource management) to another scale (such as national resource policies). We also confront proposals to transfer insights from one programme to others at the same scale (regional rainforest protection programmes in Australia to Papua New Guinea, or mine rehabilitation programmes at Weipa to Ok Tedi, or corporate strategies in one project to another). We are also expected to operate in a complex setting of resource and environmental geopolitics at various scales, and in cross-cultural settings in which different parties interpret historical circumstances affecting responses to resource management principles differently. Adopting an approach to scale that does not require a categorical separation of scale labels, but instead constitutes processes and relationships of resource management as part of the process of defining relevant scale labels is an important element in giving our model a stronger operational basis.

Scale is thus a tool for analysing and responding to circumstances. It requires theoretical and empirical endeavours, but does not lend itself to theorisation independent of other key abstractions such as space–time, environment and place. Rather than the categorical hierarchical concept that may seem natural to some readers, we need to see geographical scale as a relational matrix, in which processes and relationships constructed or manifested at one scale interpenetrate and are interpenetrated by those constructed and manifested at other scales.

Applying the conceptual toolkit

The role of theory in the conceptual framework advocated in this book contrasts sharply with its role in resource management's dominant paradigms. The purpose of the conceptual work done here is not to model objective truth, nor to set up criteria against which to test truth claims. Rather, it is to find a way to engage with the social production of knowledge and meaning in resource management systems. In large part, the task of theory building in resource management involves developing and strengthening foundations for practical responses to interaction and change in systems which are much more complex and dynamic than it has generally been acknowledged.

Our simple model of resource management systems (Figure 2.4) identifies some of the important elements of these systems. We could, of course, add more and more elements, become more and more specific, or extend the interactions with various other elements of the social totality (examples include gender and patriarchy, race, class, sexuality, other industrial sectors and systems of production). The level of generality and the spatial and temporal extension one seeks to develop depends on the specific purpose, focus and context of one's work. In our case, the purpose is not to build a universal, all-purpose model of resource management systems. Rather, the purpose here has been to develop a way of thinking about industrial resource management systems and indigenous peoples in relation to the specific values of justice, sustainability, equity and diversity. With that purpose in mind, I have advocated an approach which illustrates the broad principals of a relational approach. In this approach, readers are asked to consider how and why important categories (resources, profit-seeking industries, management agencies and diverse publics) and key concepts (such as scale and power) are constructed and used the way they are. By making the thinking behind the models used in resource management decision making more open to critical review, it is hoped that the relationships within and between the various components or elements of a resource management system and the categories used to characterise, analyse and drive them will be better understood and more open to effective intervention. In this approach, theoretical thinking (abstract, general) is not independent of empirical thinking (concrete, specific). Ideally, the two weave together, informing, supporting, challenging, testing and refining each other. The point I have tried to make earlier is that we are all involved in abstraction and in some degree of 'theorising' all the time.

Engaging in this 'theory-building' work is an important part of the wider professional literacy that has been advocated here. There is an expectation that resource managers will engage in a wider range of social theory than might be anticipated in the dominant paradigm. To understand the role of the state in resource management systems, for example, one needs to explore theories of the state. To understand the role of capital and labour in these systems, one needs a theory of political economy. Similarly, one needs to develop some competence in relevant socio-cultural theory, theories of power and theories of human–nature interaction and ecology. Exploration of this theoretical material needs to be done in ways that simultaneously acknowledge that:

- Resource management systems intersect and interact with other dimensions of social experience; and
- Each of the specific elements we choose to define within a particular system is complex and dynamic in its own right.

In other words, we need to recognise that both material reality and the discursive tools used to talk about it are contested domains. Theories, just as much as the resource localities themselves, are arenas of conflict over goals, meaning and values in resource management.

Whatever else it is, our theoretical work and conceptual tools need to be sufficiently 'robust' to tackle those elements of geopolitical realities which inconveniently intrude upon management practices at specific sites (and here, Bougainville is perhaps the most dramatic example to which we've referred), and in specific territorial jurisdictions. We must be able to handle the influences and impacts of site-specific management practices that extend well beyond the nominal boundaries of a mining lease, timber concession or fishery. Doing this requires recognition that it is not only the relationships *between* the elements of a resource management system (and between resource management systems and wider society) that are complex and dynamic, but also the relationships *within* each of those elements.

Massey (1984b: 209) provides a useful way of thinking about the concrete impact of particular histories and geographies on regional economies. She talks about regional economic landscapes containing the legacies (physical and ideological) of previous rounds of capital investment, like layers of a geological structure which continue to influence surface processes long after the forces which formed them have ceased to exist. She could just as easily have been referring to local resource management systems, and her image of rounds of investment and their persistent legacies provides an accessible way to incorporate geography and history into the basic model of resource management systems. This also highlights the importance of technological change, and the constraints placed on resource management systems by fixed investments in specific technologies. In Massey's regional economic land-scapes, it is the fixed assets of previous rounds of investment, the outdated engineering works, the ports, canals and narrow roads of previous transporta-tion technologies, the particular patterns of residential and industrial develop-ments, the legal residues of previous land tenures (in parts of Britain and Europe stretching back to feudal times – and with recognition of indigenous rights, much further), persistent structural, attitudinal and legal echoes of colonialism in post-colonial nations, and the entrenched patterns of thinking about place which constrain the ways in which old industrial spaces are incor-porated into new spatial structures of production.

Problematising the nation state in resource management systems

In all such decisions, the state – the various parts of the institutional, ideolog-ical and organisational apparatuses of government, law and political order – play enormously important roles. In most jurisdictions it is the state which claims territorial sovereignty and sovereign control of most natural resources. It is the state which defines the terms on which resources will be accessed, produced, transported and marketed. It is the state which provides the coher-ency required to co-ordinate the infrastructure (transport, communications, schools, research, health services and so on) upon which resource manage-ment systems in host nations are predicated. While there are exceptions (for example the gold rush in PNG's Mt Kare district in the early 1990s, see Jackson 1991), the role of the state is generally central in organising resource management systems. In some cases, the boundary between the state and particular corporate interests may be hard to discern. Overlapping interests of Royal Dutch Shell, the British and Dutch establishments and the govern-ments of the nations involved in developing North Sea oil and gas, for example, illustrate this. Similarly, one could point to overlapping interests of the apartheid state in South African and major resource corporations such as De Beers and Anglo-American, or to links between nationalised oil-producing enterprises such as Pertamina in Indonesia and the élites and governments of the relevant nation states. In the territories of the former Soviet Union, collec-tivised resource enterprises were often indistinguishable from the Soviet state. In many ways, the resource sector is often integrated into a military–indus-trial–state complex in which both categorical distinctions between elements of the system, and naïve assertions of conspiratorial unities are likely to be misleading. Here the need for weaving together empirical analysis and theo-retical investigations is obvious. The basic point is that, despite the rhetoric of the neo-liberal economic rationalists (who have had a devastating influence on the state apparatus in many industrialised nations and global institutions in recent years), the market is not able to organise resource management systems unassisted. If analysis and interpretation of these systems (which of course incorporate the production systems) is undertaken more holistically, we may produce more effective accounts of the complex geography of resource localities – the interacting, interdependent, mutually constitutive economic, cultural, political, totemic, biophysical and social landscapes.

At the turn of the twenty-first century it is difficult to imagine a world not organised into a number of nation states. As Anderson notes, however, nation states are a relatively new phenomenon. The rise of nationalism and such states accompanied the emergence of mercantilism and international trade; imperialism and resistance to it, and the joining of the 'national interest' and economic interest in empire, particularly in the form of the huge trading companies like the British East India Company, where national sovereignty and corporate identity were so closely linked, was the nursery of the modern nation state, with its sophisticated forms of representation, its complex bureaucratic administrative forms and its authoritative juridical structures. Throughout the contemporary world, specific forms of state apparatus shape the particular configurations of resource management practices.

Trade in resource commodities has long been a central rationale for internationalisation (access to cheaper or higher quality raw materials, access to more economic processing locations and so on) – yet international markets in this century have generally been exactly that – inter-national – between nations. The state apparatus has always been involved in shaping international trade. Despite the neo-liberal rhetoric of 'free' markets and deregulation, trade relations have never been organised on a level playing field. Market forces are rarely permitted to operate unimpeded by state intervention, particularly in those state–industrial–military complexes with the most ideological, punitive and resources power. Indeed, the 'level playing field' metaphor itself is a dramatic denial of geography, and a misreading of the forces at play.

The Australian case: the role of the state in resource development

In Australia resource industries, particularly gold mining and agriculture, were instrumental in financing the expansion of the state beyond a ramshackle colonial parody of administration (especially in NSW, Victoria, Queensland and WA). In many cases, there was dramatic interplay between the state-shaping resource industries, and resource industries shaping the state. This is particularly the case in the area of federal–state relations in Australia, where disputes over interpretation of goals, priorities and values of resource systems have often produced dramatic confrontations in which specific industry interests have successfully enlisted state and even federal governments as an advocate of sectional interests. In the specific case of relations between Aboriginal Australians, the mining industry and governments, it is possible to see the mining industry successfully marketing its sectional interest as representative of a broader national interest, excluding Aboriginal groups, which it simultaneously succeeded in representing as a parochial vested interest (Howitt 1991a).

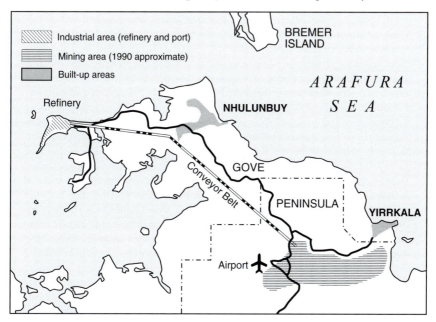

Map 3.1 Location of the Gove project

Under Westminster conventions of public administration, Australia acknowledges the importance of the separation of powers between the parliament, the public services and the judiciary. In this arrangement, the extent of interference between these sectors is intended to be minimised to avoid corruption and excessive power. It is perhaps only when it is viewed from the margins, from a point at which the world has already been turned upside down, that one can begin to see the extent to which these nominally independent or autonomous elements of the state apparatus converge. From the margins, however, it is clear that this convergence shapes resource management systems which produce political and economic power as surely as they produce resource commodities.

Snapshot: state and resources in the Gove bauxite project

The heroic efforts of Yolngu people to protect their traditional lands from mining in the early 1960s signal one of the important sources of the modern indigenous righs movement in Australia. In 1963, following unsuccessful efforts to obtain recognition, the Yolngu elders at Yirrkala sent a petition to the Federal Parliament using the traditional artistic medium of a bark painting (see Plate 8.1). Wartime shortages of aluminium, which developed as a strategically important metal during the Second World War, led the Australian government to establish a national Aluminium Production Commission, charged with developing a degree of independence in aluminium production as quickly as possible.

The AAPC oversaw construction of a smelter at Bell Bay in Tasmania, encouraged investment in aluminium fabricating plants in state capitals, and supported development of exploration for bauxite in North Australia. The transfer of personnel between the AAPC and leading aluminium and exploration interests is an example of the way in which the notional distinction between the regulatory and initiatory roles of governments and private companies can be blurred. In its original configuration in the late 1950s, the Gove deposits (see Map 3.1) were held by the British Aluminium Corporation, which also had interests in the Comalco project on the eastern side of the Gulf of Carpentaria. There were grand visions of a major industrial project linking the two bauxite deposits to the hydro-electric potential of the Purari River in Australia's colonial possession (held under UN mandate) of Papua New Guinea (see Pardy *et al.* 1978). Following acquisition of BAC by the US-based Reynolds Metals company, which was integrating backwards into aluminium from its foil-rolling requirements for cigarette packaging and taking advantage of US government anti-trust action against Alcoa and Alcan in the sale of war surplus production capacity, prospects for development of the deposit of Gove were fading. Reynolds had acquired low-cost bauxite and alumina capacity in the Caribbean and was not interested in investing in new projects that would compete with its Caribbean plants. The Australian government decided to seek new tenders for development of the deposits at Gove, and actively sought the involvement of Alusuisse, one of the Six Sisters of international aluminium. It was at this point that traditional Aboriginal owners of the region petitioned parliament not to dispose of the property which they had held in trust for thousands of years.

In 1963, Yirrkala was a Methodist mission, established in the 1930s on the northeast coast of the Arnhem Land Aboriginal Reserve in the remote Northern Territory.[4] The administration of the Northern Territory was in the hands of the national government, and the terms and conditions for development of minerals on the Aboriginal reserve would be set by the Commonwealth and enacted in an ordinance of the Northern Territory Legislative Assembly. Despite a period of sometimes ambivalent advocacy from the church that the concerns of the Yolngu at Yirrkala should be directly addressed, legislation passed in 1968 effectively handed control of a large area of mining lease over to Nabalco, a joint venture of Alusuisse and Australian finance capital.

In this case, one can see the set of interactions raised earlier in more abstract terms. In specifying the content of the categories in the model in the Gove case, one can identify the variety of linkages and imperatives that mean that no single scale of analysis that is appropriate for such a system. Within this single system (Figure 3.5), one sees the interplay of national strategic interests, global corporate imperatives, sensitivities and vulnerabilities of various ecosystems, political agendas of various players, cultural imperatives operating within the Yolngu community, and market imperatives of the international aluminium industry. Not only do these intersecting relationships overlap, contradict and reinforce each other, but they also set in train a range of other circumstances.

At Gove, the ultimate decision of a Northern Territory Supreme Court *Milirrpum v. Australia* (1971) that the Yolngu system of law and governance could not be recognised as a property right by the common law because of the doctrine of *terra nullius*, continues to echo in Australian land and resource management systems. That decision marginalised the Yolngu from controlling terms and conditions for mining and contributed to the pressure to recognise Aboriginal land rights which ultimately produced the Aboriginal Land Rights (NT) Act 1976. The impact history of the project on Yolngu communities also shaped relations between Aboriginal groups and mining companies for years to come around the nation (Howitt 1992a). And ultimately, the Australian High Court decision in Mabo *v.* Queensland No 2 (1992) overturned Justice Blackburn's view of *terra nullius*.

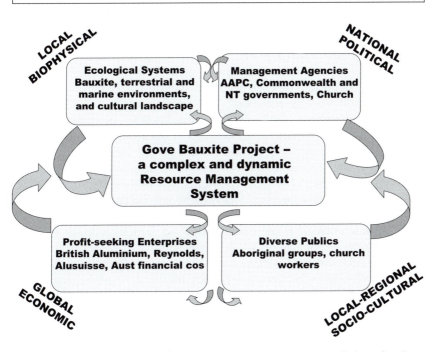

Figure 3.5 The basic model of resource management systems applied to the Gove mining case

The global arena in contemporary resource geopolitics

Globally, nation states continue to play an instrumental role in the definition and control of resources. Even in areas beyond national jurisdiction, the nation states through the UN and its various agencies influence the way in which market forces function for example, the UN Law of the Sea (UNLOS) and Antarctic Treaties. Such arrangements have dramatic impacts on the sovereignty of individual nations states, and on the rights and interests of peoples and groups whom the nation states fail to represent. The response (or lack of it) from nation states and international agencies to resource-based geopolitical issues (East Timor, Gulf War, Yugolslavia, Amazon) or the definition of certain issues as 'internal matters' for sovereign nations to resolve, often without external accountability, all shape the ways in which people are able to participate in and respond to the practical operations of resource management systems.

These points can be seen clearly in Bougainville. In 1975, shortly before Australia granted recognition to PNG, Bouganvilleans declared the Republic of North Solomons to be an independent sovereign entity. Following pressure and guarantees from the colonial power Australia the Bougainvilleans were persuaded to drop demands for independence and remain within the about-to-be-recognised Papua New Guinea, so the 1989 rebellion is not a simple reflection of recent dissatisfaction, but an element in a long-standing dispute. Yet the Papua New Guinea government successfully claimed in international forums that this dispute was an internal matter. Yet once again, many questions of 'scale' and 'vantage point' are raised. In the prolonged conflict that followed the closure of the Bougainville mine, PNG has been, *inter alia*, involved in 'hot pursuit' of rebels beyond PNG's (disputed) territorial boundary into the domain of the Solomon Islands, imposition of a strict blockade on medical and other basic supplies by the national military, peace talks with the Bougainville Revolutionary Army facilitated by the government of New Zealand and efforts to recruit a mercenary force to end the rebellion. At what point(s) do such disputes become (or cease to be) legitimately 'international' (or internal)? At what scale should accountability in these resource management systems be conceptualised? Who are the stakeholders – and what is the nature of the 'stake' each is considered to hold?

Similar questions are raised by Indonesia and Australia's claims to sovereignty over the Timor Gap. In 1975, Indonesia forcibly integrated the newly decolonised Republic of East Timor into the Indonesian nation state, following its decolonisation by Portugal. Despite historical, linguistic, administrative, and religious differences, Indonesia claimed sovereignty over the new republic. Australia did not oppose this claim, apparently feeling under successive Labor (Whitlam) and Liberal (Fraser) administrations that a small, poorly resourced and Communist nation on its northern doorstep would be destabilising. Until 1999 there were no UN interventions to protect the interests (and resources) of the East Timorese. Again, like PNG in Bougainville, prior to the fall of the Suharto regime, both Indonesia and Australia claimed that their agreement over exploration and development of the Timor Gap hydrocarbon resource was a matter of internal, sovereign concern, and not a matter for international interference or accountability. The former colonial power in East Timor, Portugal, initiated action in the International Court of Justice to challenge the Timor Gap Treaty but failed on technical grounds. Does this failure legitimate the Treaty? Does it render the aspirations of East Timorese illegitimate in any sense?

In the USA, both pre-revolutionary colonial powers and the US government signed international treaties with indigenous nations which recognised their sovereignty and independence. Churchill suggests that 'control of land

and the resources within it has been the essential source of conflict between the Euroamerican settler population and indigenous nations' since the inception of the US as a nation state, and before (Churchill 1992: 139). He goes on to suggest that:

> The United States emerged from its successful war against the British Crown (perhaps the most serious offence imaginable under prevailing law) as a pariah, an outlaw state that was considered utterly illegitimate by almost all other countries and was therefore shunned by them politically and economically ... . Indeed, what the Continental Congress needed more than anything at that time was for indigenous nations – many of whose formal national integrity and legitimacy had already been recognized by the European powers through treaties – to convey a comparable recognition upon the fledgling US by entering into treaty relationships with it.
>
> (ibid.: 141)

In Churchill's representation of the situation, one can see that the dynamics of the relationships between nation states, legal institutions, processes of government and politics, and the complex shifting relations within and between diverse interests have been fundamental to the experience of indigenous groups since the eighteenth century. Institutional structures and competing claims about control, management and regulation of resources and geopolitical processes from international trade to local rebellion, all contribute to shaping the complex resource geopolitics of such experience.

These examples could be multiplied many times over. In Indian lands in the USA and Canada treaties were signed with Indian nations as between sovereign nations. In New Zealand, the Treaty of Waitangi, at least in the Maori version, specifically excluded important resources and sovereignty from being passed over to the crown: rather than sovereignty, something called *kawanatanga* (governership) was ceded to the crown (Kaiwharu 1989). Yet the crown in New Zealand spent the next 130 years acting as if no possibility of Maori sovereignty existed. The historical denial of indigenous peoples' identity and even existence in colonial regimes, the imposition of new territorialities, new boundaries and new criteria of legitimacy, and the post-colonial empowerment of entities rooted in the colonial denial and destruction of indigenous peoples, is deeply entrenched in the ways of seeing and ways of thinking that characterise much of the dominant paradigms of industrial resource management. Using the illusion of a present and future orientation that denies links to the past, vested interests privileged, empowered and enriched by the dominant paradigm assert the need to 'move on' and 'forgive and forget'. Such arguments seek to negate the continuities of geography, history and society. They seek to detach the threads of the social fabric from their complex roots in country, culture and political economy. And they seek to assert as objective, dispassionate and unquestionable, a way of thinking that would render invisible the threads, the fabric and the stories they weave.

The triple helix of complex geography

Despite its importance in shaping the Realpolitik of resource decision making, the extent to which geography is actually understood in developing public policies affecting resource regions often seem negligible. Conventional Western wisdom continues to assume the value of industrialisation and development. Regional policies target regional development, the regions involved are rarely tackled in their social, cultural, political, economic and ecological complexity. Governments, resource companies and settler communities place industrialisation and development at the centre of regional narratives. They assert a right to make and remake history and geography in their own preferred images and to disregard and even destroy the histories and geographies of peoples who are marginalised, dispossessed and negatively affected by so-called development processes. Regional development has been conceptualised as something which regions should do – and if they won't or can't do it for themselves, then it should be done to them, with government support if necessary.

Large-scale resource projects are attractive engines for regional development outcomes for many governments. Competing notions of what constitutes regional development, however, produce competing visions of the place of resource industries in new local and regional geographies of the localities which host resource projects. Government emphasis on resource projects rather than regional (and cross-scale) dynamics needs to been turned on its head to achieve improved resource management practices and more sustainable, just, equitable and diverse regional futures. Without this, we risk continuing to treat resource regions and their populations as objects to be harnessed to the service of industrialisation and development (and to the specific service of resource-dependent governments, industries and corporations) rather than harnessing these activities to the service of humane development.

In moving beyond project-specific orientation, we need to conceptualise complex geography – at multiple scales – in a holistic and integrative way. We need to better contextualise the flows of benefits and costs from resource projects to their host localities (and wider-scale host communities). In conventional narratives of regional development, resource projects' success in economic terms alone is sufficient to constitute an appropriate goal for regional development. A more holistic regional focus, with appropriate reference to social justice, environmental and cultural sustainability, and economic viability involves a shift in focus for resource managers away from the dominant technocratic paradigm.

One can think of this shift in terms of a triple helix that winds together interdependent, ever-changing, dialectically interacting biophysical, socio-cultural and politico-economic processes and relationships at a variety of

Figure 3.6   The triple helix of complex geographies. Complex geographical totalities, such as resource localities or resource mangement systems, can be envisioned as a 'triple helix' of interweaving biophysical, socio-cultural and politico-economic systems

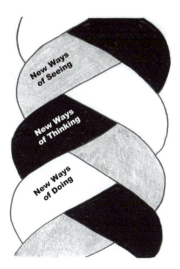

Figure 3.7   The triple helix of better resource management. There is no simple solution to the challenge of achieving 'better' resource management systems. The constant interaction of vision, theory and action provides the threads from which improvements can be woven

spatial and temporal scales (Figure 3.6). This can also be visualised as a recognition of the sorts of changes advocated in this book – changes in our ways of seeing, changes in our ways of thinking and changes in our ways of doing things (Figure 3.7). These two figures seek to show that changes in both material and discursive realities are interdependent and simultaneous. Each influences and overdetermines the other; each creates conditions for further developments in the other.

The Aboriginal notion of country parallels the situation represented in these figures. On the one hand, they weave together the biophysical, socio-cultural and politico-economic domains into a complex geography in the way that Aboriginal notions of country weave together geomorphic, mythic and social landscapes into an animate totality of known places – 'a living entity with a yesterday, today and tomorrow, with a consciousness, and a will toward life' (Rose 1996a: 7). In representing material geographies in this way, I am responding to Suzuki's call for the beneficiaries of industrialisation and development to set about creating 'a radically different way of relating ourselves to the support systems of the planet' (in Knudtson and Suzuki 1992: xxxv). On the other hand, I am seeking not just to 'spatialise' the discourses of social science and resource management, but to weave into those debates discursive geographies that are holistic, responsive and accountable. The dominant discourses in social science and resource management generally marginalise space, or rely on shallow and simplistic notions of space in which globalisation and levelling of playing fields replace living geographies with atrophied surfaces. In resource management, the technocrats silence voices from marginal places and marginal peoples and amplify voices of advocates of industrialisation and development. The dominant narratives of resource development propose that 'history' begins only when a locality is linked by industry to the wider world, and that the important speakers in such narratives are company decision makers, government policy makers and the beneficiaries of development. They encapsulate a way of thinking which is simplistic, categorical and inadequate for the task of rethinking resource management in terms of justice, sustainability, equity and diversity.

The multiple voices that characterise social, economic, political and cultural life in resource localities are silenced by the linear narratives of progress and development that subsume everything about a place and its people into the story of a resource megaproject. Like Columbus' new world, these places are 'discovered', tamed and developed. They are brought within the narrative of industrialisation and development and their meaning reduced to their part in that narrative. The dominant narrative replaces the confusing clamour of diverse voices with a generalised and homogenised monologue around the project. The people whose lives embody those marginalised voices – and those elements of the complex geography without voices that can be heard above the din of development – are displaced and devalued in favour of the common currency of jobs, revenue and trade as the measures of successful management of country. Dissident voices persist, but are easily labelled as troublemakers and undesirables whose dissent is parochial and places them outside the unquestioned 'community' whose common good benefits (by definition) from the development engendered by the project.

What happens to the complex geography of places caught up in these narratives? What happens to the people whose pasts, presents and futures are woven into the fabric of these geographies? Post-modernism, feminism and environmentalism have helped to place difference, diversity and otherness on the conceptual and practical agenda of social scientists. Real-world resource geopolitics have helped to place them on to the conceptual and operational agendas of resource managers. The polyphony that characterises the complex geographies of resource localities comes not only from their local scale diversity, but also from their multifaceted linkages to other places and other scales – the 'things that tie one locality to many others in a myriad different ways' (Massey 1993a: 144).

While its imperialist linkages must be acknowledged and their consequences addressed (Howitt and Jackson 1998; Driver 1992; Smith and Godlewska 1994), some of the traditions of geographical study provide useful foundations for developing an approach to resource management which addresses rather than avoids this complex geography, and for understanding the dynamics of the local and regional roles of resource industries. The conceptual toolkit developed here offers geographers and others a foundation for researching, judging and responding to the geopolitics of resource management – for rethinking the professional practices and implications of resource decision making in the post Cold War world. The three axes that have so strongly influenced discourses within human geography over the closing half of the twentieth century (the structure–agency axis; the time–space axis; and the integration–disintegration axis) and the discipline's five foundational concepts provide a valuable reference point for considering exemplars of resource management as both 'solution' and 'problem'.

This leads directly back to the tension between holism and reductionism in evaluating the *purpose* of investment in resource projects in host regions. It reflects the tension between bottom-up, community-centred and top-down, government- and company-centred approaches to such evaluations. At the moment, limiting the range of issues evoked for such evaluations to matters such as maximising jobs (an ambiguous notion at best in capital-intensive resource industries), maximising government revenues, maximising shareholder benefits, and serving world markets competitively, is acceptable within the dominant institutions. In other words, the dominant technocentric paradigms (and the political structures they support and are in turn supported by) assume it is politically acceptable to ignore questions of just how resource investments address the goals, values and aspirations of the host communities in the resource region. Advocates of the dominant paradigms, often using appeals to the ostensible objectivity of value-free science and value-free economics, assert the values of developmentalism and industrialisation over those adopted as 'core values' in this analysis. Their approach disguises the relationships between resource management systems and power, and the ways in which such discourses, entrenched in and reinforced by the institutional structures of regulation, investment and governance, constitute one of the critical interfaces of knowledge and power in resource geopolitics.

It is, therefore, worth restating that every aspect of resource management practice is value laden. The resource management systems in which we operate – to which we are all connected for our survival – simultaneously produce commodities *and* power. They are simultaneously production systems *and* political structures. Inevitably, these systems also produce, reflect and reinforce values. Because they produce power, they are also able to undermine the material, ideological and epistemological foundations of other value systems.

The core values of justice, sustainability, equity and diversity have been key reference points for human geography. Although they are not adopted here as 'universal', they provide a humane, generous and credible foundation for dealing with the very real challenges arising from contemporary resource geopolitics. It is important to recognise that these values are not universal, because it shapes the sort of truth claims that can be justified from them, and the sorts of propositions for changed practices that might be pursued because of them. For example, the goal of ecological sustainability is accepted (unevenly) by many national governments, and may be emerging as what might become a very widely accepted societal value. There is, however, little agreement even amongst its advocates of how best to pursue this goal. The need to use sustainability as a criterion in social and economic decision making is not accepted as urgent, or in some cases even necessary, by some sectors of society. The 'geological imperative' – if it's there dig it up, cut it down or kill it, and sell it while you can – (Davis and Mathews 1976; Howitt 1979) continues to influence Australian resource management decisions and to apply narrowly economic criteria to justify outcomes that are economically irrational at wider space–time scales.

Clarity about the personal and collective values that underpin decisions provide an important foundation for dealing with the challenges presented by resource geopolitics. There are no unambiguous answers to the difficult questions involved. There is rarely an unequivocally 'right' way of doing things. But neither are all possible choices equally good – nor are they, as some postmodernists would suggest, equally meaningless. Some choices are 'wrong', but being clear about that requires clear criteria against which to make such judgements. The conceptual framework, the resource manager's toolkit, advocated here enables one to avoid naïve relativism and being immobilised by the effort of trying to deal with positions and courses of action which all have some relative merit. In terms of the core values adopted here as the reference point for making such judgements, some outcomes are clearly better and more desirable and more worth fighting for than others – and some are demonstrably unacceptable (and worth fighting against). Our values (and the scales at which we understand them to operate), therefore, provide us with a crucial reference point for judging right and wrong, better and worse, more and less acceptable, more and less desirable. And it is these value judgements that provide a basis for framing individual and collective judgements about resource projects and their biophysical, politico-economic and socio-cultural consequences.

4 Beyond 'negotiation'
Rethinking conceptual building blocks

Negotiation is often advocated as an avenue to better resource management outcomes. Conflict resolution, alternative dispute resolution and mediation offer promise for an improvement on previous histories of exclusion for indigenous groups (O'Faircheallaigh 1996b; Ross 1999a; Lane and Yarrow 1998). Although there are pitfalls in alternative dispute resolution as a mechanism for redressing injustice (Beattie 1997), there is also a range of negotiations about country under way in Australia as a result of recognition of native title. The effectiveness of this approach in delivering the sort of integrated outcomes acceptable to Aboriginal groups remains to be proved.

Principled negotiation

It is easy to assume there are only two ways to negotiate – being soft and giving in, or being strong and getting your own way. This approach to negotiating sees power as being something that is pushed on to someone else ('power over others'). In indigenous politics and culture, self-determination has always been important and that means keeping power over oneself and resisting other people's efforts to impose their power over you. Effective negotiation is an important part of indigenous self-determination.

One approach to effective negotiation that has received considerable attention is known as principled negotiation (Table 4.1; Fisher and Ury 1991; Ury 1991). This approach was developed at Harvard University during a project dealing with negotiations at every level from the interpersonal to the global. The principled negotiation approach assumes that everybody is a negotiator because we negotiate about things all the time: whenever people come at things in different ways and work out how to deal with them, there is negotiation going on. While negotiation might be common, successful negotiation in resource management systems is hard work. In particular, cross-cultural negotiation involves setting rules that recognise indigenous people as genuine stakeholders and unsettle many taken-for-granted assumptions underpinning conventional resource management.

Table 4.1 Three views of negotiating

Power-over-you **Soft negotiation** *(positional bargaining)*	Power-over-others **Hard negotiation** *(positional bargaining)*	Self-determination *(power-over-self)* **Effective negotiation** *(principled negotiating)*
participants are friends	participants are enemies	participants are problem-solvers
agreement is the goal	victory is the goal	a wise outcome achieved efficiently and fairly is the goal
back down to keep friends	demand back-down to stay friends	separate the relationship between people from the problem(s) to be solved
be soft on the people and the problem	be hard on the people and the problem	be soft on the people and hard on the problem
trust others in the negotiations	distrust others in the negotiations	don't rely on trust but on evidence and action
change position easily	dig in to your position	focus on interests, needs and priorities rather than taking positions
make offers	make threats	explore options realistically and carefully
tell the other side what your bottom line is	mislead the other side about what you really need to get out of the process	avoid having a simple bottom line that you have to reach, but be clear about non-negotiable issues
accept unfair losses to achieve agreement	demand the other side gives up things as the price of agreement	work together to invent options that give mutual gains (win–win solutions)
search for the single answer the other side will accept	search for the single answer that you will accept	develop multiple options to choose from, decide later
insist on agreement	insist on your position	insist on using agreed criteria for assessing proposals and outcomes
try to avoid a contest of wills	try to win a contest of wills	try to reach a result that is independent of either side's will
yield to pressure	apply pressure	reason and be open to reasons; yield to principle and information, not pressure or deadlines
accept their documentation	insist on your version of documents and records	work together on a single version of important documents

Source: Based on Fisher and Ury 1991 and Ury 1991.

The persistence of indigenous rights unsettles dominant ideas of industrialis-ation and development as unproblematic goals for regional economic policy. In the United States, some Native American nations retain rights that unequivocally predate the American Constitution and were not subsumed by it. These rights include significant economic interests in sub-surface minerals, surface and sub-surface water, timber and wildlife resources. In Canada, gov-ernment efforts to discipline and extinguish common law and treaty rights have produced highly significant shifts in public policy, including political restructuring, constitutional reform and new theories of economic relations between indigenous and settler nations. In New Zealand, taking the principles of the Treaty of Waitangi seriously has contributed to the emergence of a new resource management and planning regime in which Maori values influence how regional economic decisions are made. Australia's courts and parliaments were late in entering this arena (McHugh 1996). Acknowledgment of indige-nous Australians' rights stemming from pre-colonial social formations and recognisable by the common law as 'native title', unsettles assumptions that underpin policy settings, community values and perceptions, legal and regula-tory infrastructure, and discursive communities that shape regional economic development policy and practice. In doing so, new discursive and material spaces in which different foundations for weaving economic, social and envi-ronmental justice into the social fabric are opened up. This chapter seeks to explore some of those spaces as an avenue to considering the ways in which discursive practices and the conceptual building blocks that underpin them, affect material outcomes. The chapter suggests that, pursued in isolation from a wider questioning of power relations and the conceptual building blocks of industrialisation and development, negotiation and other forms of alternative dispute resolution may have only limited success in producing 'better' resource management outcomes.

For many indigenous groups, opportunities to participate in resource-based economic activity on more equal terms are eagerly embraced. In jurisdictions around the world, diverse partnerships are emerging between indigenous groups and commercial interests. Emphasis is often placed on training, employment and production across a range of industries, particularly mining, tourism and agriculture and grazing. Indigenous economic development programmes target strengthening communities' economic base, servicing community needs and diversifying economic activity with varying success.

At a deeper level, however, recognition of indigenous rights challenges the basic building blocks underpinning regional and resource development poli-cies. Economic relations in Australian indigenous societies have always defied the conventional categories of economics. Where social relations (people-to-people relations) are ontologically embedded in ecological–economic rela-tions (people-to-country relations), categories such as 'economic base' and 'ideological superstructure' are unhelpful. And where the foundational con-cepts of 'Dreaming' can best be characterised as 'everywhere' and 'everywhen' (Stanner 1979), categories such as 'growth' and 'private profit' are difficult to grasp and operationalise. Gibson-Graham (1996) challenges the extent to which capitalist epistemology is embedded within the categories used to describe and analyse economic relations and economic processes. For indigenous peoples, the failure to incorporate even such basic elements as subsistence production into national economic statistics, or to see 'caring for country' and maintenance of indigenous cultural capital as 'productive activity' reinforces both economic and social marginalisation. And the political declaration of profit, growth and development as the singular measure of economic success entrenches environmental exploitation and cultural alienation as the fundamental basis for indigenous participation in what is widely admired as Western pluralist democracy – what Cramér refers to as the 'cleptocracy – extractive exploitation' (1994: 55). This chapter seeks to explore this discursive space in terms of resource management. It takes seriously the challenge of responding to indigenous epistemologies in the economic arena.

Challenging the conceptual building blocks in regional development discourse

Five key ideas in regional economic development discourse and resource management warrant careful interrogation. They are:

- planning;
- management;
- capacity building;
- institutional strengthening; and
- negotiating.

Much of the policy aimed at nurturing improved on-the-ground outcomes for indigenous people emphasises these strategies. Community planning and regional planning exercises are entrenched in many government, community group and private industry procedures. Planning has become the almost persistent imposition of linear notions of time (and bounded notions of space) upon social and economic activities that have previously been accountable to different values. Good management is seen as the unquestionable goal of economic planning, yet in epistemological structures that are radically ex-centric, with human affairs contextualised in sentient landscapes, management as such is almost literally unthinkable. And when it comes to those key developmentalist interventions of 'capacity building' and 'institutional strengthening', we are confronted with epistemological differences about 'capacity', 'institutions' and 'strength'. Similarly, in seeking to 'negotiate' outcomes, there is often profound misunderstanding about goals, purpose and process in even non-conflictual arrangements.

Leaving these concepts unquestioned leaves the epistemological dominance of Western liberalism (the cleptocracy) not just unchallenged, but invisible. It is part of the 'common sense' approach to resource management that reproduces injustice, inequity, intolerance and unsustainability. In rethinking the concepts that indigenous peoples might use as building blocks in shaping alternative economic futures, we really do need to interrogate the terms of engagement that set the parameters of action and debate. Strategically, in seeking to decolonise the discursive and material spaces in which indigenous peoples are implicated, we need to construct building blocks that mean something to people on the ground – we need to reconceptualise them, indigenise them and continually interrogate (and reinterrogate) them for deeply embedded colonising effects.

Planning

Planning has been a central idea in the developmentalist agenda. On both the right and the left, planning is virtually unchallenged as a basic strategic tool for achieving social, political and economic goals. Escobar (1992b: 132) suggests 'no other concept has been so insidious [nor] ... gone so unchallenged'. There is some critical literature on the role of planning in disciplining space and controlling people to rationalist visions of the future (Healey 1997; Beauregard 1989), and some effort to connect planning theory to theoretical debates about marginality, identity and difference (Sandercock 1995). But the orientation of much of this critique is more towards how to include those that planning has conventionally excluded rather than how the epistemological foundations of planning constitute some ways of thinking, some ways of being-in-place, as irrational.

Planning is fundamentally predicated on a way of envisioning the future as open to influence by deliberate human intervention. Put simply, planning is predicated ontologically on a linear, progressivist view of time. It is rooted in a view that prioritises becoming, moving towards, achieving and goal setting. It disciplines change to a singular view of what is worthwhile, valued and desirable. Using metaphors of social engineering, it universalises one version of Western experience in what Rose (1997: 4) refers to as 'hall of mirrors' where it 'mistakes its reflection for the world'. In exploring Yolngu approaches to resource negotiations, Christie and Perrett (1996) offer some insights into the ontological constraints facing application of 'planning' in other social systems. In the Dreaming, it is time's circle rather than time's arrow that provides the fundamental metaphor of change over time. Ideology disciplines social change to conform to existing patterns, forms and explanations. What might 'planning' look like in such a setting? As Escobar (1992b: 144) puts it, 'there is a need for some sort of organized or directed social change ... [but] categories and meanings have to be redefined'.

For indigenous Australians, legal acknowledgment of persistent rights has opened up prospects to challenge systems of planning and accountability that have redefined their relationship with state institutions in the 1980s and 1990s a little (Jackson 1996; Wensing 1997). One view of the negotiation and amendment of the Native Title Act 1993 is that it aimed to make the unruly pluralism of ill-defined rights and responsibilities derived from diverse systems of customary law amenable to the discipline of planning. For Aboriginal and Torres Strait Islander negotiators, the aim was to open what Pearson (1994; 1996) calls a 'recognition space': to open possibilities of allowing unruly pluralism to take root in wider Australian society, to retain space for indigenous ways of being-in-place to provide foundations for economic, social and environmental justice that do not abdicate responsibility to a depersonalised planning system, but embeds it in the lives of those who are implicated in the economic, social and environmental relationships involved.

Management

Management is perhaps an even more problematic and invisible foundational concept in the developmentalist project than 'planning'. Its absence from *The Development Dictionary* (Sachs 1992), for example, suggests that this particular technology for disciplining populations is invisible even in many critical discourses. Yet it is discourses of management that have harnessed many efforts to liberate the objects of injustice and oppression to regressive structures of discipline and power. Indigenous self-determination is reconstituted as 'community management' – and the processes of dispossession, theft and genocide (see Tatz 1998 on these terms; also Tatz 1999) that produced those settlements that the Aboriginal affairs industry reconstitutes as 'communities', the assumptions of sovereignty and identity, the aspirations of being-in-place on one's own terms are rendered invisible. Exercising the rights and responsibilities to care for (and to be cared for by) country are reconstituted as 'environmental management', or 'wildlife management' – and the ontological primacy of the human domain at the top of the hierarchical chain of being is surreptitiously embedded in the 'management systems' that are put in place to implement 'management plans'. The idea of people as kin to other species, as co-equal occupants of places, as embedded in rather than outside and above ecological relations are not just marginalised in the process but actually overruled and reconstituted.

In mission settlements and government reserves, indigenous people's lives, resources and futures were 'managed' to conform to all manner of racist presumptions. In many ways, the best that white Australia had to offer indigenous people was a well-intentioned and dehumanising paternalism that wanted to help the traumatised victims of history to manage better their post frontier realities. The tools of management – education, training, organisation, SWOT analyses, infrastructure plans, needs assessments and so on – were offered on terms that seemed generous to many. Special programmes to equip Aborigines and Torres Strait Islanders with the things they lacked were put in place, and a bureaucracy developed to manage it. The cultural alienation that success produced was seen as a temporary aberration. And the failures reconstituted as hopeless cases, or efforts to go too far too fast (with barely disguised imposition of a linear progressivist view of success).

Within this management-centred view of change, the persistence of indigenous rights is seen as simply another element to be managed, another tool in the manager's toolkit. The notion that it is not only residual rights that persist, but epistemological systems, value systems, cultural institutions, systems of customary law and deeply entrenched ways of being-in-place is only dimly glimpsed in the management-speak of the post-native title discourses of indigenous development in Australia. In some places,[1] diverse elements of indigenous society, economy and ecology continue to shape everyday life for large groups of people. The invisibility of 'management' as an ideological tool that constrains and disciplines indigenous conformity, the extent to which it actually disciplines not just the realities but also the imaginaries of being-in-place, makes it difficult to challenge. But one can begin to build an alternative vision if one considers the difference between 'co-management' arrangements for national parks or other areas (see Chapter 13), and what arrangements for organising land use, resource use and social relations might be developed by sovereign indigenous nations within wider processes of national governance. Co-operation between indigenous landowners and scientists or other experts would not be precluded by indigenous sovereignty – but the terms of engagement are likely to be extremely different to the typically paternalistic arrangements of co-management.

Capacity building

One of the fundamental lessons to be drawn from the development studies literature is the need for development programmes to target capacity building of the participants. Along with institutional strengthening (see below), capacity building is a basic strategy in development planning. Yet what is being built in these strategies? Whose capacity to do what is the focus of this work? Again, the embeddedness of profoundly powerful epistemological assumptions is difficult to escape. It is often people's capacity to plan, to manage, to participate in development opportunities, to conform to the linear trajectory of rationalist development narratives that is being built. And like so much developmentalist construction, this building is predicated on the demolition (or rejection) of the value of existing capacities. That unruly pluralism of cultural diversity is disciplined to conform to tightly controlled agendas of production, education, performance and good governance.

In achieving ownership of land or resources, in succeeding in setting up community-based enterprises, or managing community development employment programmes and so on, indigenous communities are often set up to fail. Resources are withheld, delayed or offered under strict and inappropriate conditions. Responsibilities are imposed without concomitant rights being recognised. Accountability is reconstituted in financial rather than political terms, and the intended beneficiaries of capacity building exercises and development programmes are alienated from them. Again, the terms of engagement are set externally to conform to the dominant verities of economic development discourse.

Institutional strengthening

The strategic partner of capacity building in the development discourse is institutional strengthening. Systems with unruly institutional arrangements are difficult to manage. The recognition space created by the common law's acknowledgment of native title does not extend to indigenous institutions unless they can be transformed to conform to the legal requirements of 'good governance' (accountability, transparency, efficiency and so on). In developing institutional arrangements to advance recognition of indigenous rights, the dominant developmentalist discourse strengthens institutions that it recognises. It seeks to reproduce within indigenous institutions those relationships and processes that characterise its own institutional forms. To return to Rose's 'hall of mirrors' image, much institutional strengthening is 'monologue masquerading as conversation; masturbation posing as productive interaction' (1999: 177).

It is important to make it clear that this critique of the epistemological constraints imposed by these terms and categories will not be adequately addressed by overthrowing one set of universals for another. Marginalised, traumatised, dispossessed and often dysfunctional indigenous societies are no more a source of universal truth than the flawed, dehumanised and dysfunctional systems whose smoke and mirrors approach to being-in-place has entrenched economic, social and environmental injustice as characterising contemporary social relations. In rethinking the building blocks of regional economies in ways that might entrench economic, social and environmental justice in the social fabric, we are unlikely to find concepts, categories and exemplars of what might be. Where even the imaginaries have been so deeply colonised by the dominant discourse of cleptocracy, we need to reshape not just the relationships of power, but also the concepts, language and images we use to describe, analyse and address the processes. We need to rethink the building blocks that come in the form of words, ideas and propositions as well as applying new analytical tools to the material relationships and processes. This presents multi-dimensional challenges as much to indigenous groups as to mainstream or progressive development agencies.

Jacobs and Mulvihill (1995: 9) coined the term 'viable interdependence', Rose (1999) uses 'situated availability' and Suchet (1999) suggests 'situated engagement' as a way of focusing on the task. Jacobs and Mulvihill provide an account of the need to problematise not just the institutions that derive from colonial circumstances, but also to recognise that decolonisation is an ongoing process that demands ongoing institutional change (1995: 13). Institutional infrastructures that were once part of a solution can become entrenched and insulated surprisingly quickly and emerge as part of the problem of achieving further steps along the paths of change. Similarly, it is easy to mistake employment of indigenous people to work within institutional structures that deny indigenous epistemologies for transformation of such structures into indigenous institutions (see Sullivan 1996 for Australian examples). Strengthening oppressive institutions (whether colonial or indigenous) is unlikely to provide a strong foundation for entrenching justice within environmental, social and economic relations.

Negotiating

In Australia, the post-native title period has seen negotiation become a catch-cry for indigenous empowerment. The identification of regional agreements, Indigenous Land Use Agreements, mediated settlements of claims and resource co-management solutions to land and resource-use conflicts in areas where indigenous people are asserting their claims has pushed negotiation into the strategic spotlight. Although this is essential and important, it is also a path beset with pitfalls. The importance of expert advice, legal sophistication and careful planning and strategising are factors that constitute 'negotiation' as an area in which the tension between decolonisation and deep colonisation is acute (see also Gibbs 1999). The imperative is to constantly challenge fundamental notions such as expertise and negotiation as containing the epistemological constraints that negotiation is meant to overcome. Vigilance and openness, then, are the inescapable imperatives for those engaged in processes that are meant to unsettle the certainties of developmentalist exploitation and empower indigenous interests within landscapes of co-existence.

Scale politics: regionalism, sovereignty and reconciliation

The intense localism of much of the political domain in indigenous affairs represents another challenge to the far-reaching rethinking of conceptual and political building blocks of just and sustainable regional economies. The economic reality of many remote indigenous areas is that there is a backlog of basic infrastructure and service provision (including housing, health hardware, transport and communications infrastructure) that will be overcome only by a revolutionary about-face from the neo-liberal bureaucrats who guard the public pursestrings. Governments often anticipate that large-scale resource projects may address some of these needs, although conservative political forces have opposed regulation to ensure such projects negotiate with local people as a matter of right, or invest in meaningful benefits for affected communities. At a time when Australian bureaucratic and political élites are seriously considering dismembering public health and welfare systems to facilitate greater levels of efficiency, discipline and control, allocation of massive public funds to undoing decades of trauma, neglect and abuse in indigenous settlements is unlikely. Inevitably, competition for resources (public funds, investment capital, tourist interest and so on) between indigenous areas is likely to be intense. And within indigenous groups, there is no guarantee that equity and the public good will drive successful indigenous operators in hybrid systems that continue to devalue many aspects of indigenous epistemologies.

There is, therefore, a scale politics to be considered. Remote indigenous areas are no more isolated from new globalising economic relations than the rustbelt and sunbelt industrial regions that characterise the post-modern global economy. Taking local indigenous epistemology seriously cannot involve denial of wider scale political economic processes. Indeed, one of the key challenges to remote and rural community leaders is to come to terms with complex material and ideological conditions as a basis for moving on. But neither can we pretend that the Dreaming is 'merely cultural' (Butler 1998) and without economic relevance and meaning.

In exploring new models of regional governance and economy, indigenous groups and their supporters (including those non-indigenous people whose rights co-exist with indigenous rights such as native title) must construct approaches that are capable not only of challenging the dominant terms of engagement that are derived from the operations of institutions, processes and relations that were predicated upon *terra nullius*, but also of encompassing epistemic diversity. There is no epistemic community that bridges indigenous, capitalist and socialist epistemologies. And a naïve or simplistic accommodation of diversity that denies the embeddedness of power and privilege in social, economic and environmental relations at all scales will reproduce the problems in new forms rather than open new possibilities. Rethinking resource management systems, therefore, involves not only complexly scaled political processes, but also cross-systemic conceptual processes.

In re-membering these reconceptualised building blocks into more just, equitable and sustainable regional economies, we must address the issue of multiple axes of identity, sovereignty and rights. If we revisit the metaphor of reconciliation, the effort we engage in is not an accountancy-style reconciliation, of bringing two sides together and balancing the accounts: imagining, building and refining landscapes in which multiple sovereignties, epistemological diversity and shifting identities co-exist without descent into human rights abuse and environmental or social vandalism is the hard work of reconciliation. It is not simply a matter of dealing with local antagonisms, local histories and local aspirations. It is not the imposition of another externally imposed (or even internally generated) 'correct line' or 'shining path' to liberation. It is not the devaluation of people of any description, but the hard work of working with those who are the stakeholders, in the contexts that shape being-in-place. This requires consideration of multiple scales as well as multiple stakeholders, and organising, analysing and refining engagement rather than strategic isolationism. And it is worth restating that this contextualistion is not just economic and political, but also simultaneously cultural, environmental and philosophical. This multiscale, multidimensional openness, then, is what underpins planning for, management of and negotiating about the viable interdependence, situated availability and situated engagement to which Jacobs, Mulvihill, Rose and Suchet refer.

Metaphors of change: rethinking resource landscapes

Recognition of native title, metaphors of 'reconciliation' and 'co-existence' and ideas of 'indigenous sovereignty' offer fertile ground for rethinking regional economic development strategies in Australia. In particular, admission that indigenous peoples are genuine stakeholders in the arena of regional economic activity – their transformation from marginalised victims of colonialism to active agents in the biophysical, cultural and economic landscape – demands that the unquestioned privileging of the developmentalist project be challenged at many levels in efforts to rethink resource management processes, policies and practices. This admission will not only see the emergence of negotiated settlements over specific sites, resources and projects, but will also see far-reaching challenges to institutional, legal, social and constitutional arrangements that have been predicated on assumptions of indigenous dispossession (in Australia, *terra nullius*). The discursive space created by efforts to meet the challenges involved opens up many concepts and strategies that have previously seemed settled. Ideas that were once fundamental to strategies for local or regional economic empowerment, need to be reconsidered. Ideas that might have once been rejected as anathema to local empowerment, might be amenable to appropriation, rethinking and new applications.

I have previously argued that 'recognition' of indigenous rights opens up opportunities for decolonisation of indigenous spaces (Howitt 1998b). Rose (1998) points out that most efforts at decolonisation are problematic, having embedded within them tendencies toward what she terms 'deep colonisation'. The tension between these possibilities may well be an ever-present, irresolvable reality (Gibbs 1999), but many professionals (both conservative and progressive) seek to establish certainty by reducing the dialectical complexities of new, open-ended discourses to unambiguous and singular closures. On the one hand, there is continued expansion of the racist-wedge politics of resistance to reconciliation and co-existence, illustrated most dramatically in Australia by the work of Pauline Hanson's One Nation (Langton 1996, 1997). In this view, the victimisation of rural economies by big capital (banks, telecommunications, transport, energy, agribusiness, resources and public administration) is exacerbated by pro-Aboriginal welfarism which nurtures dysfunctional Aboriginal communities to absorb public funding and restrict access to economic resources (particularly land and minerals) to which 'they' are not entitled. On the other hand, within some rural and remote areas of Australia there is a nascent suggestion that reconciliation and co-existence may offer economic salvation to depressed and marginalised communities. Funding for indigenous employment and enterprise development, land purchases and service delivery; financial flows from special legislation such as land rights and native title statutes; and negotiated agreements with development interests are all elements that are seen as mechanisms for regional economic recovery, with some flow-on to non-indigenous sectors (Rural Landholders for Co-existence 1998).

If the metaphors of reconciliation and co-existence are to offer a basis for building more equitable, just and sustainable economic relations in remote and rural communities around Australia, we need to consider how indigenous and Western epistemologies of development might differ, and what might be involved in community-level negotiation of new economic relations on the ground.

Dancing at the edge of the world

The developmentalist project has long sought to bring indigenous peoples' domains within the compass of mainstream economic relations. These areas' relationship to the economic heartlands of society are complex and ambiguous. The absence of development means that some resources remain unexploited, and this makes these areas targets for exploitation and investment (Pollin 1980; Gedicks 1993). The temptation is to rise to the challenge of securing sustainable regional economic development by harnessing the tools of developmentalism to indigenous goals. It would be easy to frame negotiation as a strategy for doing exactly that – to do something like 'moving towards sustainable regional economies'. Yet such a formulation subtly reinforces the almost invisible epistemology of developmentalism. It is oriented towards the linear narrative of development that this volume seeks to challenge and disrupt. Part of the implication of the argument presented here is that there cannot be an unambiguous movement towards a coherent strategic target. The implicit symbolism is about direction, progression and control. And it is exactly that which I seek to challenge and unsettle here. In a wonderful collection of essays, Le Guin (1989) sets about unsettling many of the conventional certainties of writing science fiction. She suggests, for example, that 'through long practice I know how to tell a story, but I'm not sure I know what a story is' (1989: 37). Under the title 'dancing at the edge of the world', she unsettles the smug assumption that, by harnessing the political, geographical, religious and artistic imagination, we can simply make the world as we wish it to be. In the idea of dancing, we see the embeddedness of one set of relationships and processes (the dance) in others (the music, the culture, the community); in the localisation at 'the edge of the world' we can begin to see that every edge is simultaneously a centre; and in the whole image, we can begin to escape the tyranny of the linear narratives of developmentalism, to glimpse the patterns of time's circle as embedded in these relationships and processes, alongside time's arrow. In such images, we may find opportunities to rethink the building blocks we use to shape and reshape regional economies so that we may weave into the social fabric those elements that the epistemology of developmentalism denies exists.

5 Reading landscapes

Cartesian geographies or places of the heart?

'Seeing' landscapes

In viewing landscapes, it is easy to revert to a naïve common sense as the basis for interpretation and judgement. For many observers a landscape can appear empty when the artefacts of one's own culture's presence cannot be seen. In shifting from visual observation to material engagement with these real-world geographies, miscues and hidden colonialism are easy to resurrect. Let us consider a series of images (Plates 5.1–5.12) from Australian resource landscapes, arbitrarily (but not categorically) classified into four types – 'natural', 'Aboriginal', 'industrial' and 'signed'. Such labels are used for discursive convenience, but they may hide more than they reveal. To some extent, what one sees reflects much of what one already knows or expects to see. Potentially, each viewer will see and understand different things in each image, and in each place. As symbolic representations of places, these images present us with a range of challenges. For example, distinguishing what is 'signed' from what is 'unsigned' depends on what signs you are adept at reading; and what is natural/Aboriginal/industrial is not always obvious.

Reading the country: resource management systems as new geographies

We have already discussed the idea that resource management systems simultaneously produce both commodities and power. These economic, social, cultural and environmental systems are as deeply implicated in cultural landscapes as those in which human identities are shaped – the nourishing terrains referred to by Rose (1996) and the worlds turned upside down referred to in Leon Rosselson's song. Resource management systems, and the actions (and omissions) of resource managers, also create new geographies (and histories). These new geographies consist of new places; new relationships within places and new relationships between places; new relationships between people; and between people, places and ideas. This is as true in the case of the location of a waste management facility in suburban Los Angeles as it is in the case of a large-scale mine near a remote indigenous settlement, or forest clearances in tropical homelands of tribal people. It is as true in the planning of urban infrastructure in Sao Paulo or Bangkok, as planning the resettlement of people displaced by dams in China, India and Laos or administrative decisions in the Navajo–Hopi area of Arizona, or developing management plans for conservation areas in Africa or Australia's Great Barrier Reef. In all resource management systems, the perceptions, attitudes, values, ethical standards and aspirations of those involved are fundamental to its structure and operation. One observer suggests 'the most important element of an ecosystem is the state of mind of the persons who use it' (Blay 1984: 130).

Such matters have currency in the difficult language of post-modernist social theory. They are, however, not of only academic interest; the issues and their implications are much too important for that! Different perceptions and values – different senses of place – underlie many of the geopolitical conflicts that have shaped and continue to shape social experience at all scales. Yet in the worlds of technocratic and scientistic dreaming which characterise so much of the Realpolitik of resource management, there is little room for 'sense of place' beyond the application of sophisticated Geographical Information Systems to document exactly what is there to be utilised. Such systems aim to capture local geographies (at whatever scale) in a tight Cartesian framework, where grid references, physical descriptions and quantitative measures of vectors, direction and size suffice for most purposes.

If we define resource management as a technical task, there is little room for the geographical imagination, and little reason to shift from the certainties of Cartesian space to the vagaries and uncertainties of complex geographies. Yet it is exactly this that this book has argued is a crucial element of a resource manager's toolkit:

> The distinctive quality of the geographical imagination is that it aims to grasp personal, social and environmental processes in the interrelationship. For the person who has developed the geographical imagination, no individual actions are without environmental and social consequences, and nowhere is remote, for the entire earth is implicated in each of its places.
>
> (Relph 1989: 158)

Exercising responsible judgement as resource managers[1] requires us to develop many skills, much knowledge and deep understandings. But, perhaps above all, it requires 'an act of geographical imagination' (Relph 1989: 158). This includes an ability to 'read' landscapes – not simply as if they were texts, but as complex records of interaction, interrelationship and change over time and space. To some extent, using the word read in this context may be too constraining, as knowing a place, developing a multidimensional sense of place, involves all the senses and facilities of human experience.

In the case of indigenous peoples, for example, where a strong relationship often exists between physical, totemic and cultural landscapes, many of the

Plate 5.1   Finke River, near Palm Valley, NT. The world's oldest riverbed in a landscape shaped by Aboriginal land management practices over many generations

Source: R. Howitt 1989.

Plate 5.2   Palm Valley, NT. There is always some ambiguity about landscapes. Is this 'natural', 'Aboriginal' or industrial?

Source: R. Howitt 1989.

Plate 5.3 Palm Valley, NT. 'Wilderness' tourism? 'Natural' landscape? Or tourism-based regional development?

Source: R. Howitt 1989.

Plate 5.4 Harding River, near Roebourne, WA. 'Natural', 'sacred' or just a great site for a dam?

Source: R. Howitt 1980.

Plate 5.5 Nature or culture? Where is this place? – A 'Namatjira' landscape in WA's Eastern Goldfields near Laverton

Source: R. Howitt 1990.

Plate 5.6 Andoom, western Cape York, Queensland. Open-cut mining on this vast scale reshapes entire landscapes, affecting biodiversity, drainage and culture

Source: R. Howitt 1994.

Plate 5.7   Mission River Estuary, Weipa, North Queensland. The 'edges' between 'nature' and 'industry' are often not clear-cut at all

Source: R. Howitt 1995.

Plate 5.8   Weipa, Queensland. World Heritage listed shell mound in Uningan Nature Reserve. These massive middens confirm a long cultural history in the area

Source: R. Howitt 1992.

Plate 5.9   Gove, NT. There are many ways of 'signing' a landscape. What might one read into this signage near Yirrkala in northeast Arnhem Land?

Source: R. Howitt 1990.

Plate 5.10   Who controls access? Access to land has been a key conflict between indigenous Australians and the mining industry since the late 1970s. At these sites at Comlaco's Weipa bauxite mine (left) and a WMC site near Kalgoorlie (right), there is no ambiguity about who controls access

Source: R. Howitt 1993, 1990.

Plate 5.11   Red Beach near Weipa, Queensland. What can be read into this 'order'? The sign is authorised by Comalco's Town Manager and might easily be seen as an effort to control Aboriginal camping. In fact it protects an area of Aboriginal land and a popular Aboriginal fishing spot from unauthorised tourist camping

Source: R. Howitt 1994.

Plate 5.12   Leaf litter at Weipa. What can you 'read' about the country from this? Even at the micro-scale, a new literacy of landscapes provides avenues for improved understanding. This leaf litter on a beach near Weipa on western Cape York Penninsula reveals much about biodiversity (mangrove leaves and seeds dominate), environmental controls (note the presence of burnt materials, suggesting fire as an important element of the landscape) and human actvity (the absence of plastic and other rubbish from this tidal detritus is revealing)

Source: R. Howitt 1979.

reasons for emphasising a professional literacy for resource managers which extends to this skill in 'reading' landscapes (and an awareness of the limitations of these skills), are particularly clear. When dealing with cross-cultural relationships, miscues in reading cultural information, including sense of place, are often easier to recognise than when we think we are 'at home'. Yet even when you think you are working in your home culture, miscues are common. For example, in more familiar urban landscapes, elements of the city's basic infrastructure bear very different messages for people from similar cultural backgrounds but different class, political, gender or age contexts. The rapid transit system treasured by commuters may have displaced residential communities; the luxury warehouse apartments treasured by international investors may have replaced inner-city industrial employment; the global standard sporting facilities that attract international media attention to the city's 'quality of life', might have destroyed remnant habitat of endangered species, community recreation space, or cultural heritage materials. Similarly, the exciting post-modern landscape of an international financial centre in a sophisticated downtown area may be hostile to local homeless people, or groups of teenage boys who get labelled as gangs. Treasured nature reserves can be interpreted as threatening and unsafe for women or children. Similarly, in rural settings, a city dweller's rural idyll might be a displaced agricultural producer's private hell. In no circumstances can a single reading be universally authoritative.

Many metaphors have been used to try and capture this notion of a multiplicity of dynamic meanings of place. Davidson talks of Australian landscapes as narrative (1987); Kobayashi draws parallels between landscape and dance (1989: 164–5); Myers talks about 'the country as story' and 'geography as code' (1986: 59, 66); Soja talks of the difficulty in matching the historical sequence of texts and narratives with the spatiality and simultaneity of maps and geographies (1989: 1); Duncan and Duncan consider the transformation of texts into landscapes, and vice versa (1988) (as does Myers in his study of Pintupi lives).[2] Young puts it this way:

> In observing and interpreting the landscape we are often immediately aware of the human use of resources within that particular environment … . However, … the landscape also consists of 'layers', reflecting historical processes which have resulted in its continuous transformation, and which stem from changing economic, political, cultural and demographic factors.
>
> (Young 1992: 255)

Much is inscribed into and recorded upon the landscape – either physically or symbolically – which affects resource management practice. Yet remarkably little of this information is subjected to critical analysis and interpretation. Resource managers are rarely held publicly accountable for the sometimes dramatic consequences of their demonstrable illiteracy in cultural landscapes.

The power to turn worlds upside down carries ethical imperatives that should weigh heavily on decision making.

Because of the ethical implications, it is important to tackle the tension between conventional images and metaphors used in the dominant paradigm of resource management, the Cartesian geographies and these more complex, dynamic and culturally referenced spatial metaphors. We need to challenge the image of resource management as a technical task, in which places can be reduced to dispassionate, and to some extent interchangeable grid references in Cartesian space. Resource managers, whose decisions have substantial power in people's lives, need to allow resource localities, the real settings of our work, to become *places* – to see them as imbued with multiple cultural meanings, diverse human experience, and ecological dynamism. The places in which resource management systems are embedded are objects of contested interpretation and uses. They are cultured places. They are *places of the heart*. They are not reducible to statistical descriptions of their 'resources' (as if resources are things and not relationships), nor to grid references on maps or cells in spreadsheets.

Physical/totemic/social landscapes: cultural geographies as 'places of the heart'

For many Aboriginal people, the landscape in which they live is a seamless fabric of physical, spiritual and cultural threads. The geomorphic landscape reflects and confirms the same cosmological truths that shape the relationships within the currently living community of people. Stanner's description of the Dreaming as 'the everywhen' (1979: 24), for example, points us to a fundamental ontological reference point – how cultures conceptualise the passage of time. In most Western philosophy, the passage of time has been conceived of (imagined) as characterised by sequential linearity. This leads Western cultures towards ideologies of development which imagine growth as development; more as better; past as discontinuous with the present. This is the metaphor of time as an arrow – always constructing a trajectory towards (or away) from something. It leads many non-Aboriginal people to imagine the Dreaming as a time long past, a point 'back in the beginning of time'. In contrast, many Aboriginal ontologies emphasis the circularities of time; the passage of time as a cycle, reflected in seasons, in lifecycles, in daily cycles, in complex interacting, mutually constitutive cycles in which interaction and change confirm and renew relationships. We find metaphors of breath, tide and season here. This is the metaphor of time as a circle, in which limitless growth involves disruption rather than development. Development, understood in lifecycle rather than arithmetic terms, becomes a process of realisation, not accumulation. And the Dreaming becomes an ever-present reality; a touchstone of everyday life. It cannot be conceived as a moment in a distant past.

For many Aboriginal groups, it is as if the social fabric itself is woven from a geographical weft (the land) and a historical warp (the creation narrative) and that it lies snugly over the land, simultaneously accommodating and respond-ing to (as well as shaping and being contained within) every feature, every place, every time and every past, present and potential person, in a diverse and complex ontological unity.[3] In 'reading' such landscapes even the most clever and learned outsider is reduced to illiteracy. As Muecke points out in relation to Mr Roe's ability to 'read' his own homeland (in Benterrak *et al.*,1984: 63), the notion of literacy needs to be redefined:

> [Mr Roe's] culture has insignia which represent everything of importance to it: clans, families, movements of people, classical myths and recent events, animals, seasons, plant life, the layout of the country. Do we fail to call it writing because it is kept from white or because it is erased and redrawn during the telling of stories? Must a trace endure to qualify as writing? A better word for [Mr] Roe than 'illiterate', with all its bad connotations would be the French word *analphabète* – someone who doesn't know one particular Western system of writing.
>
> (Benterrak *et al.* 1984: 63)

The teaching of such reading requires what Suchet (1999) terms 'situated engagement'. There can be no simple, singular fix to teach literacy in reading landscapes constructed in other cultures. Yet as this book has shown, the absence of literacy in complex multicultural environments is a common source of misunderstanding and conflict in many resource management systems.

Woodley (1992) examines provision of interpretative materials about local culture to tourists in remote parts of Canada and discusses the interplay of landscape, meaning and identity. In reviewing the experience of the Inuit community of Baker Lake, NWT, she notes that there was a contradiction involved in the task of developing such materials:

> Tourists are motivated to travel to remote parts of the world by a fascina-tion with different cultures. However, cultural differences between hosts and guests create communication barriers that can lead to negative interactions.
>
> (Woodley 1992: 45)

In assisting the small Inuit community develop a visitor centre, Woodley strove to find the common ground between hosts and visitors. She aimed to develop materials that might provide a basis for communication to take place between cultures. In an approach which has some parallels to the participa-tory, empowering and interventionist approach to social impact assessment (SIA) discussed below, Woodley argues for a highly participatory and ongoing interpretive planning process within the appropriate communities.

Tabula rasa to terra nullius to terra mater

Cross-cultural communication, of realising the implications of a multicultural definition of environment, is fundamental to most resource management settings. In the case of indigenous stakeholders in resource management systems, we have previously considered what indigenous groups might have that is of value to resource managers. We identified two substantial contributions. First were the philosophical principles of holism, integration and ethical responsibility – principles that are clearly embedded in ontologies such as the Aboriginal approaches discussed above. Second was the specific ecological knowledge of particular local resource systems. There is also an ethical dimension in regards to the intrinsic importance of individual and collective rights of indigenous peoples, and the value of cultural diversity.

Western ontologies risk treating the earth as some sort of blank slate, a *tabula rasa* on which to inscribe, and from which to trace, the aspirations and achievements (and mistakes) of the most powerful, the most arrogant, the most violent, the most greedy (Wolf 1982; Berger 1991; Blaut 1993). The theological imperatives of the Judaeo-Christian traditions used by some resource managers to frame their principles (for example Morgan 1987, 1991) rely on metaphors of conquering, subduing and taming the world of nature. The world of man, and I use the masculine deliberately, is separate from and superior to the world of nature, and the world of business is the highest form of civilisation: its economic and geological imperatives justifying even the most unsustainable solutions to problems so long as they are economically justifiable (see Suchet 1999 for a critique in relation to wildlife).

Treating the cultured landscapes of Australia as *tabula rasa* – a blank slate on which to compose the wonderful narratives of the story of Australian mining, or fishing, or forestry, or tourism – is fundamentally unacceptable on many grounds, yet many resource systems have done exactly that. In Australia, the metaphor of the *tabula rasa* is completely unsustainable because of the obvious presence of indigenous people. Yet, until the early 1990s, non-Aboriginal law was sufficiently arrogant to assert its singular superiority in a disputed jurisdiction, a contested landscape. Until the Mabo decision in 1992, non-Aboriginal law simply decreed that the land was *terra nullius* – as if the settled systems of law, the consistent and continuing relationships between land, myth and people – simply did not matter in the face of the British conquest of nature and people. It is such contrasts and tensions between the conventional Cartesian geometries of resources and the complex topographies and topologies of places of the heart that contextualise the ethical imperatives faced by contemporary resource managers.

6   Ethics for resource managers

What does ethics have to do with resource management?

The professional field of resource management needs to be understood in its social, political, cultural, economic and environmental contexts. There are many reasons why resource managers should be able to recognise and respond to a wide range of interactions and dimensions of change in their professional work. From the outset, it has been argued that a professional education that isolates resource managers from understanding the wider contexts of their actions would be completely inadequate. The need for clarity of values, intellectual rigour, flexibility and openness, and professional literacy which includes a degree of both technical *and* philosophical sophistication has been emphasised. The aim, in other words, has not been to advocate a particular method or set of methods of resource management, but to nurture the philosophical means of choosing and refining the most appropriate available methods – the most logical, the most effective, and the most ethical possible. As Relph put it:

> Method in the absence of philosophy opens the door for confusion and even violence because it is detached from its logical and ethical contexts.
> (Relph 1989: 150)

This approach to resource management has deliberately dissented from the dominant paradigm, in which methods and techniques are often emphasised in ways which make philosophical and ethical issues appear remote, irrelevant and slightly comic, in order to challenge and unsettle it. In exploring a range of ways of doing resource management, it has been shown that every aspect of resource management practice is value laden. The resource management systems in which we operate – to which we are all connected for our survival – simultaneously produce commodities and power. They are simultaneously production systems and political structures. Inevitably, these systems also produce, reflect and reinforce values, meaning and identity. Because they produce power, they are also able to undermine the material, ideological and epistemological foundations of other value systems.

The four core values used as a reference point for our exploration of resource management systems (ecological sustainability, social justice, economic equity and cultural diversity) have been key issues in the discipline of geography for a long time. Although they are not universal, they provide a humane, generous and credible foundation for dealing with the practical issues we face. Clarity about personal and collective values provides an important foundation for dealing with the ethical challenges we all face in working at the interface of resource management systems with other aspects of society. As glimpsed in the case studies considered above, there are no clear and unambiguous answers to the difficult questions raised in the ethical domain. There is rarely an unequivocally right way of doing things: but neither are all possible choices equally good – nor are they, as some post-modernists would suggest, equally meaningless. The approach developed here should not reduce us to naïve relativism, immobilised by the effort of trying to deal with positions and courses of action which all have some relative merit. In terms of the core personal values referred to, some outcomes are clearly better and more desirable and more worth fighting for than others – and some are demonstrably unacceptable (and worth fighting against), in the context of these values. Our values, therefore, provide us with a crucial reference point for judging right and wrong, better and worse, more and less acceptable, more and less desirable.[1] And it is these value judgements that provide a basis for dealing with ethical issues.

Resources, power and values: traditional and industrial systems

In the resource management systems developed by small-scale traditional societies, where local needs were the driving force for the use of local resources, common values are central to the identification, use, management and replenishing of the resources which provided the means of survival. Like large-scale industrial resource management systems, these systems reinforce power structures – some of which are far from equitable, and some of which are unsustainable. The position advocated here is no nostalgic romancing of small-scale, indigenous and traditional systems, but what is significant about them is that:

- These traditional systems generally produced use values – things which were used in social life;
- They relied directly on successful management for seasonal survival;
- The scale of production was such that even in the event of a catastrophic failure of management practices, damage was geographically limited.

The development of larger scale industrialised systems of resource management focused on production, and built on the international stage created by colonialism and the commodification of resources as industrial raw materials, changed each of these points of reference. In the process of alienating these systems from those who participate in them,[2] large-scale industrial resource management systems have relied on:

- Commodification – production of exchange values rather than use values (trade rather than local use);
- Geographical spread of risk – mismanagement in one part of the global system does not threaten the survival of the total system, because places themselves have become commodified and interchangeable;
- The scale of production and impacts is of sufficient scale that catastrophic failure of management practices has widespread consequences, with planetary scale impacts an acknowledged possibility.

The overlap between the professional practice of resource managers and the construction of social, economic and political power in industrial resource systems is inescapable. Even a brief consideration of the interplay between state and corporate power in industrial systems, and their impact on indigenous people and other local resource management systems, will confirm that the patterns of empowerment and enrichment produced by the industrial systems are accompanied by parallel processes of disempowerment and marginalisation (Howitt 1993b).

Inevitably, resource management can be constructed as both part of the problem and part of the solution. What is clear, however, is that the professional practice of resource managers involves constant and substantial engagement with issues to do with values and ethics. All resource management demands value-laden choices, and the constraints on those choices involve notions of ethical standards (among other things such as costs, quality, risk and so on). In responding to these issues, we are inevitably required to deal with the overlap between ways of seeing and ways of doing. In other words, it is not sufficient to consider these questions in abstract. They are concrete questions of practice – what do resource managers *do*, rather than how do they think about what they want to do.

Ethics in professional practice

In the real world, issues related to values and ethics are much less straightforward in many ways than they appear within the comfort of an academic critique. There is a substantial difference between talking about making decisions and actually making them. Recognising that decisions are also shaped not only by technical understanding but also by the (interacting and dynamic) influences of personality, education, culture, class, responsibilities, relationships and so on, the picture becomes even further complicated. Where one's ethical 'bottom line' is drawn in practice may be quite different to where one would like to think it will be drawn in abstract. Some situations may be sufficiently extreme to warrant some specific action. For example, a covert proposal by an employer to dispose of nuclear waste in an uninformed community would probably provoke all of us into becoming whistle-blowers. Such circumstances, however, are rare. Most situations are many shades of grey rather than clearly black-and-white. Competing readings of circumstances, divergent vantage points and differential understanding of imperatives all place different interpretations on ethical concerns and consequences of actions and omissions by resource managers. What of the situation that involves a marginal increase in public risk, in return for maintenance of employment and local incomes in recessionary times? What of a situation which compromises a specific minority interest in order to produce a substantial benefit to the majority population? Consider the following brief examples.

Values and ethics in practice 1: to leak or not to leak?

In 1976, soon after Aboriginal Land Rights (Northern Territory) Act was passed, one of Australia's leading mining companies undertook a confidential assessment of the mineral potential of every Aboriginal Reserve in remote Australia with a view to obtaining pre-emptive exploration titles prior to any legislative move to recognise land rights. An alternative view is that the company was seeking to establish a land bank of exploration titles that would give both political leverage and a long-term exploration base. Given the likelihood that existing interests such as mining and exploration rights would act as a restriction on Aboriginal people receiving full recognition, what should a company insider have done when they became aware of the company's strategy?

The company involved employed an Aboriginal Liaison Officer, who had grown up on a reserve in southern NSW. Although part of the 'stolen generations', this man's cultural and social background clearly provided a different set of constraints on his action in this situation compared with virtually everybody else inside the company. Faced with this dilemma, the Aboriginal Liaison Officer chose to leak the documents to Aboriginal and environment groups. His decision had substantial personal consequences (he was sacked), as well as much broader implications. The company's reputation as an antagonist of Aboriginal rights at that time was reinforced in the activist community – but the company's internal culture of mistrust of and hostility to Aboriginal people was also consolidated by this incident. For many company officers, the man's action confirmed the company's dilemma in dealing with Aboriginal issues: Aboriginal people simply could not be trusted, and were impossible to deal with. The mindset became further entrenched, and this no doubt contributed to the sort of strategies adopted by the industry in the late 1970s through to the 1990s, characterised by dealing as 'good neighbours' rather than recognising indigenous interests as genuine and legitimate stakeholders (Howitt 1998b).

In this case, the direct and indirect consequences of the decision made by the particular resource manager reverberate across wider geographical and temporal scales. There was no simple 'best answer' to the question of what should be done in such situations. Personal ethics and values shape the professional standards for all of us.

Values and ethics in practice 2: sustainability

A second example for consideration is the question of sustainability. For many resource management students, opportunities for professional employment as resource managers will involve participating in the utilisation of non-renewable resources. In the context of debates about sustainability, this presents difficulties. By definition, non-renewable resource management systems are not sustainable in the longer term. It is not just the operational phases of resource management systems that need to be managed equitably, but also the closure stage. What is to be left behind? In the past, many local interests have received little more than the dust left behind by the mines and mills after closure. Whole communities are often expected to move on at the convenience of the producing companies (Thomas 1975, 1979). What is the ethical responsibility of the resource planner in that sort of situation? How does one balance loyalty to shareholders, to employer, to colleagues, to personal preferences and so on? What are the ethical imperatives in this sort of situation?

Norgaard (1992) notes that the institutional sources of values (church, science, state and education) are linked. The combined values of and institutional support for scientism, developmentalism and statism for most of the nineteenth and twentieth centuries effectively sanctioned resource managers to act as they saw fit in the public interest:

> The public sanctioned technocrats, engineers, agricultural scientists, foresters and planners to act – to combine publicly held values with scientific knowledge – on their behalf. This sanctioning was rooted in a common vision of progress and a shared faith in how Western science and technology could accelerate development.
>
> (Norgaard 1992: 85)

In effect, the debate about sustainability represents a crisis in common values. There is no longer a common vision, if indeed there ever was, if one tries to include the visions, values and aspirations of those who were excluded or marginalised from the 'common visions' of the past. Nowhere is this clearer than in the field of environmental issues, where the failure of science and technology to identify problems where others are already seeing a major crisis has emphasised the diversity of values driving current debates. As Norgaard puts it, the late twentieth century has seen:

> A pastiche of dialogues between people of different economic, environmental, and ethical understandings working in international agencies and academic institutions. Joined by leaders of national governments, nongovernmental organisations and traditional cultures, from industrial and developing nations alike, this discourse is steadily transforming our understanding of the desirable and the possible.
>
> (ibid.: 89)

While Norgaard might assert that 'sustainable development is accepted as policy', the reality is that putting such a policy ideal into practice is far from straightforward or uncontested. But what we have considered in this course, particularly drawing on the criticism of conventional resource management systems by indigenous peoples, is the need for a new ethic.

This same issue echoes in the crisis in modes of representation of human experience; of the inadequacy of traditional models of that experience; of the need for management systems which are cognisant of wider implications of certain actions – of cumulative impacts, indirect impacts and overlapping consequences. We have heard of the desperate need for indigenous groups to achieve renewal and recognition, to have autonomy:

> Among all these Indian groups, people so mired in the interminable process of fighting against or lobbying for programmes, applying for funding, and wrestling with the demands of those who insist that theirs are the social programmes that provide the single, best agenda for being Indian, that they scarcely have time to meet with one another and determine how they themselves might nourish their own culture.
>
> (T. Johnson 1991: 26)

We hear the same echoes in much of the post-modernist and feminist critique of dominant industrial and social practices in the West. The idea of a totalising discourse that could proclaim a definite set of true answers to the fundamental questions – whether derived from neo-classical economics, science, Marxism, religion or political doctrine – is strongly contested in the post-modern period. In most resource management systems, and the overlapping political, economic and social systems in which they are embedded, faith in scientism, developmentalism and statism continues to dominate. The dissident voices, however, now come from much closer to the mainstream than ever before. Dissident scientists such as Suzuki, dissident economists such as Shumacher (1973), Max-Neef (1992) and Ekins and Max-Neef (1992), provide pointers to the sea change under way. They also open paths for different futures.

Values and ethics in practice 3: to publish or not?

Closer to home, for me, was a dilemma faced in my work for Aboriginal people at Weipa in the early 1990s (Howitt 1992b, 1994). This work enmeshed me in the webs of relationships between Aboriginal groups at Napranum and the mining company Comalco. In 1992, I was commissioned by the executive of Weipa Aborigines Society (WAS) to review the previous twenty years of WAS's operations. WAS had been established by Comalco in 1973 as a vehicle for funding community development projects in Napranaum (then Weipa South) without inflaming the politics of Aboriginal land rights in the area. This consultancy has become central to my role at Weipa and has involved an effort to co-construct – with a number of Aboriginal people involved in the Napranum Aboriginal Corporation, the Weipa Aborigines Society and the Napranum Aboriginal Community Council, and with active involvement and support from

Comalco staff – a narrative of the area which challenges the pre-eminence of mining in people's view of it. In challenging this 'imagined centre' (Howitt 1995), my Aboriginal colleagues and I tried to overturn deeply entrenched views of development in which local Aboriginal people have felt alienated from and victimised and marginalised by development processes constructed by and for the metropolitan, corporate centre, not the remote periphery of Aboriginal Weipa. At the same time, we actively asserted a range of alternative 'centres' for a narrative of local and regional sustainability and justice. These focused on a range of Aboriginal priorities, including such diverse concerns as improved employment in mining and related industries, better training and educational opportunities, language and cultural maintenance programmes, land claims and land care issues, and improved cross-cultural programmes within the mining company.

This work presented a range of ethical dilemmas. As anybody working closely with any community organisation knows, the line between providing explanation, advice and direction is difficult to draw, and always hard to negotiate in practice. I found myself playing many roles as counsellor, mediator, interpreter and so on, as the people involved in long-running organisation-building and cross-cultural negotiations drew me into their decisions, discussions, desperation and hopes. In my previous theoretical discussion of this work (Howitt and Douglas 1983), I had emphasised the danger of engaging with mining companies because they seemed capable of appropriating even the most well-intentioned work to their own rather than community means. More recently, however, I have argued that recognition of both local and wider scale fragmentation within mining companies can provide a valuable way of challenging corporate-centred developmentalist narratives of Aboriginal communities such as Napranum (Howitt 1995; 1998b). But engaging with the company inevitably means that key action cannot proceed according to community timetables, because they are subject to negotiation with the corporate partner in the process.

The practicalities of dealing with the ethical domain in this situation can be illustrated with reference to a book manuscript prepared during research at Napranum in 1993. Publication of this manuscript was proposed by the Aboriginal people involved in the process[3] as a way of giving their negotiations over transformation of the relationship between the community and company more credibility. The book manuscript provided a wider view of the issues involved in the relationship between Comalco and Aboriginal people at Napranum which had been discussed in the original review. Under the title *Part of the Damage?* – using a quotation taken from an interview with a past chairperson of WAS – the manuscript argued that Comalco had clearly been part of the damage done to Aboriginal people at Napranum in the period since mining began in the early 1960s, recent developments had provided strong foundations for the company becoming part of the healing of this damage, through moves towards at least symbolic recognition of and respect for the continuing interests and concerns of the traditional Aboriginal owners of the mining areas, and acceptance of and support for the need to Aboriginalise paternalistic structures such as those of WAS. Like all such manuscripts, there was room for improvement, but, given the demands of ongoing discussions with the company, the perceived need to achieve some breakthroughs prior to departure of key staff, and the time constraints placed on both fieldwork and writing time by teaching and family commitments, I was reasonably pleased with the draft manuscript which was given limited circulation for discussion and approval in community and company circles in January 1994.[4]

Despite the generally positive response of Aboriginal people to the book version of the report on WAS, many of whom felt it told much of the 'real' story of their experience over recent decades, wider circulation and publication had to be delayed when it became clear that there was a high level of unanticipated hostility to the document at senior levels in the company, and that this antagonism to the book might well derail, rather than reinforce, the whole process of Aboriginalising WAS and healing the damage experienced at Napranum.

This situation faced me with many dilemmas. On the one hand, the manuscript did, I believe, present a reasonable version of a story that needs to be more widely known. My efforts to ensure that the people whose voices had so often been silenced at Weipa were included in my account of the story meant that many people felt that the book would be theirs as much as mine. Considerable Aboriginal effort and excitement had already been put into planning what photographs to include in a published version. Yet the purpose of the book was not just to bring a particular local story to public attention, but to advance a process of local Aboriginal empowerment and recognition. In a covering letter to Comalco's managing director, I put the dilemma in the following terms:

> I recognise that some of my interpretations and conclusions are unlikely to be well-received within Comalco and CRA. I hope that our different perspectives can be addressed constructively ...
>
> It is certainly not our intention that my conclusions and comments should become destructive of the work underway at Weipa. Therefore, I hope we will be able to discuss any matters of continuing concern directly, and that publication of this manuscript can play a constructive role in increasing understanding within Comalco and CRA, and the wider community, of the ways in which (Comalco's) operations affect, and are perceived by, indigenous people.
>
> (Howitt to Managing Director, Comalco Ltd, 28 January 1994)

After considerable discussion and a meeting between myself and company managers in Sydney and telephone discussion with NAC, it was agreed that publication would be delayed, and that Comalco sponsor continued research to improve the manuscript. It was also suggested that I should be prepared to provide some input into the company at higher levels, in part to increase understanding of the process under way at Weipa.

Of crucial importance in NAC's acceptance of this outcome was the Executive Committee's concern for the fate of the NAC Executive Committee's request for the Comalco trustees to wind-up the old WAS, and the keenly awaited response from Comalco to a submission to finance the transition costs involved in transforming a paternalistic WAS into an autonomous 'NAC'.[5] In this situation, I inevitably found myself confronted with conflicting concerns and aspirations. On the one hand, I recognised that the purpose of the piece was to support, not destroy, the changing relationship between the community group and the company; on the other hand, I'd put a lot of effort into the writing and explanation of the piece.

It would be possible to tell this story as if the company were seeking to silence the research. It is more appropriate, however, to see the decision as reflecting the delicate balance between the various agendas evolving around the study, and the relationships and processes into which the study was woven. Central to my understanding of these is an assumption that change is possible – that it is worth pursuing a reorientation of Comalco's local activities towards producing more favourable and sustainable outcomes for Napranum Aboriginal people – and that many aspects of the processes and changing relationships are precisely the things that are necessary for reconciliation between indigenous and non-indigenous Australians at the grassroots level. Reverting to archetypal caricatures of a censorious and conspiratorial mining company at that stage would risk negating much of the real reconciliation that is already under way. So the purpose of the manuscript, and its place in the process became valued over its concrete form as a potential book. The irony of this decision some years later is that despite further research, and extended negotiations over a native title agreement, and despite significant achievements in negotiations with another mining company, it was not until April 2001 that Comalco finalised the negotiations anticipated in 1994. In mid-1999, NAC faced financial and managerial problems that threatened their survival because the support for managerial training requested from Comalco never appeared.

One important aspect of the ethical difficulty involved here is that other people would probably draw the line somewhere different to my own decisions. For people outside the specific relationships and processes woven around the research, and into which the research has been woven, the rationale influencing our decisions may be less convincing. For people not holding our assumptions, particularly those more cynical than we have been about the potential for change in a large mining company, the outcomes to date might reflect unacceptable compromises and failures. At a professional level, what needs to be considered is the extent to which my behaviour was appropriately accountable and ethical.

New values and new ethics for new regional geographies of resources

For resource managers dealing with non-renewable resources, the challenge is how to respond to the implications of these value debates in such systems. In these resource systems, the balance of costs and benefits has clearly been weighted in favour of large institutions (nation states, companies etc.) and distant markets rather than local interests. In many places, remote from the mainstream of social and economic life, non-renewable resources provide the vehicle for connecting to the world at large. An ethic which requires implementation of management practices which put what is contributed to local people on centre stage, along with what costs are imposed locally (environmental, psychological, cultural, social and so on), and what is left behind when the non-renewable resource is exhausted, would substantially change what is acceptable, and what is possible.

For example, Rose (1988: 379) advocates a mind shift, a 'change in perception' towards wider acceptance of a land ethic in Western thought which:

> is not human-centred [and which] must involve knowledge of other living species and other living systems [and in which] (r)esponsible action can only be based on a sound understanding of what is going on in all parts of the system.
>
> (Rose 1988: 386)

In some situations, resource managers will be constrained by professional codes of ethics or codes of conduct. For example, most universities have ethics guidelines which constrain research. Some professional bodies, such as the Australian Association of Anthropologists, have codes of ethics. In these documents, general guidelines are provided to suggest how one should respond to a range of circumstances. For example, it is clearly improper to quote personal details of an informant (for instance in an anthropological study) without permission; it is clearly desirable to avoid conflicts of interests in professional domains, and to disclose them when they occur.

While such codes may provide broad guidance, and may in some circumstances carry considerable legal as well as professional weight, they cannot provide incontrovertible guidance in all circumstances and situations. In the end, it is in the complex and dynamic interface between personal values, professional standards and institutional values and cultures that most of us have to face these difficult decisions.

Part IV

Case studies

Solemn Declaration
World Council of Indigenous Peoples

We, the Indigenous Peoples of the world,
united in this corner of our Mother the Earth
in a great assembly of men of wisdom
declare to all nations:

We glory in our proud past:
    when the earth was our nurturing mother,
    when the night sky formed our common roof,
    when Sun and Moon were our parents,
    when all were brothers and sisters,
    when our great civilizations grew under the sun,
    when our chiefs and elders were great leaders,
    when justice ruled the Law and its execution.

The other people arrived:
    thirsting for blood, for gold, for land and all its wealth,
    carrying the cross and the sword, one in each hand,
    without knowing or waiting to learn the ways of our worlds,
    they considered us to be lower than the animals,
    they stole our lands from us and took us from our lands,
    they made slaves of the Sons of the Sun.

However, they have never been able to eliminate us,
    nor to erase our memories of what we were,
    because we are the culture of the earth and the sky,
    we are the ancient descent and we are the millions,
    and although our whole universe may be ravaged,
    our peoples will live on
    for longer even than the kingdom of death.

Now, we come from the four corners of the earth,
    we protest before the concert of nations
    that, 'We are the Indigenous Peoples, we are a People
    with a consciousness of culture and race,
    on the edge of each country's borders and
    marginal to each country's citizenship.'

And rising up after centuries of oppression,
    evoking the greatness of our ancestors,
    in the memory of our Indigenous martyrs,
    and in homage to the counsel of our wise elders:

We vow to control again our own destiny and
    Recover our complete humanity and
    Pride in being Indigenous peoples

                        Port Alberni 1975

7 Case studies

A research tool for resource management

Even a brief review of the resource management literature reveals how important the case study is as a method in resource analysis. O'Faircheallaigh, for example, notes that 'a substantial amount of research has now been conducted into the effects of resource development on indigenous peoples, but the existing literature is overwhelmingly empirical and case study in nature' (1991: 228). In some books, brief case studies suffice to make a general point (e.g. Burger 1990; Ekins 1992; Knudtson and Suzuki 1992; Bodley 1982; Moody 1988). In others, more detailed case studies (Connell and Howitt 1991b; Maybury-Lewis 1992; Cant *et al.* 1993; Howitt 1996) are collected to demonstrate aspects of an argument or set of arguments expressed in a general introduction. In still others, a single detailed case study forms the core of a book that seeks to contextualise a particular case and generalise from it (Brody 1981; Gedicks 1993), or an idea or process becomes the 'case' to be examined from different perspectives and at different scales: for example Blaut (1993) provides a case study of geographical diffusionism and Jacobs (1996) tackles ideas of empire and identity. It is easy to think that by 'doing a case study' we have learned something. A lot of professional education is driven by the goal of acquiring new information, new 'facts' and new content. Content-led curriculum development remains an enduring feature of far too much professional education in this field, and case studies provide an unequivocal information base for content-led curricula. In setting up a series of case studies to facilitate rethinking resource management, we have highlighted:

- The importance of interaction and change in resource management systems;
- The complexity of relations within and between the elements of resource management systems and wider scale (historically, socially and geographically) processes;
- The value of diversity and holism as principles in approaches to resource management;
- The tension between bottom-up and top-down approaches to dealing with resource management issues;

- The importance of vantage point (among other things) in understanding what it is we are looking at, participating in and responding to;
- The importance of linkages between resource management systems and between elements which operate or are constructed at different scales.

A strongly practical or applied orientation that acknowledges the importance of seeing and thinking about both the purposeful focus and wider context of resource management rather than simply collecting and organising 'facts' has been developed here. Our practical orientation has been not been treated as independent of the need to organise ideas about key concepts, core values and foundational arguments. Having given some attention to these issues, it is now appropriate to turn to the case study method as a way of tackling the relationship between thinking and action. Rather than simply 'doing a case study', then, the chapters that follow aim to demonstrate the value of the case study method as a tool in pursuing improved resource management outcomes.

In tackling case studies, both in terms of reading the literature and undertaking one's own research, several important questions must be considered. What is it that one might learn or seek to learn from a particular case study? What might one learn from a case study approach in resource management? What *is* a case study after all? Where might this method of organising information fit into concerns about seeing and thinking differently and doing resource management better? These are important questions if one is to avoid the 'passing-parade-of-case-studies' syndrome. It is far too easy to find oneself drawn into the philatelist's approach to simply collecting case studies as objects, rather than engaging with the material and discursive context and implications of the issues under discussion in any particular study.

It is unhelpful to reduce a case study to a collection of 'facts' to be documented and discussed. This risks representing case studies as disengaged description of material realities not requiring any engagement with discursive realities and theory. Within geography, for example, much attention has been given to locality studies as a particular form of case study – a case study of local relations. In reviewing the efficacy of these studies Massey explained why case studies cannot be reduced to 'mere description'. After all, she wrote:

> There is no such thing as totally neutral description uninformed by a world view of what is significant and how phenomena are linked together.
>
> (Massey 1993a: 147)

In other words, whether it is implicit or explicit, whether or not the author critically reviews it, or develops it in ignorance, even the most descriptive case study has a conceptual framework which affects its content, meaning and value. It is also worth emphasising that we are talking about *case* studies – studies which illustrate a specific case of something more general. The best

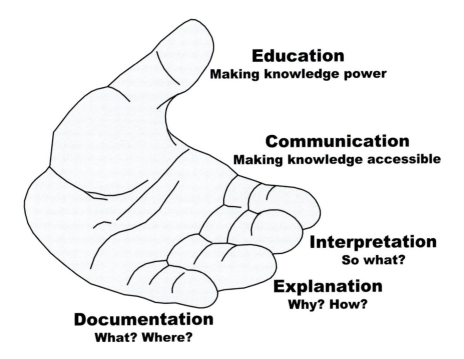

Figure 7.1   A 'five finger' model of academic responsibility

case studies make their readers specifically and critically aware of what the more general case illustrated by the particular study is.

There are several reasons for using a case study approach to resource-related research:

- To provide knowledge as a basis for understanding specific circumstances
- To provide an empirical basis for developing generalised models
- To identify common ground in reaching policy directions across a range of situations
- To provide a basis for making decisions.

In synthesising insights from cultural, social, economic and/or biophysical domains case studies offer a common research method for academic and industry research. The role of an academic researcher in resource management extends to five responsibilities – documentation, interpretation, explanation, communication and education (Figure 7.1). My own efforts have targeted these responsibilities in several ways (see Box).

Illustrating the five key areas of academic responsibility

Documention: Documenting specific circumstances and their meaning(s), providing useable documentation to Aboriginal groups about how mining companies operate, why they operate that way, what imperatives they face, why government departments operate the way they do, what environmental, social and economic consequences may arise from various courses of action and so on. In an age in which the World Wide Web gives instant access to more information than can possibly be processed, the idea of 'documentation' may seem outdated. Yet there is still a need to carefully, honestly, rigorously document and witness events, relationships, consequences and experience.

Explanation: In many academic quarters, 'explanation' is deemed the highest form of analysis. In positivist epistemologies, principal components of analytical methods may be found which purport to measure the amount of 'explanation' contributed by various components of an analysis. Explanation in resource management is rarely like that. If one's explanation is insufficient or unconvincing in certain circumstances, can it be considered scientifically adequate? Is it reasonable to continue advocating an explanation in the light of hostile responses from people (whether mining industry or Aboriginal community) whose 'reality' is being explained?

Interpretation: Analysis and interpretation is, in many ways, the bread-and-butter work of the academic. Constructing meaning from known facts is not a simple task. For students who have come to resource management from a biophysical sciences background, this volume aims to highlight the challenges that are presented in the social domain. It should have also unsettled assumptions that the nature of the task of resource management is to identify and manage causal relations. In many circumstances, the task involves 'creating meaning' rather than 'identifying causes'. In my own work, I have been involved in providing analysis and interpretation to all sorts of people. I have tried to do so for Land Councils, for local government groups, for parts of communities, for mining companies, for government departments, for the Royal Commission into Aboriginal Deaths in Custody, and for various inquiries. In emphasising meaning rather than causation, the imperative for analysts to exercise judgement is highlighted.

Communication: A lot of academic work remains in the realm of description and interpretation. Intellectuals are obliged to communicate to society about the work that society enables us to undertake. Some peer-to-peer communication is essential, but to limit our communicative efforts to academic journals is to miss something important. In communicating about our work, our thinking, our conclusions, our arguments, we hold a mirror to those we work with. As one Land Council lawyer said to me once, we are paid to think (and 'they' don't have time to) and if we don't 'think' then what?

Education: In terms of academic responsibilities, education is a more profound (and specific) process than 'communication'. Education is not about acquiring new facts. It is about developing new understandings and the means by which to generate new knowledge, and new insights. Following Freire (1972a, 1972b, 1976) it is also the case that when we 'really learn' something, we are different people.

Although case studies involve description and documentation, they also involve interpretation of information, comparison with other situations, making judgements about relevance, meaning and significance and intervention to achieve particular goals in response to conclusions from a range of cases. Research should also aim to provide an explanation of why and how things are as they are, and some interpretation of the meaning and implications of the explanations proffered. Beyond that, researchers have some responsibility to make information accessible in a variety of ways to a range of audiences other than academic peers in learned journals and books. The task of communication is one that often drops off the agenda of busy researchers who are pushed on towards the next funded project rather than making sure the people who would benefit from knowing about the last one actually get access to it. It also needs to be said that in the context of discussions of power and empowerment, academic researchers' responsibilities for educating people about their insights and understandings do not stop at the border of the university campus! It is all too easy for research-funding bodies to overlook the extent to which spending time 'on the ground' educating research participants about the implications and meaning of research conclusions is integral to case study research.

Inevitably, case studies are 'partial' in both senses of this word. No single study can hope to provide an exhaustive representation of all the elements of a particular set of circumstances, its contextual links and its historical and geographical development. So all case studies are partial in the sense of incomplete. Neither can any case study escape the implications of positionality and abstractions of vantage point that shape the way information, relationships and events are seen, interpreted and represented. Research that reaches justifiable conclusions will inevitably be partial in the sense of advocating a particular view and set of outcomes as preferable. This is not a retreat from objectivity, but recognition of the relational nature of research in the highly charged context of resource management.

One of the problems, however, is that once one recognises that everything is related to everything else, it is easy to lose focus by trying to be exhaustive and encyclopaedic in cataloguing things that are related and interesting. Whether reading or researching a case study, one needs to critically consider one's purposive focus and informative context. In other words, case studies should be framed to emphasise relevance over relatedness, and significance over interest. The aim of case study research is *not* just to collect whatever information is available on a chosen topic. Rather, a case study should use transparent and rigorous methods to illustrate, substantiate and explore the implications of a *significant* argument about a defined topic. If this is done, case studies can be a valuable method for identifying, linking and comparing issues in resource management.

This notion of comparison is taken up by Jull (1992b), who suggests that:

> The purpose of comparative studies in socio-politics [for which we could read 'resource geopolitics'] is to understand better the workings of social and political practices [to which we could add resource management practices] by examining different approaches taken in different (but somehow comparable) situations in order to find better ways to solve problems at home. Away from home we can be less blinkered by habit and prejudice, by our upbringing and commitments and we can see more clearly.
>
> (Jull 1992b: 4)

Jull also notes the importance of personal travel in comparative research:

> Only a personal visit makes sense of a place, of a context and of a situation to be compared with one's own. Without that context it may be misleading, even dangerous, to think one knows what one is talking about.
>
> (ibid.)

This raises the important issue of fieldwork in professional education (see also below). While there has been a proliferation of university courses on resource and environmental management, the prospects for including compulsory international field experience in the curriculum have become more remote in most institutions. Most student learning, and much comparative study must rely on the literature – on other people's case studies and data – to shape arguments that identify, link and compare important issues.

This limitation makes it even more important to exercise critical reading skills. This is particularly true where one finds material one agrees with. It is very easy to be ruthlessly critical of material that one opposes in terms of basic values and conclusions. For example, I find my students are capable of delivering withering attacks on a provocative piece of right-wing propaganda from *Readers Digest* published as an article of 'enduring significance in condensed permanent booklet form' under the title 'Time to stop the war against mining' (Heilbuth and Raffaele 1993). Exercising the same level of ruthless critique against David Suzuki (Knudtson and Suzuki 1992) or Al Gedicks (1993), however, typically proves much more difficult for them.

Doing resource management research

Applied research in resource management draws on diverse disciplinary backgrounds and value positions. Many university courses in this field bring people together in ways that dialectically marry the strengths of scholarship, activism

and production-centredness. Many fields of professional resource management involve applied research in various guises. In undertaking investigation,
analysis, interpretation of various aspects of complex and dynamic resource
management systems, one is inevitably drawn into the task of making sense of
diverse, often contradictory information from a wide variety of sources. Information is rarely reducible to a singular set of facts open to just one interpretation. The researcher quickly learns that relevant information sets are
constructed from a variety of vantage points, with a variety of purposes, with
different assumptions about how the world fits together, and a different scope
and level of generality. The particular sense that we make of the information
available to us (or created by us in our research efforts), will depend on our
answers to a range of questions such as:

- What is the purpose of our research?
- What is (are) the source(s) of our information?
- How might we make sense of it?
- How might we recognise and deal with entirely new information?
- What is our position (vantage point) within the particular resource man-
  agement system? What other positions exist within it?
- Who are the (critical) audience(s) for our efforts to make new sense of the
  world?

Many manuals on social and environmental research are available that discuss
specific research methods (for example Bernard 1988; Bouma 1996; Denzin
and Lincoln 1998; Hay 2000; Perry 1989; Stake 1995). Establishing which
problems are amenable to quantitative investigation, and which require application of qualitative methods is important (Dowling 2000). As we established
earlier, many important issues are not reducible to measurements that can be
analysed statistically. In other situations, the research problem is to learn to
listen to information that comes from beyond one's frame of reference. In any
setting, the effective researcher quickly learns to question the nature and
meaning of information. Reading available documents, reports, opinions and
so-called 'facts' is always a matter of applying critical skills rather than just
reading for information. In dealing with information and interpretation in
multicultural environments, the interpretation, analysis and presentation-
through-writing are often deeply embedded within each other. Interpretation
of social meaning is not something that is produced through manipulation of
a computer software program, but involves a conversational process with
informants, checking, cross-tabulating, rechecking the sort of sense one is
making of information. While the various moments of a research process may
be deeply implicated in each other, it is nevertheless, possible to discuss some
general issues of research preparation and planning. I have long advocated a
'Five P Approach' to research – **p**reparation, **p**atience, **p**ersistence and a **p**en
and **p**aper (now being displaced by a la**p**top and a **p**rinter!). Research is not a
random process. It is worth considering what one thinks is the difference

between 'applied research' and 'investigative journalism'. One important characteristic of scholarly research is the contextualisation of proposed work within an existing literature or discourse (the literature search). Opening the discursive spaces created by existing work relevant to the proposed research topic often opens up avenues for investigation, comparison and discussion. Framing responses to these questions in the form of a research proposal helps to see research in terms of engagement with real activities, places and people, rather than as an abstract set of problems, techniques and locations (see Box). While each research proposal will face particular challenges, common questions need to be addressed in developing research that is consistent with an applied people's geography approach.

Developing a research proposal

Preparing a proposal for field-based research on resource management requires consideration of many issues. Although the specific information included will vary depending on the details of the topic, some general guidance can be offered.

1   Project title

This should be no more than 10 words and should capture the 'big idea' that your research is about in straightforward terms.

2   Brief statement of the research problem and its context

In this section you should situate the particular problem/topic in relation to broad issues or concerns either related to a particular site or situation, or perhaps your course of study if you are a student. This should identify the theoretical discourse or professional debates that your research will engage with or address, and explain where any proposed case study fits in. This section of your proposal should also clearly identify the purpose(s) of the research. Explaining why a piece of research is worth doing, or worth approaching in a particular way, is important in framing questions of proposed methodology, theoretical orientation, research timeframes, resource requirements etc.

3   Principal information sources to be used (data)

You need to make clear just what information sources you will be trying to access in the field (in other words who will you interview and why; what documents or statistics etc. you will collect, and how). You also need to identify key documentary sources and the existing scholarly and other literature that will be significant in your research.

4   Other information sources to be consulted

Any research question will involve you in looking at sources of information other than those derived from fieldwork. You need to think about material in academic libraries, in government departments and in private sector locations and other places that might help you tackle your topic. You might also identify specific people and organisations whose work might be relevant.

5   Research plan (strategy)

In this section you should outline the tasks that are necessary to complete your project, what priority you will give each task, how you will tackle them, and how you will evaluate your success in each. This section must identify research methods (what you will do to collect or create relevant data), and provide a basis for thinking about ethical procedures for your proposed work (for example, how will interview participants provide informed consent?).

6   Analysis and interpretation

In this section you should indicate any particular analytical and interpretive methods and issues which you think will be important in your project.

7   Output and significance

In most projects, specific attention needs to be given to proposed output. For students, this might initially be a straightforward task in the form of a set essay or thesis. In most work, however, there is a need to consider the most appropriate forms of presentation for different audiences and the resource requirements for various alternative formats and their accessibility. You should also give some preliminary consideration to the likely significance of your research.

Dealing with fieldwork

In most aspects of resource management the compelling reference point remains the complex bundles of real-world relations – the material realities of real-world resource management systems, real decision makers, real affected communities and ecosystems, real commodity markets – rather than disembodied theoretical abstractions of these things. Many of the relevant information sets are simply inaccessible without fieldwork. The intersection of these material spaces and the discursive spaces of theory and debate that occur in research is fundamental in shaping understanding of the operations of resource management systems. Even in circumstances where we give priority to a theoretical or conceptual agenda, the management emphasis of resource management leads us towards an applied, realist, focus. The emphasis on management involves a concern about intervention – about affecting influence towards 'management' goals – whether they involve sustainability, justice, impact minimisation, profit maximisation, or some other set of issues. This is a long way from the naïve representation of geographical fieldwork as simply looking around and collecting whatever comes to hand. It is certainly not the process of seeking to identify and measure spatial causes for spatial patterns within resource management systems.

As Massey compellingly argues (1984a, *inter alia*), complex and dynamic geographies cannot be reduced to a quantifiable dimension of space. Complex and dynamic resource management systems cannot be reduced to summary statistics of production, reserves, prices, costs and so on. The 'field' is a crucial arena for developing, refining, evaluating and implementing our ideas and understandings. It simply cannot be avoided. It is also the case that 'book learning' changes shape when it is confronted with the grounded realities of the 'field'. This was certainly my own experience as a young researcher (see Box).

A formative field experience

In the late 1970s, after reading of social injustices in far North Queensland (Roberts 1975; Roberts *et al.* 1975; Roberts and McLean 1975; Stevens 1969), I planned to undertake my undergraduate honours thesis on issues of Aboriginal land rights on Cape York Peninsula. I knew there were large mining companies involved, but because I wanted to 'help the Aborigines', I thought I needed to study them! En route to the 'field' at Weipa, I was taken aside by a couple of Aboriginal activists who expressed their concern that I wanted to study the victims rather than what they saw as the cause of the problem – the mining companies. I had already read Laura Nader's influential papers (1964, 1974), and after a long discussion, my research topic changed emphasis, to focus on the strategies of the mining companies. But, again, the 'field' confounded my student book learning. Despite the acknowledged problems at the mine at Weipa where I was studying, evil people did not run the company with malice towards Aboriginal people. In managing the mine–community interface, and balancing the demands of shareholders, landowners, markets and governments, the corporate strategies I was studying were shaped by a wider range of forces and events than I had realised (Howitt 1978, 1979). Had I limited my study to company reports and existing materials, it would have been easy to continue as an ignorant critic. My field experience pushed me to become a more informed critic, and shaped my work to be more useful in strengthening Aboriginal understanding of the circumstances they faced. Its informed criticism also made it less easily dismissed by the mining companies, and ultimately opened further avenues to pursue improved outcomes in Weipa (Howitt 1995).

Of course, many challenges face field-based research in resource management. Vested interests in resource management systems create barriers for researchers. Activist communities and private interests alike try to capture researchers for their own purposes. Equipment failure, unexpected personal responses to loneliness, violence, and a host of other circumstances, cultural miscues, misunderstanding, natural hazards and unexpected scheduling problems can all disrupt field research disastrously. But of course, all this raises the interesting and deceptively simple question: 'What, and where, is the "field"?'

In human geography there is a tradition that can be caricatured as the *Boys' Own* tradition, where the field is a remote, hostile and exciting place; and fieldwork becomes an heroic and macho undertaking that tests the mettle and quality of young geographers. This tradition is linked with geography's imperial links to exploration (see Howitt and Jackson 1998; Hooson 1994; Smith and Godlewska 1994), to extensions of Europe's frontier, to the mapping (and acquisition) of new worlds, to the collection of trophies, trinkets (and land titles). In this tradition, the field is contrasted with home: 'there' is contrasted with 'here', and 'they' are contrasted with 'us'. The field, in this perspective, is inevitably constructed as, and responded to, as Other – as entirely different and disconnected from the 'non-field', as somehow alien, unfamiliar, perhaps threatening, certainly exciting and unusual. In anthropology, this sort of construction of an exotic field as the location in which fieldwork is done has produced a series of crises: the crisis of representation (Marcus and Fischer 1986; Fothergill 1992; Sardar 1992–93; Kaliss 1997), crises of authority (Crang 1992; McDowell 1994), and challenges to the privileged status of various sorts of knowledge (Kanaaneh 1997; Jacobs 1997).

In geography, the centrality of the field to the discipline has not yet produced pervasive critiques seen in anthropology since the mid-1980s, but there are serious questions at issue. Issues regarding the construction of knowledge through research, the ethical implications of the research, the epistemological implications of certain research methods and the nature of cross-cultural research transactions are now in debate within and beyond geography. One of the implications of these debates is that it is increasingly clear that virtually all research is 'cross-cultural', and that relations between researchers and the people who are the subjects of research have important methodological, philosophical and ethical implications. In other words, the field in which research is undertaken does not need to be 'out there'. Home is as much a research field as 'away'; 'in here' is a cross-cultural field equal to 'out there'. Insider fieldwork, phenomenological fieldwork, studying up into the structures of power in our own familiar world – the world of the family, the neighbourhood, the university, the everyday world of our own lives – is not somehow outside the scope of 'the field', while fieldwork that involves travel, cross-cultural research and an outsider status is (see for example Ellis 1998). Nast (1994) argues that feminist field methods emphasise research as a collective activity. She notes that:

> fieldwork allows 'fields' of everyday bodies and problems 'out there' to be incorporated into and thereby subvert what has historically been the preserve of the white, the masculine, the abstract – the ivory tower.
>
> (Nast 1994: 57)

This renders the field in which research is undertaken as politicised, gendered, classed, and always problematic. As Katz observes, this weaves together the research moment and the other domains of social life:

> I am always, everywhere, in 'the field'. My practices as a politically engaged geographer ... requires that I work on many fronts – teaching,

writing, and non-academy based practice – not just to expose power relations, but to overcome them.

(Katz 1994: 72)

So 'the field' that so richly symbolised the separation of geography's research
object from 'the home' comforts of the researchers, turns out to be far less
exotic and separate than many might presume. And as Kobayashi observes,
this interweaving of place, research and power embeds the field within the
academy:

> It has resulted in the voices of the marginalized being taken seriously, if
> only in limited contexts. It has thus empowered many to be more politi-
> cally effective.

(Kobayashi 1994: 74)

For resource management, the implication of such discussions (see *Professional Geographer* 1994; Gibson-Graham 1996) is that those elements often
treated as externalities – outside the field of immediate competence or relevance – need to be reconceptualised as internally related to professional
practice.

Even though the context may be more familiar and comfortable, a student
interviewing an academic, or surveying a fellow student/fellow shopper/
fellow resident will face many of the same issues of principle, method and perspective that need to be addressed as a resource management professional
doing the same things with an Aboriginal elder, a political decision maker, a
corporate manager or a trade union organiser. In each case, we are inescapably
involved in the construction, interpretation, testing, verification and communication of knowledge.

This leads us to deal with the cycle of research–action–reflection that characterises the approach of action research or participatory-action research
(PAR) (McTaggart and Kemmis 1988). This is a common approach in education, where there is an interventionist intention in the research itself – the
research aims to affect outcomes rather than simply document them. This
approach is also relevant to much social and environmental research; it is akin
to the interventionist intention of 'management' itself. It emphasises the dialectical relationship between research and its application and refinement,
between theory and action.

This does, however, return us to precisely those issues of epistemology and
ontology that were considered earlier in terms of 'ways of thinking'. The
knowledge and understanding produced by research is rarely in a form that
allows it to be 'discovered' as if it were simply waiting 'out there'. Knowledge
is socially constructed through social processes such as research: it is always
contextualised and situated, and it is always produced in response to questions
such as: What do we know? How do we know it? What do we want/need to

know? How might we get to know it? What are the consequences of knowing it? (What are the responsibilities of 'knowledge'?)

Writing about complexity: presenting a case study

One of the most difficult tasks that all resource managers face in tackling these issues is doing justice to their own understanding of the complexity they face when writing a report or essay. One quickly realises that the big challenge in dealing with socio-politico–economic-cultural aspects of resource management is to deal with complexity. In particular, the need to address 'polyvocality' and the 'multiple voices' that provide alternative rationales, alternative readings and alternative foundations for understanding events, processes and relationships (McDowell 1992, 1994; Rodman 1992) requires consideration. It is precisely this complexity that often makes conventional narrative styles difficult to sustain in writing adequately about resource management. In almost any example one can think of, it is essential to identify a diversity of 'players' and 'positions', to identify multifaceted and dialectical links between them across time and space. This means that in writing about complex situations, planning your writing – what to include, what to leave out – is more important than many people realise. In advising my own students about writing essays and theses, I emphasise the need to sit down and work out what they are trying to say. In particular, I insist on the need to frame the argument in terms not of the case study, but in terms of the bigger questions that the case is intended to illustrate. I also urge them to be clear about what they absolutely have to have in their essay to support what they are trying to say – to be clear about what constitutes evidence, and what it allows and doesn't allow them to say. Next, it is essential to think about what order things need to/might be able to come in. What different sort of sense of the topic will their reader make of the material presented if it is presented in alternative orders, or with alternative emphasis? What sense do they want their reader to make of the material? And crucially, how important is each section of writing or each aspect of the topic in relation to the overall word limit, audience skills and so on? Careful planning will assist a writer to strike an appropriate balance between description, explanation and interpretation in a piece of writing. Most writing has to do all these things, but you need to get the balance right. In my experience, the most critical issue in thinking and writing about complexity is time. Whether it is a student writing an essay, or a consultant writing an impact study report, leaving inadequate time for the writing and the thinking required to do it is a recipe for inadequate results. In contrast, those who give themselves time to actually think, plan and review their writing not only write better, but also understand more of the complexity they confront and will eventually have more to offer in terms of insight and practical suggestions.

8 Recognition, respect and reconciliation
Changing relations between Aborigines and mining interests in Australia

Since the early 1990s, Australia has confronted several nasty legacies of its colonial heritage in a rather blunt form. Overtly racist politicians, including the 'Independent' federal politician Pauline Hanson,[1] have peddled a 'no natives, no exotics' approach to defining Australian cultural identity. A great deal of media attention and public debate throughout this period focused on issues of indigenous rights and Australian identity that came to prominence in the wake of the High Court's 1992 decision in Mabo and its 1996 decision in Wik.[2] Community polarisation over indigenous rights has overlapped with divisions over questions about Australia's constitutional monarchy, multiculturalism as a policy framework, migration and industrial reform. Ten years earlier, when the West Australian and Federal Labor Party governments backed away from a policy that committed the ALP to a national framework to grant Aboriginal land rights (Libby 1989), the mining industry was united in its vehement opposition to such recognition. Indigenous Australians faced a reprehensible campaign of misinformation, misrepresentation and scaremongering, from which there was no public dissent among mining and mineral exploration companies.

Many things in the 1990s tempered the naïve and arrogant myopia of corporate excess in the 1980s. For example:

- Complex corporate responses to CRA Ltd's losses in the Bougainville rebellion;
- Incarceration of senior corporate figures in the notorious political–industrial alliance that became known as 'WA Inc.';
- Negotiation of effective mining and exploration agreements by Aboriginal traditional owners in the Northern Territory under the Aboriginal Land Rights (Northern Territory) Act 1976 (McLaughlin and Niemann 1984; Howitt 1991c; Teehan 1994);
- Considerable effort in reconciliation between indigenous interests and some sectors of the mining industry (see for example APEA 1988; Council for Aboriginal Reconciliation 1993);
- Recognition of the inevitability of change and the opportunities for

competitive advantage from developing good relations with indigenous groups (Howitt 1997a; Davis 1996; Wand 1996).

The interaction of these changes, along with significant changes in technologies, markets, management, regulation and all the other dimensions of 're-structuring' (Dicken 1998; Fagan and Webber 1994) produced a resource management system in which it has been possible to consider the way in which the core arguments of this book are played out on the ground.

   This chapter presents a case study of why indigenous rights and the core values advocated in this book are not appropriately considered 'externalities', but must be addressed as an integral component of the decision making landscape of resource management. The chapter focuses on the turbulent period of legislative, public and corporate debate in Australia through the 1980s and 1990s, and draws on field-based research and secondary materials. It argues that recognition of native title and indigenous identity, respect for indigenous people and their values and experience, and reconciliation through negotiated agreements about mining projects and indigenous rights in resource regions of Australia is a realistic framework for achieving more just, sustainable, equitable and tolerant outcomes in mining communities throughout Australia.

Terra nullius: the legacy of colonialism

Before 1992, Australian governments operated land and resource management systems as if Aboriginal and Torres Strait Islander peoples had not existed or held any rights prior to British settlement. Although it was a contested notion in international law at the time (Reynolds 1996), the British argued that they acquired sovereignty over Australia by right of discovery and settlement because it was legally *terra nullius* – land belonging to no one. The Australian colonies established systems of land titles and property rights that ignored the possibility of indigenous Australians holding any legal status or enforceable rights. The doctrine of *terra nullius* was not finally entrenched as the underlying principle of Australian property law until the late 1830s and early 1840s. In 1832, for example, Tasmania's Governor George Arthur urged the Colonial Office to avoid repeating the 'fatal error in the first settlement of Van Dieman's Land, that a treaty was not entered into with the natives' (cited in Reynolds 1996: 115). In further correspondence with the Colonial Office in 1935, Arthur continued to urge the conclusion of a treaty before settlement commenced in South Australia, and a similar policy was advocated in 1836 by the military commander at the fledgling Swan River colony of Western Australia (Reynolds 1996: 115). In 1840, the British Crown concluded the Treaty of Waitangi with Maori chiefs in New Zealand, where the crown faced military defeat and expulsion from its dominion (see McHugh 1991). Despite the 'air of unreality surrounding the prevailing legal and constitutional pretensions [which was] apparent to clear-eyed settlers', colonial policy in Australia contrasted with policy across the Tasman. It was also in contradiction to the longstanding paternalism of the British Crown's:

> protestation of goodwill towards tribal peoples and its assumption of a protective role in its relations with them, particularly their land rights. This paternalism is manifest in the formalities of Crown–tribe relations throughout most of the 350 years of imperial activity. For the most part Crown authority over tribal peoples and their land was established not by usurpation or conquest but by treaty. Even where tribal people were forcibly vanquished by the British their subjugation was secured formally by treaty rather than reliance upon the fact of enforced submission.
>
> (McHugh 1996: 308)

There is no simple explanation of why the British failed to heed the advice of people such as Arthur, and why they persisted with the fragile myth that crown sovereignty in Australia was established neither by settlement nor conquest but by occupation. Reynolds suggests that it was, perhaps, simply too hard:

> The vast size and the nature of the Australian continent made nonsense of theories of sovereignty which emerged in the British Isles during the early modern period. At any time in the nineteenth century there were many sovereigns in Australia and many systems of law. The fiction of settlement by occupation allowed this reality to be overlooked, something which couldn't have occurred if it had been accepted that the British established themselves by cession or conquest. If colonial government had sought to have Aboriginal communities cede sovereignty they would have been required to do the hard work of negotiation over a long time in all parts of Australia. Treaties or other formal agreements would mark the spread of British sovereignty and pin it down in time and space.
>
> (Reynolds 1996: 117–18)

The crown's self-serving paternalism has not protected indigenous peoples in the Commonwealth from the burdens of the colonial and post-colonial hunger for resources. In Australia, the legacy of the legal fiction of 'occupation' was a system of property law and resource rights that denied that 'there were many sovereigns in Australia and many systems of law' (ibid.: 117) and entrenched and enforced the prerogative of resource developers, particularly miners. State laws submerged pre-colonial sovereignties, identities, laws and rights of indigenous peoples throughout the continent (Bartlett 1993). Indeed, indigenous Australians were categorised as occupying the bottom of the 'great chain of being' (Maybury-Lewis 1992:38–9). The doctrine of *terra nullius* was confirmed in the Gove Land Rights Case (Blackburn 1970).[3] Even the action of establishing Aboriginal Reserves, an act which Reynolds (1987: 133) argues was intended to create islands of native tenure remaining above the flood tide of settlement was seen by Blackburn as a confirmation of the absence of indigenous rights, because:

> it implies not that the sovereign recognizes rights in the natives, but that it has the power to dispose for their benefit of any lands, irrespective of what the natives claim.
>
> (Blackburn 1970: 253)

Anti-Aboriginal racism has been as deeply entrenched in Australian thinking as the structural racism of *terra nullius* was entrenched in Australian law. The expansion of agricultural and pastoral settlement in Australia created a huge hunger for land and water; the discovery of gold from the 1850s whetted the colonies' hunger for resources and capital (Howitt 1993b). Industrialisation and developmentalism dominated as the twin pillars of public policy in shaping land, resource and property law.

The Mabo decision and native title

On 3 June 1992, however, the Australian High Court delivered a judgement which transformed the geopolitics of resources in Australia. The decision was the culmination of an action commenced by five residents of Murray Island (Mer) in the Torres Strait, including Eddie Koiki Mabo, in May 1982 (Map 8.1). The claimants sought a declaration that annexation of their traditional lands and waters to Queensland had not extinguished their pre-existing rights. By 1992, Mr Mabo and two other plaintiffs had died, but the High Court confirmed the existence and persistence of their right 'as against the whole, to possession, occupation, use and enjoyment of the lands of the Murray Islands' (Orders in *Mabo* [*no. 2*], see Bartlett 1993a). The key conclusions of the judgement can be summarised as follows:

- Indigenous Australians had rights that predated the acquisition of sovereignty by the British Crown;
- The common law was capable of recognising these rights and obliged to do so;
- Following settlement, the crown could extinguish native title by granting interests in land with a clear intent to extinguish native title (for example, interests which involved exclusive occupancy such as freehold and some leasehold titles);
- Unless specifically extinguished by a valid act of government which expressed a clear and deliberate intent to do so, native title persisted as a common law property interest in contemporary Australia, with the content of the native title interest determined according to the law and customs of the indigenous people connected to the land in question;
- Native title could also be extinguished in the event of loss of connection

between the people and the land, by extinction of the descent group, or
by loss of law and custom;
- Native title could only be surrendered to the crown and was otherwise
  inalienable;
- Action to extinguish native title must be consistent with the terms of
  overriding Commonwealth legislation (such as the Racial Discrimination
  Act 1975);
- The legal principle of *terra nullius*, long assumed to underlie Australian
  land and resource management systems and to exempt Australian governments from dealing with pre-colonial interests, was not supportable.[4]

Despite confirming the massive extinguishment of native title, and refusing to
consider the possibility of residual sovereign rights of indigenous Australians,
the Mabo decision was widely seen as overturning colonial relations between
governments and indigenous Australians – a judicial revolution (Stephenson
and Ratnapala 1993). The recognition of native title in Mabo [no. 2],
however, was limited. The High Court identified passage of the Commonwealth's Racial Discrimination Act 1975, as the point at which discriminatory,
arbitrary and immoral extinguishment of native title was rendered illegal.
Prior to that Act, titles created by virtue of colonial violence, theft, dispossession, removal and marginalisation were legal. The crown had a peremptory
right to extinguish native title derived from its claim to radical title on acquisition of sovereignty. As Aboriginal and Torres Strait Islander Social Justice
Commissioner Michael Dodson observed:

> In confirming the Crown's right to claim sovereignty and gain the power
> to extinguish native title, the source of that right was not scrutinised to
> any substantial degree. The High Court appeared conscious of the flaws
> in the theories by which the Crown claimed sovereignty over new territo-
> ries and 'acquired' for itself the right to extinguish Indigenous titles but it
> deemed these issues non-justiciable. The Court restated the view that the
> acquisition of sovereignty gives rise to the right to extinguish native title
> … It is the exercise of this 'paramount power' rather than the claim of
> sovereignty itself to which the dispossession of Indigenous people is
> attributable …
>
> The assumed power to extinguish the property of Indigenous people
> was sanctioned by the common law in Australia in Mabo [no. 2] … [an
> act which] amounts to the entrenchment of the legacy of colonial racism.
>
> (Dodson 1995: 78–9)

Despite its limited recognition of persisting native title at common law in Australia, governments and industry saw the decision in Mabo [no. 2] as a source
of great uncertainty and threat. The validity of titles created without regard
for native title since 1975 was brought into question, as was the status of
indigenous Australians' rights and interests in areas throughout Australia. The

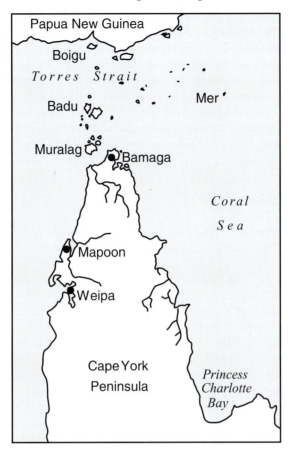

Map 8.1   Torres Strait showing Mer (Murray Island)

Mabo decision dealt only with specific claims of Torres Strait Islanders on the island of Mer, in the eastern Torres Strait. It left untested the nature and extent of native title that might remain on mainland Australia. And it left untested the question of how continuing indigenous interests in lands, seas, waters and resources were to be accommodated in specific places where a system predicated on the absence of indigenous rights purported to create new and inconsistent interests for non-indigenous interests.

In 1991, in the wake of indigenous protests during national celebrations of the bicentenary of British settlement in 1988, and protests at the Brisbane Commonwealth Games, and prior to the decision in Mabo [no. 2], the Commonwealth established a Council for Aboriginal Reconciliation. The Council was established under Commonwealth legislation, with a charter to promote a formal reconciliation between Aborigines and Torres Strait Islanders and other Australians, and a term that would expire on 1 January 2001 – the centenary of federation in Australia (Council for Aboriginal Reconciliation 1994: ch 3). Reconciliation, like the High Court's belated recognition of native title, was greeted with ambiguity amongst indigenous Australians and their supporters. The risk was that entrenched racism would see native title reduced to a restricted bundle of use rights with no contemporary benefits[5] and the reconciliation process reoriented to become a reconciliation of indigenous people to their dispossession.[6] The early 1990s, however, saw a number of events which further transformed the cultural politics of identity in Australia and interacted with the Mabo decision in ways that directly affected the geopolitics of resources.

Mining and Aborigines

> The history of the mining industry in Australia, particularly in recent years, has been punctuated by episodes of conflicts between interests of the industry and those of Aboriginal people. Several remote locations, such as Gove, Noonkanbah, Coronation Hill, Yakabindie, McArthur River, Rudall River, have been scenes of disagreement and misunderstanding ... .
>
> On the surface, it is hard to find two sets of interests which, while co-existing and interacting in contemporary Australian society, are more culturally different than those of miners and Aboriginal people.
>
> In general, the differences are stark. It can be said that Aboriginal society is locally oriented, while mining companies are increasingly international. Aboriginal people generally place a higher value on social and cultural concerns than economic ones. Traditional Aboriginal values emphasise religion, family and co-operation. By contrast, companies emphasise the marketplace and its competitive economic environment.
>
> (Council for Aboriginal Reconciliation 1993: 3, 5)

Conflicts over resources between Aboriginal groups and mining companies have punctuated Australian political, economic and social affairs on many occasions since the 1960s. The persistent pattern of 'dispossession, displacement, marginalisation and alienation in periods of rapid change precipitated by minerals-based industrialisation' (Connell and Howitt 1991b: 198), the stark cultural differences, the considerable antipathy each group has expressed for the other at various times and until recently the invisibility of indigenous rights to Australian lawmakers all underpin the history of relations between indigenous groups and the Australian mining industry.

Aborigines and Bauxite at Weipa and Gove[7]

The bauxite mining operations at Weipa, on the west coast of Queensland's Cape York Peninsula, and Gove, in northeast Arnhem Land, epitomise many aspects of both the worst impacts and best prospects for positive outcomes from mining on Aboriginal land in Australia. The identification of significant resources of the strategic aluminium ore at both sites came at a time in the post-war reconstruction of the Australian economy when development of self-sufficiency in aluminium metal production was a Commonwealth government priority. Globally, the aluminium industry was undergoing rapid change. In North America, anti-trust action was weakening the market dominance of Alcoa and Alcan, and post-war sale of government surplus smelting capacity, constructed to meet the demands of military production, enabled major new players (Kaiser, Reynolds) to enter the industry. In Europe, primary metal production was dominated by the French corporation Pechiney and the Swiss Alusuisse. Post-war industrialisation in North America and reconstruction in Europe and Japan, the expansion of aluminium-hungry militarisation on both sides of the Iron Curtain during the Cold War, and a wide range of new applications in transport, communications and other areas all underpinned the drive to identify new bauxite resources. In Australia the Australian Aluminium Production Commission was actively developing smelting and processing capacity in Australia, constructing the Bell Bay smelter based on Tasmanian hydro-electric resources.

Initial identification of large-scale bauxite resources in north Australia occurred in the 1950s, at precisely the time when this restructuring of industry structures and company strategies was very active. At the same time, Aboriginal people in north Australia were experiencing a reimposition of colonial relations in the wake of wartime relaxation of some of the strict controls affecting their lives. Christian missions at Gove (Anglican) and Weipa (Presbyterian) were actively involved in implementing government policies of removing Aboriginal children from their families and their cultures. In Queensland, where highly racist legislation imposed extraordinarily high levels of intrusion and control over the everyday lives of Aboriginal people, children at the Weipa mission were inculcated into a dormitory system where missionaries rather than parents defined basic standards, values and aspirations.

Despite being reduced to the status of non-citizens and wards of the state, Aboriginal people in the 1950s were not able to secure legal protection of their rights. In the Wik case, the claimants argued that the Queensland government had a fiduciary duty not to detrimentally impinge upon their existing rights, and that included the Comalco Act 1957, which created the bauxite mining lease granted to Comalco on western Cape York Peninsula. Damages claimed by the Wik were not accepted by the High Court:

> To permit a party to attack the validity of the Comalco agreement on the basis of alleged default or impropriety in the steps leading to its execution would be to frustrate the clear purpose of the legislation … . The fact that other persons (such as the Wik) may thereby have lost rights previously belonging to them is simply the result of the [normal] operation of the legislation which is not impugned … . the Comalco Act had the effect of giving legislative force to the Comalco agreement. To permit the Wik to

now question the validity of the Comalco agreement is contrary to the plainly intended effect of the Comalco Act. Inherent in this conclusion is the further one that damages and other relief cannot be obtained for alleged breaches of duty resulting on, or constituted by, the making of the Comalco Act or flowing from the Comalco agreement. This is so because, once executed as parliament provided, the Comalco agreement itself took the force of legislation. *This was not the usurpation of legislative power. It was the exercise of it.* The suggested injustice of the Comalco agreement and of its consequences for the Wik is not then a matter for legal but only for political redress.

(Kirby in 141 ALR 129 at 289–90, emphasis added)

Similarly at Yirrkala, where the Alusuisse-led Gove Joint Venture (Nabalco Ltd) was granted rights to mine and process bauxite and to develop a town and port, the rights of traditional Aboriginal owners counted for very little. In 1963, the Yolngu clans at Yirrkala sent their famous bark petition to Canberra, seeking acknowledgment that neither the federal government nor the Northern Territory was entitled to dispose of their country without consultation and agreement (see Box and Plate 8.1).

To the Honourable the Speaker and Members of the House of Representatives in Parliament Assemblies

The Humble Petition of the Undersigned aboriginal people of Yirrkala, being members of the Balamumu, Narrkala, Gapiny and Miliwurrurr people and Djapu, Mangalili, Madarrpa, Magarrwanalinirri, Gumaitj, Djambarrpuynu, Marrakulu, Galpu, Dhalnayu, Wangurri, Warramirri, Maymil, Rirritjinu, tribes, respectfully sheweth

That nearly 500 people of the above tribes are residents of the land excised from the Aboriginal Reserve in Arnhem Land.

That the procedures of the excision of this land and the fate of the people on it were never explained to them beforehand, and were kept secret from them.

That when Welfare Officers and Government officials came to inform them of decisions taken without them and against them, they did not convey to the Government in Canberra the views and feelings of the Yirrkala aboriginal people.

That the land in question has been hunting and food gathering land for the Yirrkala tribes from time immemorial; we were all born here.

That places sacred to the Yirrkala people, as well as vital to their livelihood are in the excised land, especially Melville Bay.

That the people of this area fear that their needs and interests will be completely ignored as they have been ignored in the past, and they fear that the fate which has overtaken the Larrakeah tribe will overtake them.

And they humbly pray that the Honourable the House of Representatives will appoint a Committee, accompanied by competent interpreters, to hear the views of the Yirrkala people before permitting the excision of this land.

They humbly pray that no arrangements be entered into with any company which will destroy the livelihood and independence of the Yirrkala people.

And your petitioners as in duty bound will ever pray God to help you and us.

Certified as a correct translation, Kim E. Beazley

Bukudjjulni gong'yurru napurrunha Yirrkalalili Yulnunha malanha Balamumu, Narrkala, Gapiny, Miliwurrurr nanapurru dhuwala mala, ga Djapu, Mangalili, Madarrpa, Magarrwanalinirri, Gumaitj, Djambarrpuynu, Marrakulu, Galpu, Dhalnayu, Wangurri, Warramirri, Maymil, Rirritjinu, malamanapamirri djal dhunapa.

Dhuwala yulnu mala galki 500 nhina ga dhiyala wananura. Dhuwala wanga Arnhem Land yurru djaw'yunna naburrungala.

Dhuwala wanga djaw'yunna ga nhaltjana yurru yulnungunydja dhiyala wanga nura nhaltjanna dhu dharrpanna yulnu wlandja yakana lakarama madayangumuna.

Dhuwalanunhi Welfare Officers ga Government bungawa lakarama yulnuwa malanuwa nhaltjarra nhuma gana wanganaminha yaka nula napurrungu lakarama wlala yaka lakarama Governmentgala nunhala Canberra nhaltjanna napurrungu guyana yulnuyu Yirrkala.

Dhuwala wänga napurrungyu balanu larrunarawu napurrungu näthawa, guyawu, miyspununwu, maypa;wu nunhi napurru gana nhinana bitjarrayi näthilimirri, napurru dhawalguyananadhiyala wänganura.

Dhuwala wänga yurru dharpalnha yurru yulnuwalandja malawala, ga dharrpalnha dhuwala bala yulnuwuyndja nhinanharawu Melville Bathurru wänga balandayu djaw'yun nyumulunin.

Dhuwala yulnundja mala yurru nhämana balandawunu nha mulkurru nhämä yurru moma ga darangan yalalanumirrinha nhaltjanna dhu napurru bijarra nhakuna Larrakeahyu momara wlalanguwuy wänga.

Nuli dhu bungawayu House of Representatives djaw'yn yulnuwala näthili yurru nha dhu lakarama interpreteruy bungwala yulnu matha, yurru nha dhu djaw'yun dhuwala wängandja.

Nunhiyina dhu märrlayun marrama'-ndja nhinanharawu yulnuwu marrna-mathinyarawu.

Dhuwala napuru yulnu mala yurru liyamirriyama bitjan bili marr yurru napurru hha gonga' yunnna wangarr'yu.

(Wells 1982: 127–128)

In his judgement in the Gove Land Rights case (17 FLR 10: 141–294), Justice Blackburn confirmed the principle of *terra nullius* as a legitimate basis for an agreement between the Australian government and an international

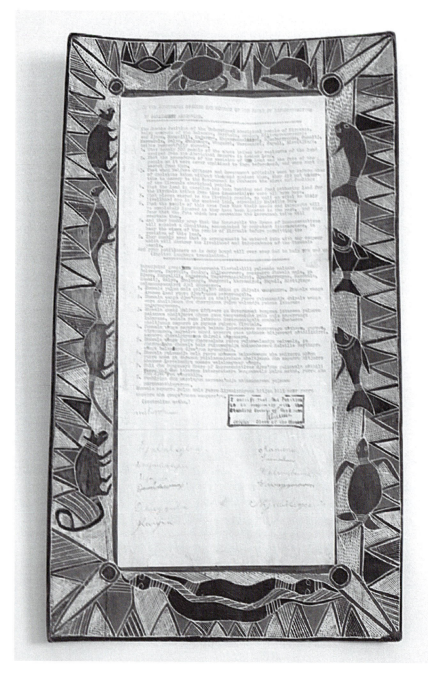

Plate 8.1 The Yirrkala Bark Petition. Presented to the Australian House of Representatives by the people of Yirrkala (NT) on 28 August 1963

Source: Reproduced with permission of the Speaker of the House of Representatives, Australian Parliament, Canberra.

mining company to develop a large-scale mining and refining operation against the expressed wishes of the native title owners. In opposing the Yolngu claims in the Gove Land Rights case, the Commonwealth asserted 'in the aboriginal world there was nothing recognisable as law at all' (17 FLR 10: 265). While Blackburn ruled that the Yolngu clans could not establish proprietary interests in their traditional estates (17 FLR 10: 273), he rejected the Commonwealth's assertion:

> I am clearly of the opinion, upon the evidence, that the social rules and customs of the plaintiffs cannot possibly be dismissed as lying on the other side of an unbridgeable gulf. The evidence shows a subtle and elaborate system highly adapted to the country in which the people led their lives, which provided a stable order of society and was remarkably free from the vagaries of personal whim or influence. If ever a system could be called 'a government of laws, and not of men', it is shown in the evidence before me.
>
> (17 FLR 10: 267)

Not only did the government refuse to acknowledge the Yolngu leadership as representing a legitimate interest, they also simply assumed that 'development' was not only what the nation needed, but also what Aboriginal people in the remote Arnhem Land Aboriginal Reserve needed. In 1952, when the bauxite reserves of the Gove Peninsula and the Wessel Islands were first identified, 'national development' was a priority, and an independent aluminium industry a strategic imperative. The Northern Territory's Crown Law Officer was quoted as dismissing very early expressions of Aboriginal concerns about development of bauxite in the region in the following terms:

> It would be a pity if natives who went to the Wessel Islands once a year to hunt turtles were protected at the cost of the Nation's ability to supply aluminium for aeroplanes.
>
> (quoted in NT News, 4 September 1952)

As a concession to these concerns, however, the Commonwealth did agree in 1952 to forego royalty receipts from mining on Aboriginal Reserves in the Northern Territory and earmark those monies as a form of compensation for the collective benefit of Aboriginal people in the Northern Territory (Altman 1983: 5–6; see also Williams 1986a).

The Yolngu actions over the Nabalco mining and processing project coincided with and contributed to a shift in societal values in white Australian society. The bark petition, the land rights case, the Gurindji walk-off, the 1967 Referendum and many other actions shifted the policy context away from simple confrontation and conquest and post-colonial paternalism towards (however ineffectually) self-determination. Legal debate gave way to policy debate, and by the early 1970s, land rights for Aboriginal Australians had become a policy platform of the major political parties. Parliamentary inquiries established that social conditions at Yirrkala were appalling (Australia 1963), and continued to be distressing despite resource-based development (Australia 1974). The 1968 Gove Agreement (Mining [Gove Peninsula Nabalco Agreement] Act 1968) entrenched Nabalco as virtual sovereign over 13 496 acres of Yolngu land for forty-two years, with a right to undertake any mining or associated activity even to the extent of polluting, diverting or otherwise interfering with the Yirrkala water supply. Despite recommendations from the 1963 Parliamentary report (Australia 1963), the later Parliamentary review reported:

> Until 1973, no action had been taken ... [and] in the meantime many of the matters about which the Yirrkala people had expressed apprehension in 1963 became realities.
>
> (Australia 1974: 6)

Aluminium-based development: success story or chimera?

At both Weipa and Gove, church authorities sought to maximise benefits of development and minimise harm, but proved ineffectual. In both cases, the strategic imperatives of the rapidly changing international aluminium industry governed company decisions. At Weipa, for example, the task of identifying appropriate joint venture partners became a major concern to the Australian partners, as some international companies were already well supplied with bauxite and sought partnership simply to delay development of this vast new deposit. At Gove, the project was never likely to be wildly profitable in its early years, but it provided Alusuisse with a secure captive source of low-cost alumina for its European and Icelandic smelters, and attractive taxation prospects. In both cases, Australian investors brought a range of strategic demands, and state and territory governments emphasised industrial development as an end in itself. At Weipa, the Comalco Act made virtually no provision for Aboriginal interests, and the Queensland government took a number of actions which contributed dramatically to the negative impacts of the Comalco mine on the lives of Aboriginal people in western Cape York Peninsula. At Gove, despite some provisions to protect Aboriginal interests, a mechanism to secure some financial benefit to Aboriginal people and, from 1976, transfer of the land to Aboriginal ownership under the Commonwealth Aboriginal Land Rights (Northern Territory) Act 1976, Yolngu people have suffered considerable disruption and ill-effects from the project.

In many ways, the Australian aluminium industry is a stunning success story. In three decades, it has evolved from nothing to a globally significant activity; Australian producers such as Comalco, Nabalco and Alcoa of Australia have integrated forward from low-value high-bulk production of bauxite for direct export, to establish major value-added processing of alumina and primary aluminium for export. In the process, the industry has developed or contributed to the development of coal mines, power stations and infrastructure, new towns and ports and a hydro-electric scheme in New Zealand, and has also greatly expanded downstream processing for domestic markets. In other words, here is a resource industry that has successfully pursued value-added processing and delivered industrialisation and development to Australia in a big way. Yet, in the context of greenhouse gas debates, aluminium's status as congealed electricity has become one of the features of Australia's out-of-step policy proposals at the Kyoto Environmental Summit in November 1997. Following the massive restructuring of the industry in the mid-1980s, which produced a global shift in smelting from Japan to Australia, Australia took on a new role in the industry, destabilising the International Bauxite Association and becoming a major trader in aluminium metal (Howitt and Crough 1996). And not only have the 'benefits' of this industrialisation and development success story emerged in recent years as double edged: it is also clear that the distribution of benefits of development to indigenous Australians most affected by the mining projects that underpin the industry has simply failed. In the 1990s Aboriginal people at both Weipa and Gove sought to redress their grievances through negotiation. Their difficulties in succeeding provide an exemplar of the challenges facing resource managers in dealing with the fundamental issues of justice, sustainability, equity and diversity.

Part of the damage: Comalco and Nabalco's responses

At both Weipa and Gove, bauxite operations were developed with strong government support on lands previously gazetted as reserved for the use and benefit of Aboriginal people. In the wake of the Wik decision in 1996, and the special legislation granting mining leases to Comalco (Queensland's Comalco Act 1957) and Nabalco (the Northern Territory's Mining (Gove Peninsula Nabalco Agreement) Ordinance 1968), it seems clear that the failure to acknowledge and protect the pre-existing rights of Aboriginal people in those reserve lands was sufficient expression of a clear intent to extinguish native title to have negated the legal standing interests of Aboriginal people in these areas. In both cases it was argued by governments that the development of the mines would bring the benefits of industrialisation and development to the previously remote and isolated communities. At both Weipa and Gove the new towns for the mining operations were built within a short distance of existing Aboriginal settlements established by Christian missionaries with government support. At Napranum (previously Weipa South) and Yirrkala, Aboriginal people raised concerns about the damage the mines and their employees and contractors did to their country, their people and their communities. Both mines (and at Gove a major alumina refinery to process the bauxite) were developed without formal environmental and social impact reviews. The absence of statutory support for Aboriginal rights in both jurisdictions meant that Aboriginal people relied on political rather than statutory avenues to achieve recognition of their concerns.

The mine at Weipa was one of the first of a new generation of remote northern resource projects which were playing a crucial role in integrating Australian raw materials into emerging global markets in the early 1960s. The Queensland Aborigines Act 1936, which was in operation with relatively little amendment until the mid-1970s (see for example Rosser 1987) controlled every aspect of life for Aboriginal people and prevented them from taking legal or political action to protect their rights. Mission paternalism as well as government antagonism made it difficult for Aboriginal leaders to raise concerns without facing extreme sanctions, including removal from the area. Despite their efforts to participate in the development boom Comalco's presence thrust upon them, local Aboriginal people were marginalised and pauperised by the process of development at Weipa (Howitt 1995; Stevens 1969). Many of the negative impacts of the mine and related developments were well entrenched before they received academic, government, church or company attention. In the early period of mine expansion, the small mission-based settlement at Napranum faced:

- The need to accommodate people displaced from the neighbouring settlement at Mapoon, when in 1963 it was closed at gunpoint by state officials and its residents forcibly resettled at Napranum and a shamefully named new settlement at 'Hidden Valley' (now New Mapoon) in the Northern Peninsula area of Cape York (Roberts *et al.* 1975; Wharton 1996b);
- Direct dispossession through the granting of mining leases over former Aboriginal Reserve lands;
- A rapid influx of construction and operational workforces dominated by young, unaccompanied non-Aboriginal men;
- Massive environmental damage as open-cut mining operations were undertaken literally alongside the boundary of the village;
- Greatly increased access to alcohol;
- Increased pressure on a wide range of culturally and economically significant resources within the immediate area of the Weipa Peninsula;
- Changing patterns of accessibility to traditional lands around the area;
- A dramatic juxtaposition of their own poverty and the developing wealth of the new mining community, exemplified by the contrast in accommodation in the two settlements following a contribution from Comalco to the church for housing (Plate 8.2), and the development of government funded infrastructure (schools, a hospital, recreational facilities, etc.) in the mining community, while infrastructure in Napranum remained non-existent or vastly inferior;
- A range of social, cultural and psychological impacts which rendered community members increasingly vulnerable to internalised violence, alcohol abuse and damage.

Plate 8.2   Aboriginal housing at Weipa. Aluminium houses provided by the Presbyterian Church for Aboriginal people at the Weipa Mission in the mid-1960s using funds provided by Comalco. The contrast with the housing provided by the company for its employees was never lost on Aboriginal people

Source: R. Howitt 1996.

As one community member put it to me in the early 1990s during fieldwork at Napranum aimed at redressing this history of neglect and abuse:

> Comalco can't ignore us and we can't ignore them ... They've been part of the damage done to Aboriginal people. They're taking part of our self-respect and telling us what to do ... I wouldn't have talked like this five years ago – this talking comes from oppression and I think how oppressed my older people were ... They're mining the land of Aboriginal people. To understand the link between that land and our people is important for Comalco.
>
> (Fieldwork interview, ex-chairperson,
> Weipa Aborigines Society, Napranum July 1992)

The Aboriginal families who bore the brunt of the damage imposed by the mining at Weipa faced a systematic response that was unjust, unbalanced and unsympathetic from state governments and ambitious corporate processes, and a complicit public silence.

Similarly at Gove, the Yolngu people living at the mission settlement at Yirr-kala expressed concern and direct opposition to mining because of their fears about its potential effects on the social and cultural life. Unlike the Aboriginal community at Weipa, Yirrkala became a focus of some national attention following their petitioning of the federal parliament, advocacy of their concerns by the local missionary (Wells 1982) and two inquiries into conditions at the settlement (Australia 1963, 1974). Despite this elevation of their concerns, however, there was no systematic monitoring of or attention to negative impacts as they developed. In a community in which personal relationships were dominated by strong sanctions from a system of customary law that was largely unchallenged on the ground, the rapid growth of the mining town at Nhulunbuy with alien values, unfamiliar social behaviour and no knowledge of or respect for Yolngu values and priorities, even the improved access to the 'benefits' of industrialisation and development such as education and health facilities, became a source of discomfort and doubt as Yolngu people came face-to-face with the demands of the powerful domains of balanda culture.[8]

In both Weipa and Gove, the juxtaposition of relative wealth in the new mining towns and the deepening poverty of the Aboriginal settlements exacerbated the sense of alienation and powerlessness amongst Aboriginal people. Government and company expenditure produced improved health, education, recreation and service infrastructure in these localities, but spatialised racism severely restricted Aboriginal access to these facilities. Although the success of the 1967 Referendum provided some recognition of indigenous Australians as part of Australian society, it reduced community controls over access to and abuse of alcohol, and combined with restructuring in pastoral employment, mission administration and other matters to decrease rather than reinforce indigenous self-determination in many places. The combination of dispossession, alienation, resentment, fear and violence produced a tragic cycle of self-destructiveness in both places throughout the 1970s (see for example Wilson 1982; more generally Hunter 1993).

The level of alcohol abuse and community violence provoked some change of direction among Aboriginal, company and government agencies. At Weipa, Comalco established a new organisation, Weipa Aborigines Society (WAS), in co-operation with state and Commonwealth authorities, to help direct and apply community development funds at Napranum. At Gove, Nabalco co-operated in the community's development of Yirrkala Business Enterprises (YBE).

Yirrkala Business Enterprises

YBE was originally developed as a paternalistic business undertaking in 1968 by the Methodist Church at Yirrkala, with little obvious emphasis on Aboriginal priorities:

> A white man ran it then ... I don't suppose there was any idea then that an Aborigine would ever run the company. The emphasis for the first twenty years was on Balanda control of Yolngu community affairs.
> (Gatjil Djerrkura quoted in *Australian Business*, 28 February 1990: 45)

By 1987, the company's non-Aboriginal administration had provided the company with a legacy of debt of close to $700 000 following an unsuccessful attempt to use YBE to penetrate the real estate and construction market in Darwin. A community meeting accepted a proposal to revise the structure and expand the board of directors to include representation from each of the Yirr-kala clans. Gatjil Djerrkura, a member of the Wangurri clan, a former Regional Director of the Department of Aboriginal Affairs and the Chairperson of the Commonwealth's Aboriginal and Torres Strait Islander Commission (ATSIC) since 1996, was appointed as General Manager and he adopted a very different approach to dealings with Nabalco:

> I guess that instead of using confrontation I've taken a strategy of co-operation ... [The large debt burden on YBE was] a big cost to us and it was something that should never have been set up. But that was something that came from a manager that believed in confrontation. He would fight Nabalco, and so he thought that since he wasn't doing any winning here, he'd look for other pastures – and it was a bit more competitive out there.
> (Gatjil Djerrkura, fieldwork interview, Nhulunbuy, May 1990)

YBE now operates as an incorporated Aboriginal organisation with each of the Yolngu clans having an equal interest in the company, and each clan providing one director to the YBE Board. It provides contract services to Nabalco (for example, earthworks, contract mining, rehabilitation), to Nhulunbuy Corporation, the company's local government authority on the town lease areas (beautification, garbage collection, tip maintenance and so on) and the NT Department of Transport and Works (for instance maintenance of the Gove–Bulman road).

Under Gatjil Djerrkura's leadership, YBE developed an innovative and successful approach to securing some economic and developmental benefits for Yolngu from Nabalco's operations in pursuit of its broad objectives:

- That YBE Pty Ltd endeavours to promote social and economic advancement of Aboriginal people of the Yirrkala region;
- That training and employment of Aboriginal individuals be regarded as a priority;
- That Aboriginal self-management and self-determination must be encouraged and implemented according to their pace, style and development;
- That non-Aboriginal employment be minimised in order to carry out training programmes and allow permanent Aboriginal employment within YBE Pty Ltd.

(YBE business papers, no date).

With a strong emphasis on training, minimum contracts between YBE and casual employees of one full working day, new arrangements for payment of staff and greater encouragement of participation by Yolngu women, YBE presented a new image to Nabalco and other employers in the region. Nabalco's engineering manager, responsible for employing casual Aboriginal labour through YBE 'found that YBE's crews for Nabalco were turning up in full strength – often over strength' (*Australian Business* 1990: 46). In early 1990, YBE had a payroll of 129 men and 42 women, mostly casuals. This compares with Nabalco's total workforce in April 1990 of 827 (Howitt 1992a: 40)

YBE's emphasis on training, contract employment and labour pooling rather than seeking to increase direct employment of Yolngu within Nabalco is widely supported among Yolngu. Gumatj clan leader and NLC chairman Galarrwuy Yunupingu indicated that Nabalco's initial response to increased levels of contracting to YBE was not positive:

> We tried direct employment with Nabalco, but there were many disagreements between the company and Aboriginal people because of the company's insistence on strict conditions. So it's better to direct the work through YBE and Nabalco can fit in with the local people ... This way, work can be directed towards all thirteen clans. But the company had to be forced to use YBE ... The YBE setup allows people to do what they can do and what they are good at.
>
> (Fieldwork interview, Nhulunbuy, 24 May 1990)

YBE's approach has been very different to the Comalco-dominated approach of Weipa Aborigines Society (WAS) in the 1970s and 1980s.

Weipa Aborigines Society

In 1972, following substantial public criticism of its employment practices and in the context of escalating political debate about Aboriginal land rights in Australia, the Comalco Board sought an initiative to depoliticise Aboriginal demands for recognition at Weipa. The board recognised that Comalco could not rely on the Queensland government to address the needs of Aboriginal people at Weipa. Regardless of the legal situation, the board concluded that Comalco and not the government would be blamed for the fate of local Aborigines. The board approved funding of the Weipa Aborigines Society (WAS) to sponsor community development projects for Aborigines at Weipa. Trustees included representatives from Comalco and the state and federal governments. The WAS Executive Committee added a group of five local Aboriginal leaders to this group of trustees.

In its first ten years of operations, WAS emphasised 'bricks and mortar development'. Its projects included paving roads, building and running a pre-school, water, sewerage and drainage work and so on. It was run as a highly paternalistic organisation, with the Aboriginal executive members having little influence over priorities and decisions, and no power at all over management of the money provided by Comalco and the governments. Despite its shortcomings, the community appreciated the work of WAS. Involvement in WAS's operations also exposed a group of company managers and community leaders to long-term personal relationships with each other; this laid the foundations for less ideological responses to some issues later on (see below). The structure of WAS, unlike most other Aboriginal organisations, was firmly rooted in the culture of corporate Australia rather than the local community. This is significant because it allowed the company personnel to feel relatively comfortable in what was, for them, a very alien situation.

In the mid-1980s, following changes in government, company and community personnel and wider changes in the Aboriginal affairs arena, the paternalism of WAS began to be challenged by a new generation of Aboriginal leaders. The emphasis of the group's work changed from 'bricks and mortar' to 'people development', with considerable investment in development of pre-employment and vocational training, and later development of a training centre in Napranum. By the early 1990s, this had moved to development of joint enterprises between WAS, Comalco and the Napranum Aboriginal Community Council providing contracted services to the company for cleaning, maintenance and parks and gardens. Parallel to these operational changes, WAS also began devolving decisions away from the Melbourne-based trustees towards Weipa-based company officers and the Aboriginal executive members.

Renegotiating relationships at Gove and Weipa

At both Weipa and Gove it is possible to quantify the value of mineral production, and the value being returned to the affected Aboriginal communities and native title interests. The two operations have developed under different legal arrangements, with the Nabalco operation having to deal more directly with Aboriginal interests since the passage of the Commonwealth's Aboriginal Land Rights (Northern Territory) Act in 1976. This legislation and existing arrangements have seen a direct accumulation of royalty-type payments through the Aboriginal Benefits Trust Account (ABTA) for distribution to Aboriginal interests in the NT, including traditional owners. While the actual receipts of the ABTA are small compared with the funds received by governments and corporate participants at Gove, the absence of any discretionary funds accumulating to Aboriginal interests at Weipa is glaring. Yolngu pressure to renegotiate the more onerous provisions of the 1968 mining agreement at Gove, and a strategy of not negotiating expanded mining or tailings disposal sites for the project until Nabalco would agree to reconsider and renegotiate the project conditions, has been matched by a determination to revisit the terms and conditions of the Comalco project.

At Weipa, the management of WAS commissioned a strategic review of its organisation and operations in 1992.[9] During this review it became clear that neither WAS nor Comalco had yet come to terms with many of the longstanding Aboriginal grievances about Comalco's mining operations and presence in the region and their underlying concerns of Aboriginal people, including the

Aboriginal people involved in the work of WAS. It was also clear in the review that the failure of WAS to be accountable to the community, and particularly to the traditional Aboriginal landowners of the mining areas, was a source of continuing resentment. And perhaps most fundamental was the different views of just what WAS's 'people development work' could and should involve. In response to the review, the Aboriginal members of WAS decided to establish a completely Aboriginalised organisation to take over the work of WAS, and to wind up the outdated, paternalistic and unsatisfactory existing organisation. In 1993, under the leadership of Chairperson Sandy Callope, a new organisation, Napranum Aboriginal Corporation (NAC), was incorporated to take over the Comalco-dominated operations of WAS. Comalco was urged to wind up WAS and hand over its assets and operations to NAC. NAC's membership and decision making processes were completely Aboriginalised. In response to recommendations of the strategic review, it added a Cultural Programmes section to the existing Pre-School, Training Centre, Enterprises and Administration sections.

In shifting from the old-style paternalism of WAS, NAC faced many difficulties. In particular, the funding base provided by the company meant that there were still 'strings' attached to money, and rapid development of programmes more in tune with Aboriginal perceptions and priorities was hampered; winding up WAS required approval of the Comalco Board and was very slow in coming. At the same time, Comalco's involvement in litigation of the Wik claim meant many company managers were antagonistic to funding 'good neighbour' work which had not kept them out of court!

Establishing management and accountability structures that allowed NAC to maintain existing operations, including promising enterprises in brickmaking, flyscreen production and sawmilling, at the same time as moving the organisation in entirely new and unfamiliar directions oriented to cultural maintenance, protection of native title and resource rights presented challenges to all involved. Despite requests since 1994 for support for training for the members of the NAC executive in their legal obligations and for developing management and policy skills, Comalco has failed to provide effective support for the Aboriginalisation process. Adoption of a constitution in late 1994 has entrenched the area's traditional owners and elders as central to the structure and operations of the NAC[10], and clarified the balance between the corporate-oriented work in enterprise development and some aspects of training, and the community/consensus-oriented work in education and cultural programmes. Maintaining that balance, and widening the financial support for NAC's work remains, however, a continuing challenge for the organisation.

At the same time, following the Mabo decision and with support from the Cape York Land Council, the Wik and Thayorre peoples, whose traditional territories include lands south of Weipa covered by the Comalco mining lease, commenced action to determine their common law native title and to seek rulings on the extent to which the Queensland government had failed in its fiduciary duty to native title interests to protect those rights against extinguishment and diminution. The Wik claim placed considerable pressure on Comalco and threw its previous paternalistic approach to managing relations with the Napranum community into doubt. In late 1995 the company gave a public undertaking to negotiate a settlement of grievances and concerns with the affected communities and native title interests regardless of the outcome in the Wik claim. As part of those negotiations, Comalco funded a major review of the economic and social impacts of the thirty-year-old mine and related activities, the environmental performance of the mine and cultural resource management concerns.[11]

In parallel with Comalco's efforts, the Canadian-based Alcan sought to activate a bauxite mining lease just north of Weipa at Ely. While Comalco stalled and failed to grasp the opportunities presented in negotiation processes, producing doubt and suspicion amongst the affected communities, Alcan rapidly moved through environmental and social impact assessment processes to a negotiated settlement with native title and community interests. While the original Comalco negotiations remained incomplete at the time of writing (mid–2000), Alcan's were finalised in 1997. Comalco and Alcan negotiated an arrangement for Comalco to take over its obligations under the Alcan Agreement of 1997.

Part of the healing?

At both Gove and Weipa, Aboriginal efforts to renegotiate legal relationships underpinning resource production from their traditional lands have proved inconclusive. The Wik decision confirmed the legal extinguishment of native title by the Comalco Act in 1957, but also that the moral landscape had changed. Similarly, the validity of the rights created by the Mining (Gove Peninsula Nabalco Agreement) Ordinance in 1968 has been confirmed. In both cases, the mining companies' insistence on a strictly legalistic approach to the complex process of intercommunity and intercultural relations has proved inconclusive. The ostensible commitment of both companies to an agenda of reconciliation has yet to produce an equitable or sustainable outcome on the ground in either community. In both places, there is a strong community effort to pursue healing of the damage caused by projects imposed on them in an era when paternalistic colonial practices sidelined them from decisions about their own futures. The ability of the personnel in the companies to deal with these changes, however, will be tested not in words, but in actions as the shift from management to negotiation of relations with local communities proceeds.

Aborigines and oil exploration at Noonkanbah

For six days in early August 1980, Australians watched as a paramilitary force organised by the state government escorted an oil drilling rig in a convoy of fifty vehicles up the west coast of Western Australia and on to Aboriginal land at Noonkanbah. The convoy was the culmination of months of dispute, confrontation and tension over oil and mineral exploration in the West Kimberley region. The Noonkanbah dispute reflected deep divisions in Western Australia over issues of land rights, human rights and political ideology. It became an important focus of media attention and political debate and the convoy's dramatic escort through Aboriginal, community, church and trade-union protests was a symbolic expression of the state government's hostility to addressing longstanding Aboriginal grievances over land and human rights. As the convoy approached Noonkanbah, a last-ditch blockade was organised at Mickey's Pool, a few kilometres from Noonkanbah on the only road into the property. Twenty-two people, including five Uniting and Anglican Church leaders, were arrested as police broke up the blockade. This confrontation came at the end of a long campaign by Aboriginal people at Noonkanbah to regain control of their traditional territories and to restrict mineral and oil exploration while they re-established a viable economic and cultural foundation for their communities through pastoral activity.[12] The Yungngora community at Noonkanbah was ultimately unable to prevent drilling on their land. Western Australia's hostility to recognition of indigenous rights and its determination to protect the notion of *terra nullius* as the foundation of land and resource law was at the forefront of conservative opposition to negotiation and implementation of the Native Title Act 1993. Western Australia's campaign against national recognition of native title was spearheaded by the Conservative Premier Richard Court, son of Sir Charles Court who had been the Premier who sanctioned the assault on Noonkanbah in 1980. The complex historical roots of conflict at Noonkanbah provide a compelling demonstration of the need for negotiations about the future to come to terms with complex histories.[13] The interaction of corporate, government and indigenous politics in the 1980 dispute demonstrates the ways in which nominally distinct domains interact in resource management systems and need to be reconciled in quite practical ways to achieve better resource management outcomes. Finally, in this brief case study, the lessons that the Noonkanbah dispute holds for more recent discussion of indigenous rights in Australia confirms the importance of multilateral, regional approaches to reconciliation and resolution of conflict rather than reliance on state mediation. Noonkanbah demonstrates the extent to which the developmentalist state is an unambiguous protagonist of resource-based development, even where there are no resources to be exploited!

The colonial pastoral frontier in the West Kimberleys

The pastoral station at Noonkanbah was established in 1886 as the first permanent non-Aboriginal presence in the region. By the mid-1880s, settler–Aboriginal relations in the Kimberleys were violent and poised for further deterioration (Pederson and Woorunmurra 1995: 33–50). Pastoral colonisation of the Kimberleys was critical to Western Australia's quest for independence from

Britain, and frontier violence elsewhere in the state had already raised political criticism in London. The government in Perth relied on increasing revenue from pastoral leases; pastoralists relied on unpaid Aboriginal labour. Sexual abuse of Aboriginal women was widespread in the region, as was misunderstanding and mistrust between pastoralists and Aboriginal people. Law and order, then as now, was a politically sensitive issue, with protection of settler lives and property a higher priority than protection of Aboriginal people's rights as British subjects. On the pastoral frontier, settlers, police and administrators faced the dual problem of acquiring control of land and resources and transforming the regional Aboriginal population into a compliant workforce without further exciting the moral and legal concerns of the imperial government in London.[14] The leaseholder at Noonkanbah, Isadore Emanuel, was active in agitating for a stronger police presence and succeeded in securing establishment of a police post at the station in the late 1880s. On the eve of West Australian independence in 1889, changes in the administration of police and justice in the Kimberley region laid the foundations for a frontier war between Bunuba people in the King Leopold Ranges north of Noonkanbah and pastoral interests and police (ibid.: 49–50).

Conventional historical accounts written by the colonial victors of this frontier paint a stark picture of brave settlers and police confronting wild, uncivilised and unpredictable savages in a hostile and difficult landscape, focusing on the deaths of a single policeman and three stockmen. As Pederson and Woorunmurra note, this conventional history 'fails to mention the slaughter of hundreds of Aboriginal people during the period 1894 and 1897' (1995: 197). Like all frontiers, circumstances in the West Kimberley reflected myriad influences. Pearling, sheep and cattle oriented to different markets, within and beyond Western Australia, all attracted investors. A short-lived gold rush at Halls Creek to the east attracted some interest, but soon faded away with richer discoveries further south. Race relations at the colonial frontier north of Noonkanbah in the 1890s were affected by a number of factors: the passions and prejudices of individual settlers and administrators; the personal histories of individual Aboriginal people (notably, for example, the Bunuba resistance leaders Ellemarra and Jandamarra)[15]; political and economic contingencies at various scales and places; changing technologies related to fencing, transport, communications, firearms and mining; and economic conditions including the international depression of the period.

At Noonkanbah, tensions between Aboriginal workers and station managers simmered until mid-1896, when 'station workers and "bush" Aborigines joined forces with ... [Aboriginal] prison escapees to embark on a full-scale attack on white settlement along the Fitzroy [River]' (Pederson and Woorunmurra 1995: 159). Using fire as a weapon, the traditional owners sought to engulf the whole valley in flames. A firefront fifty miles wide roared down the valley towards the colonial outpost of Derby, destroying hundreds of miles of fences and terrifying the settlers. This act of defiance was met with brutal force and most of the rebels were killed by police immediately afterwards.

Enslavement, exile and return: pastoral conditions at Noonkanbah

The Aboriginal people who survived the violence and disease of the invasion of the West Kimberleys were offered 'protection' and 'employment' on the pastoral stations that occupied their traditional homelands. By the 1950s, living conditions for Aboriginal people on Noonkanbah station were notoriously squalid and degrading. Citing the reports of Native Welfare Officers, Kolig reports 'a picture of exploitation, negligence and callousness *vis-à-vis* Aborigines' (1987: 43; see also Hawke and Gallagher 1989: 61–74). Basic facilities and services were simply not available to Aboriginal people at Noonkanbah. Even when a single water tap was provided in 1970 it was of only limited value 'because it was shut off by the management whenever there was a dispute with the Aboriginal residents' (Kolig 1987: 45). Prolonged tensions and disputes (ibid.: 45–49) came to a head in August 1971, following the dismissal of one Aboriginal woman and the death of a second. On 18 August 1971, the entire Aboriginal population on the station left, commencing a period of exile, social disintegration and pain:

> Only those who know how much Aborigines have always been prepared to suffer just so long as they could stay on what they consider their home-country, can imagine how strong the community's disgust must have been to finally reach boiling point.
>
> (Kolig 1987: 50)

Many people feared reprisals such as those recounted in oral traditions from earlier periods of invasion and occupation. Again, one sees the overlapping of causal processes at work here, with displacement of Aboriginal people from pastoral stations accelerating in the late 1960s following an industrial court ruling that Aboriginal pastoral workers should receive equal pay for equal work;[16] changing markets and technologies were also at work in the industry; new societal attitudes towards Aboriginal people (reflected most compellingly in the 1967 Federal Referendum)[17] were reflected in new policies and ambivalent support for struggling communities; and changing geographies through displacement, ecological change and social relations all affected the experience of Aboriginal people from Noonkanbah in their period of exile as fringe dwellers on the outskirts of Fitzroy Crossing.

Even before leaving Noonkanbah, the community had sought return of some of their land to their official care. From the turmoil of the Fitzroy Crossing camps, that struggle intensified. Initially under the umbrella of the Kadjina Community Incorporated, and later, following a split within the group, under the umbrella of the Yungngora Community Association, the community sought government support to purchase Quanbun and Noonkanbah stations, in their traditional heartlands (Hawke and Gallagher 1989: 78–85). A change of government federally had produced dramatic policy changes in Aboriginal affairs, with the Whitlam Labor government

(1972–5) clearly committed to addressing many long-festering problems in indigenous Australia. Following the return of land at Wattie Creek to the Gurindji people by Whitlam himself, an Aboriginal Land Fund Commission was established (see Palmer 1988), and in 1976 it purchased Noonkanbah and Waratea pastoral leases on behalf of the Kandjina and Yungngora communities. Within months:

> The people were home and free. Ninety years after the arrival of Isadore Emanuel they were once again masters of their own land, and owners of a cattle station to boot. They now had a future to believe in.
>
> (Hawke and Gallagher 1989: 85)

Returning to Noonkanbah, however, the Yungngora people found the rich country of the Noonkanbah lease had been plundered by the pastoral managers. Key infrastructure and plant was in poor condition, the country was in poor condition from overgrazing, fences were neglected and bores unusable. Despite the problems they faced, including tensions between the two community groups, tensions between state and federal departmental staff, appalling conditions of the properties, and chronic lack of capital and support, both groups struggled to forge new foundations. A bilingual community school was established with generous support from the Nomads Educational Foundation, a group established by the fiercely independent 'Strelley Mob' of Aboriginal people from the Pilbara region to the south.[18] By late 1978, Noonkanbah was emerging as a confident and optimistic community moving towards economic, educational and cultural autonomy on their own lands. Their respite from outside pressure was, however, short-lived. In late 1978, diamond fever gripped the Kimberleys, and prospecting at Ellendale just north of the Noonkanbah lease seemed to be amongst the most promising areas. At the same time, ancient coral reefs and shallow marine deposits of an ancient sea in the area were seen as highly promising indicators of oil and gas. In November 1978, the Yungngora lodged 95 objections to mineral claims made by CRA Exploration Ltd, a subsidiary of the major British–Australian mining group CRA Ltd. The Noonkanbah dispute was on its path to rapid escalation and drama.

Diamonds, oil and the search for justice

> Despite the spectre of miners, Noonkanbah was a confident and happy community at the close of [1978] … . There were nearly two hundred people in the Community. Most were living in camps that were not much better than those of the old days or the fringe camps, though some families occupied the buildings formerly reserved for the white staff and the shearers … .
>
> They were the custodians of the land, steeped in its Law and Dreamtime stories; and now the owners of the white man's pastoral lease as well.

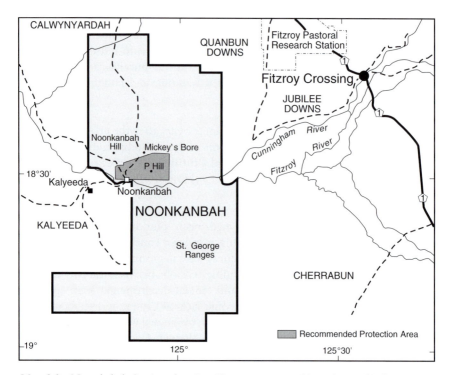

Map 8.2   Noonkabah Station showing Umpampurru and its sphere of influence

> They had a proud reputation ... as 'properly strong Lawmen'. They were
> proud Aboriginal people and they planned to stay that way, and bring up
> their children in the same way.
>   And they were strong. They had won the fight for their land. They had
> won the fight to run and manage their land their way, and were doing it
> well. They had won the fight to teach their children in their own school,
> in the way they believed to be right. They believed in themselves and their
> own power. And there was a spirit abroad in the Kimberley communities
> that what the Noonkanbah mob had done could be done by others, and
> that they had the capacity to change the present and build for the future.
> The pendulum was swinging back their way after ninety long years.
>
> (Hawke and Gallagher 1989: 89–90)

Hawke and Gallagher's description of the situation at Noonkanbah in late
1978 captures a widespread feeling of optimism and strength in indigenous
politics in the region at that time. Yet the 'fight' they refer to was really only
just beginning with the transfer of property titles. Recognition of Aboriginal people as stakeholders and participants in their own right was (and
remains) anathema to the ideologues of the developmentalist state and their

beneficiaries. Oil and diamond exploration activities had both caused problems for community and cattle operations at Noonkanbah in 1977–8. By May 1978 there were nearly 500 mineral claims pegged on the property (Hawke and Gallagher 1989: 103); the onshore oil exploration permit EP97 covered the whole property, and was held by a consortium that included Amax, Whitestone Petroleum, Pennszoil and others (Howitt and Douglas 1983: 61). Following damage to sacred sites, roads, fences and a series of blunders and misunderstandings in community consultation between the community and both CRA and Amax, the WA Aboriginal Legal Service commissioned an anthropological report from Dr Kingsley Palmer on the general situation at Noonkanbah, and the location of sites of significance as part of the preparation of a case in the Mining Warden's Court. This limited report was based on very brief fieldwork.

In his judgement on Yungngora objections to CRA's diamond exploration activities and tenures on the Noonkanbah pastoral lease, the Broome Mining Warden acknowledged that some of the CRA claims to which the Yungngora community objected to were within the spheres of influence of recognised sacred sites, and that knowledge of the sacred geography of the area was far from complete. Although the Mining Warden found there was no legally sustainable objection to the granting of CRA's claims, he did not simply dismiss the Aboriginal concerns, tackling their objections seriously and sounding a warning that was heeded by neither the developmentalist state nor Amax in the later dispute over oil-drilling at Noonkanbah:

> It would be insensitive not to recognise the sincere and deep interest of these Aboriginal people in the land they see as theirs. It is clear that they are deeply worried and, to a degree, feel threatened by the mining development in the area. This concern and worry has manifested itself in the objections made to these claims. It is a matter of comfort that this manifestation has taken lawful, as distinct from illegal and hostile, form.
>
> If only as a matter of self-interest, the Government, the Mining Companies and the community at large would do well to look at the issues raised in these proceedings and take positive steps to attempt to abate the concern expressed by the Aboriginal people.
>
> (quoted from the judgement of Magistrate and Mining Warden David McCann, in Hawke and Gallagher 1989: 23)

Although the Broome Mining Warden accepted that site recording in the Palmer Report was far from complete, Amax went on to rely on the map accompanying this report as an authoritative indication for the location selected for a wildcat well to be drilled near Pea Hill (Umpampurru) on Noonkanbah in the 1979 dry season (see Map 8.2). Umpampurru was well known as a significant site in the region, linked to major Ngarranggani (Dreaming) figures associated with goannas, lizards, frogs, kangaroos, turkeys, snakes and other reptiles. The influence of the Ngarranggani at

Plate 8.3 Pea Hill, Noonkanbah, WA

Source: M. Gallagher 1980.

Umpampurru were understood to act as 'power spreading out under the ground into the surrounding country', bring fertility and wealth to the locality (Hawke and Gallagher 1989: 123). Umpampurru, like all such sites throughout Australia, was part of a cultural and mythic landscape, woven into a rich tapestry of society, myth and country. Dancing grounds and ritual sites associated with it were part of a system of sites and rituals linking people throughout the region. According to the government anthropologist who investigated the dispute over Amax's proposed drill site, there was no doubt that the location was within the 'sphere of influence' of Umpampurru and another Ngarranggani site (Bundarra Goodun), and was itself used in preparing initiates for rituals (quoted in Hawke and Gallagher 1989: 123). The anthropologist Peter Bindon concluded that any possible site from which to drill into the target geological formation would be within the sphere of influence of the scared sites, that the sites had considerable religious and economic significance to the community, and that any drilling activity within the zone he identified would be detrimental to the site complex (Hawke and Gallagher 1989: 126–8) (Plate 8.3).

Amax's strategic context

Amax, the company in the lead of the joint venture on EP97 at Noonkanbah, was an innovative US-based transnational resource company. Australia was a priority target for Amax investments, with interests in iron ore in the Pilbara and bauxite prospects in the northern Kimberleys. Despite claiming a record of good environmental performance and relations with First Nations in the USA and Canada during the Noonkanbah dispute, Amax was in fact involved in disputes with First Nations and environmental groups on a range of issues (see also People's Grand Jury 1977) that included disputes over:

- Water usage with the Papago nation in Arizona;
- Personal injuries and deaths of Navajo uranium miners;
- Damage to the sacred mountain Mount Tolman following legal challenges from the Collville Confederation;
- Dubious contract negotiations for coal rights with the Crow and Northern Cheyenne nations of Montana.

In the late 1970s, Amax was one of the leading strategic innovators in the international resources sector. Under the leadership of Ian McGregor (Chairman 1967–77) and Pierre Gousseland (Chairman 1977 onwards), Amax built out from a dominant position in the molybdenum industry to a varied resources portfolio, emphasising diversification, research and development, containment of environmental conflict, targeting investment to politically 'stable' areas (including Australia), developing tight management controls, innovative financial arrangements and developing strategies to deal with indigenous peoples (*Business Week* 1976; Howitt and Douglas 1983: 54–60; Howitt 1990). Despite its strategic strengths, Amax experienced dramatic slumps in sales and earnings from 1980–85, along with escalating debt, falling asset values and lower share prices. The company faced simultaneous problems at the level of specific projects, the enterprise as a whole, industry sectors and systemically. The timing of the Noonkanbah crisis affected the company's plans to obtain 'naturalising' status under Australian foreign investment guidelines and offer a public float of its Australian operations. These were eventually largely disposed of during the 1980s as the company contracted to its core businesses and core areas of operations.

Towards conflict and confrontation at Noonkanbah

The ancient Devonian reefs of the Canning Basin had been assessed for oil and gas in the 1950s and 1960s. A geologist now working with Amax's joint venture partner Whitestone Petroleum had been involved in these early efforts, and maintained considerable optimism about specific sites he felt had not been properly evaluated. Although he felt the area was not 'dripping in oil', he acknowledged that the area was seen in the industry in the mid-1970s as 'one of the most promising onshore plays in Australia' (fieldwork interview, P. Purcell, Exploration Manager Whitestone Petroleum, Perth, 8 May 1980). A WA Mines Department spokesperson suggested that this optimism was unlikely to be fulfilled, but that the government was determined that it and not Aboriginal people would make the decisions about oil exploration in the state:

> There's a company here that wants to drill a well to look for petroleum. Now they could drill it some other time, but the Mines Department wants them to drill it now, because this is when we said they should drill the well. So if there is any opposition to the drilling of the well, the Mines Department are forcing this drilling; and it's opposition not to a company; but it's opposition to the Government.
>
> (M. Johnstone, supervising geologist, Sedimentary
> and Oil Division, WA Department of Mines
> speaking at a meeting at Noonkanbah 30 May 1980,
> quoted in Hawke and Gallagher 1989: 138)

In June 1979, the Yungngora community confronted government and Amax staff at the locked gate to Noonkanbah, refusing them access to begin work. Injunctions against further work were sought and obtained by the WA Aboriginal Legal Service, and the debate over the nature of anthropological evidence and the concept of 'spheres of influence' around sacred sites began to rage. For the bureaucrats drawing lines on maps to define rights and interests, sacred sites protected under the WA Aboriginal Heritage Act 1972 were, at best, locatable as points on a map. They were to be protected if development permitted, but they were seen by the politicians as quaint relics of a stone age culture. The developing state retained for itself power to overrule the recommendations of anthropological experts at the WA Museum Aboriginal Sites Department. As trustee of the lands trust holding the lease over Noonkanbah, the state also claimed ultimate rights over the granting of permission for mining-related activity. Finally, as the arbiter of land and resource management in the state, the WA government claimed an ultimate veto over impediments to 'development'. Anthropologists were accused of inventing the idea of 'spheres of influence' in order to stop Amax's activities, and Aboriginal people were accused of inventing sacred sites with the same motive (Kolig 1987: 136–7).

As the tension over the proposed drilling within the sphere of influence of Umpampurru increased, Amax sought to be released from the timetable approved by the WA Mines Department as the basis for granting EP97. The WA government was, however, determined not to give ground (either figuratively or literally) to the land rights movement that it saw as a Communist-inspired, left-wing anti-developmentalist programme aimed at undermining WA's economic and social prosperity. Developmentalism, paternalism, racism and greed combined in government policies and strategies to create Noonkanbah as the front-line against radical opposition to the rule of law in

Plate 8.4   Amax staff speak to Aboriginal people at Noonkanbah
Source: M. Gallagher 1980.

the state. In the state elections in February 1980, land rights, law and order and freedom of access to resources for development were key issues. Despite winning a further term in office, Conservative Premier Sir Charles Court's Liberal Party was overwhelmingly rejected by Aboriginal voters in the Kimberleys (Hawke and Gallagher 1989: 169–70). By early March 1980, plans were in preparation to commence drilling at the controversial site near Pea Hill during the 1980 dry season. A new ministerial team confronted the problem. The government refused to transfer any further pastoral leases to Aboriginal ownership while the drilling was delayed. Minsters insisted the site was not shown as a sacred site on the WA Museum's register of sacred sites, and therefore was not protected by the Heritage Act. Under police protection, contractors were on site setting up drilling equipment for a water bore by the end of March 1980, confronting continuing Aboriginal opposition (Plate 8.4). Trade union bans were placed on drilling equipment, street protests in Perth criticised the state government, there were calls for federal intervention, including proposals from the ALP opposition for a Royal Commission into Aboriginal land rights in Western Australia. At Noonkanbah itself, Aboriginal lawmen gathered to draw on the power of the Ngarranggani. On the afternoon of 2 April, the contractors and their equipment was being withdrawn from the site (Hawke and Gallagher 1989: 208–19).

The eviction of Amax from Noonkanbah in April 1980 made national headlines and infuriated the West Australian government. In this dramatic moment, Noonkanbah loomed as a watershed in indigenous affairs in Australia. Against the explicit demands, pressures and intervention of a developmentalist state, Aboriginal people and Aboriginal law had prevailed in a highly public way. Ideological battlelines were drawn. Sir Charles Court talked of 'ideological penetration and manipulation' of Aborigines at Noonkanbah, and labelled the community's stand as an 'insurrection against legitimate authority' and trade union support of the Yungngora as aiming 'to destroy the authority of the Government and the law' (from press releases in April 1980, quoted in Hawke and Gallagher 1989: 225).

Court visited Noonkanbah in May 1980 and demonstrated his inability to listen to or understand what Aboriginal people were saying. He suggested they were in breach of the conditions of their pastoral lease over Noonkanbah. The community responded strongly in a letter signed by eight elders:

> We cannot agree to Amax mining in our Sacred Areas because we would be breaking our Law. We cannot break our Law. If you force the drilling on our Sacred Areas, we cannot help you and you will be held responsible for the consequences.
>
> This is rich country and the Government is living off land that belongs to the Aboriginal. It is riding on the back of the Aboriginal.
>
> (Elders of Yungngora Community, letter to Sir Charles Court
> dated 9 June 1980, quoted in Hawke and Gallagher 1989: 250)

In July, using contractors and state government contracts to circumvent union bans, earth-moving and water-drilling equipment was moved on to the site at Noonkanbah. A blockade at the Noonkanbah gate was broken and Amax established their site camp. This period of confrontation at Noonkanbah was also a period of intense confrontation and negotiation over diamond discoveries in the East Kimberley (see below). In late July, as Amax established their camp at Noonkanbah, CRA announced the signing of a secret agreement with a small group of traditional elders at the Argyle diamond site. While the government was hostile to CRA's direct negotiations, the pressure on traditional owners and the 'divide and rule' tactics involved at Argyle put even more pressure on the Yungngora at Noonkanbah. To circumvent union bans on the oil drilling rig, which had been working at Enneabba on the central coast of WA, the state government had secretly employed truck owners on contract and organised an operation of military precision to transport the drilling equipment, accommodation modules, and other materials to the Noonkanbah site. At 1:00 am, these vehicles left Perth for Enneabba under police escort. On the morning of 8 August the fully-laden convoy commenced its journey up the coastal highway to Noonkanbah. Aboriginal protesters at Roebourne and Tabba Tabba Creek near Strelley confronted the convoy. An official trade union picket line confronted the trucks near Port Hedland. At the turnoff to Broome, 200 protestors confronted the convoy. For my own part, I was directed by the Kimberley Land Council (with whom I was

Plate 8.5   Police break up the blockade at Mickey's Pool near Noonkanbah, 1980

Source: M. Gallagher 1980.

working as a PhD student) to stay behind in Derby to avoid compromising my contact with Amax and Whitestone personnel. I was left in my tent writing protest songs. National protests supported the Yungngora, and at Noonkanbah, senior law people from places around the Kimberley, the Pilbara and the Northern Territory joined with Christian churchmen to make a last stand at Mickey's Pool, a strategic point on the road that the convoy could not avoid. Police broke up the blockade (Plate 8.5) and the dispersed protesters and police awaited the convoy. At the last moment, the drilling crew, who had recently been recruited into the Australian Workers Union, announced that they would not work on the equipment while it was required to work behind a barbed-wire fence because of Aboriginal protests. The convoy rolled on to Noonkanba without a crew to operate it.

On 27 August the chairman of the WA Aboriginal Land Trust, the nominal holder of the pastoral lease at Noonkanbah, announced that the ALT had no objection to drilling at Noonkanbah, and called on the union bans on drilling to be lifted. With this dramatic development, on the eve of the departure of an Aboriginal delegation to address the United Nations in Geneva, drilling at Noonkanbah commenced. Most members of the community were away at the Annual Races at Fitzroy Crossing. Court travelled the next day to Noonkanbah to inspect the drilling operation, which was now proceeding under new contractors employed by a shelf company set up months earlier by the WA government. The Aboriginal opposition to the Court government's anti-land rights policies was in disarray. International criticism was met with statements from the Australian government that Noonkanbah was an exceptional case. At Noonkanbah, a meeting of the Federation of Aboriginal Land Councils called on all land councils to suspend talks with mining companies until meaningful discussion about removing the drill at Noonkanbah commenced. In the face of this turmoil, the community returned to its pastoral operations and the long slow task of healing themselves not only of the hurt of recent dramatic events, but the pain of the previous ninety years. The echoes of this dispute, however, lingered long afterwards.

The lessons of Noonkanbah

More than any previous local dispute over Aboriginal rights in Australia, the Noonkanbah dispute mobilised public attention to the substance of Aboriginal grievances over the past, the present and the future. In this dispute, the interaction of spiritual, historical, cultural, environmental and economic issues was flagged. The depth of antagonism between ideologies of justice and ideologies of developmentalism in wider Australian society was highlighted. The burdensome depth and weight of Australia's legacy of colonial racism, paternalism and fear was revealed. The lengths to which those privileged by that inheritance would go to avoid addressing Aboriginal concerns was also sketched out. Although this dispute was eventually resolved in favour of the developmentalist state, the wildcat well drilled at Noonkanbah between August and November 1980 was dry. The Devonian reefs of that part of the Canning Basin revealed no treasures, and the well was plugged with only a crude iron plate to mark the site of this political drama. Yet this dispute placed indigenous rights in Australia on the regional, state, national and international agendas in new ways. It also deeply affected the personal agendas of many of us whose lives were touched by it. Despite the politicians' criticism of the idea of 'spheres of influence', the Ngarranggani at Umpampurru extended their influence very widely in struggle for recognition of and respect for indigenous rights in Australia.

The influence of Noonkanbah was in many ways contingent on unrelated events – the return of a displaced community to its homeland; the re-election of a radical right-wing government; the election of an ineffectual conservative government federally, the continued national commitment to securing oil self-sufficiency in Australia; the antagonism between the union movement and the WA government; the corporate strategies of Amax and its partners; and the personalities of many of the individuals involved. Many other factors could be added. Yet it is also clear that certain lessons emerge from the Yungngora experience. The marriage of indigenous law and the development of intercultural skills among younger leaders provided the community's protests with credibility in both Aboriginal and wider media settings. The co-ordination of community protests with trade union, church and wider community support was also critical in shifting the protests upscale to national and international forums. The emergence of institutional forums like the Marra Worra Worra (local) and Kimberley Land Council (regional), and institutional support from other indigenous organisations was also critical at various times in empowering particular community actions. Ultimately, however, I feel that the most telling lesson of this dispute is the extent to which the Yungngora community and its Aboriginal supporters wove together a cohesive response to political and economic pressure that was rooted in traditional values, knowledge and law.

For resource managers the Noonkanbah story confirms the integration of resource landscapes into those of more mythic and more conventionally political proportions. For Amax, the debâcle at Noonkanbah contributed to a major revision of strategies that led to a contraction back to US-based investments. Despite prioritising politically stable investment areas, and the company's sophisticated public relations, its reputation in Australia was severely tarnished by this dispute. The need to pursue reconciliation rather than simply compromise for its own sake was also reinforced at Noonkanbah. Misunderstandings and incomprehension multiplied in this dispute as key participants from industry and government simply failed to listen to or understand what was being said by community leaders. In following a 'win-at-all-costs' approach, they escalated the dispute rather than moved it towards resolution. In failing to acknowledge the underlying grievances of the Yungngora mob, they ensured that any resolution would avoid dealing with the critical issues in dispute. Without such reconciliation, such issues will continue to disrupt developmentalist efforts to narrate a new landscape from which indigenous people are displaced.

Negotiating without power? Aborigines and the Argyle Diamond Mine

At the same time as the Yungngora at Noonkanbah were becoming embroiled in disputes about diamond and oil exploration, Aboriginal people in the East Kimberley were also facing greatly expanded mineral exploration activity. In 1979 Aboriginal people involved in mustering cattle in and around Smoke Creek in the East Kimberley region of Western Australia discovered damage to a major sacred site complex. The damage was caused by CRA Exploration's diamond exploration programme which had identified a major kimberlite pipe and potential diamond resource. Amidst the great secrecy of the frantic search for diamonds in the Kimberley region in the late 1970s, CRA Exploration was involved in exploration on Ellendale Station near Noonkanbah (see above), at Oombulgurrie on the Forrest River Aboriginal Reserve north of Wyndham and on Lissadel Station near the Aboriginal settlement of Turkey Creek. In a report released on 21 October 1979, CRA announced discovery of diamonds in the Argyle/Lissadel area.

The CRA annual report for that year outlined its policy on exploration in areas where Aboriginal people had interests in the following terms:

> Aboriginal land holders are consulted before exploration is commenced on their land and care is taken to safeguard sacred sites. The assistance of the relevant museum authority is also sought in order to avoid unwitting disturbance of sacred sites.
>
> (CRA Ltd Annual Report 1979: 19)

This policy was contradicted by the company's practice at Argyle. Conflict over development of the Argyle mine produced one of the first voluntary agreements between Aboriginal people and a major mining company outside the provisions of the Aboriginal Land Rights (Northern Territory) Act. Although the Western Australian government was strongly opposed to the concession CRA made to Aboriginal people in the Argyle Agreement (see below), the bargaining position of Aboriginal negotiators was undermined in many ways by the strategies developed by CRA as well as the hostility and intervention of the Western Australian government. The conflict over resource development in the East Kimberleys represents a critical moment in the development of relations between Aborigines and mining in Australia, as well as an important element in the thinking of mining companies, government and the general public in Australia about these issues. In terms of the impacts of the mine on the affected Aboriginal groups, the history of this project is also instructive.

Discovery, dispute and negotiation: roots of the Argyle Agreement

The Argyle Diamond Mine was established in a setting of tension and controversy over its effects on Aboriginal people in the area.[19] The major diamond resource at Argyle, a kimberlite pipe located at a gap in the Carr Boyd Ranges known to local Aboriginal people as *tayiwul*, was an important sacred site. The Argyle Diamond Mine was discovered and developed by participants in the Ashton Joint Venture (AJV), which was managed by CRA. Rio Tinto (57 per cent) currently manages the mine, with other investors comprising Ashton Mining (38 per cent), and the WA Diamond Trust (5 per cent). The Western Australian government gave permission for destruction of *tayiwul* in circumstances which seemed deliberately aimed at undermining the ability of local Aboriginal authorities to influence or benefit from regional development. In the same period, use of force against Aboriginal protests at Noonkanbah in August 1980 (see above), reinforced the notion that Aboriginal people were secondary to the interests of developers and their efforts to set the agenda on or become participants in and beneficiaries of exploitation of the region's resources would not be tolerated by the state government.

The identification of diamonds in the East Kimberley in 1979 represented a major change to the regional economy. It was the first major resource project to link the region to global commodity markets. Earlier pastoral activities, development of a large-scale irrigation programme in the 1960s and spasmodic tourism had failed to secure a major boost for economic activity in the region – although it produced massive disruption for Aboriginal people (Coombs *et al.* 1989; Pederson and Woorunmurra 1995). For local Aboriginal people, the actions of the mining company threatened the integrity of sacred sites within and around the mining lease area. In particular, the *tayiwul,* which was both the centre of the identified diamond source material and an important women's sacred site, became a focus for dissatisfaction, controversy and dispute.

Despite the mining company's stated policy of community consultation and site protection, it failed to make any meaningful contact with local Aboriginal communities before starting excavation in the area. When Aborigines working as helicopter musterers in the area saw damage to sites in the area, they notified the Kimberley Land Council (KLC). The KLC was a fledgling community-based organisation advocating the rights and interests of Aboriginal people in the Kimberley region, which was facing a massive boom in resource, tourist and agricultural development. CRA was notified on 7 November 1979 of the existence of sacred sites in the areas they were working and asked to stop further work. The next day, CRA replied that the company was seeking advice on sacred sites from the WA Museum. On 5 December, CRA received a report from the WA Museum locating 58 sites in the work area, including three sites within the main area of activity:

> The Museum requested that CRA repair the damage, [and] recommended that CRA not do further work without proper evaluation of the sites by the Museum, and [also suggested] that CRA apply to the Minister to continue work and employ a Warmun Aborigine to help in identifying the boundaries of the sites.
>
> (Howitt and Douglas 1983: 48)

In the following months, the scale of the potential resource at Argyle began to emerge – and to submerge Aboriginal community concerns. The company's main concerns were commercial secrecy and continuing to work in the prospective area to prove its significance. The Western Australian government was ideologically committed to a strongly developmentalist agenda, and during 1979 had already demonstrated its willingness to overrule its own heritage protection legislation to secure resource-based development. Despite their increasing profile in political debates, Aboriginal people in the region had little experience or practical support in meeting the developmentalist behemoth of the international mining industry and the state government. Contrary to its co-operative rhetoric, the company adopted quite blunt divide-and-rule tactics that secured reluctant endorsement of its project from one section of the regional Aboriginal community. Consistent with its prodevelopmentalist policies, the government intervened at several points to limit

Aboriginal people's access to legal redress, and opposed even the limited concessions the company's divide-and-rule strategy offered to Aboriginal groups.

By April 1980 a meeting of traditional Aboriginal owners of the affected sites had informed the CRA and its joint venture partners that they would need to avoid working within two miles of the crest of the mountains at the Barramundi Gap (*tayiwul*) to avoid infringing the terms of the Aboriginal Heritage Act. The developers, however, ignored this advice because:

> That boundary would have covered the whole of the best diamond area ... (instead) the companies decided that a more appropriate boundary, which they felt would protect the site and still allow them to mine the major Kimberlite pipe, would be a distance of 1 km from the centre of the site.
>
> (Howitt 1989c: 235, emphasis in original)

Aboriginal opposition to activity which threatened this site met with vehement criticism and direct intervention by the Western Australian government. Prosecuting CRA for damage to the site under the WA Aboriginal Heritage Act required action from a committee of the WA Museum. This became impossible to secure when the Minister directed the Museum not to act. The Minister eventually approved an application to destroy the site. The government also amended the Heritage Act to ensure that future Aboriginal actions would not be able to disrupt orderly progress towards production at the Argyle mine. Private legal action proposed by some Aboriginal people was apparently blocked by these amendments.

In an attempt to secure some of their interests, one group of the area's Aboriginal traditional owners negotiated a private agreement (the Argyle Agreement) with CRA. The group involved was focused on the Manadangala outstation on the Glen Hill pastoral lease north of the exploration area, and was led by Mr John Toby, who had been a vocal critic of the company's actions and a leading agitator for prosecution of the company. Mr Toby had dropped legal action against CRA in early July 1980, when the WA Museum commenced its own prosecution. When ministerial intervention prevented continuation of the museum's action, Toby seems to have made a judgement that negotiating some benefits, however small, with CRA was a better outcome than facing the inevitable development of the diamond mine without any support at all (Dixon and Dillon 1990). Following a series of meetings at Glen Hill, a small group of traditional owners was flown to Perth and provided with legal advice at the company's expense, before signing the Argyle Agreement in late July 1980. This agreement committed CRA to a very modest annual expenditure on capital works on the group's outstation at Glen Hill. Generally characterised as Argyle Diamond's 'Good Neighbour Programme', these funds were administered in ways that precluded Aboriginal control over priorities and decisions. Terms of the Argyle Agreement itself limited the applications for which the funds can be used. It was also explicit that this was a good neighbour arrangement, not a legal obligation or royalty-type arrangement.

The terms of the agreement were confidential between the parties. Aboriginal people from Turkey Creek (Warmun) and elsewhere, who had not been consulted, expressed strong disapproval of the Mandangala group's giving in to the pressure from the mining company and the government and declared that the agreement was invalid. It is clear that although some of the signatories were senior traditional owners of sites affected by the project, the negotiation of the Argyle Agreement excluded a large number of people whose consent and involvement in discussions about the affected sites was required by Aboriginal law. Despite its limitations, the WA government strongly opposed the Argyle Agreement. WA Premier Sir Charles Court expressed grave concerns to CRA Chairman Sir Roderick Carnegie, saying that:

> The financial arrangements were 'so specific that it must be interpreted as compensation and payments in lieu of royalty'. In other words, he saw the arrangement as implicitly recognizing Aboriginal rights in land and thus creating a precedent for the establishment of a land rights regime in Western Australia.
>
> (quoted in Dillon 1991: 142)

The legitimacy of concerns expressed by the Warmun and Doon Doon communities and the KLC was eventually acknowledged, at least partially, when the company's good neighbour programme, established under the Argyle Agreement, was extended to include them in 1983.

The value and effectiveness of the good neighbour programme contributions to these communities, although significant, is limited by the enormous back-log in provision of even the most basic infrastructure such as water, housing, health hardware and so on. A long history of frontier hostility, bureaucratic and missionary intervention, reprehensible neglect and deliberate marginalisation of indigenous interests in Western Australia has imposed massive human and environmental costs on Aboriginal people. In facilitating primitive accumulation on the scale represented in the Argyle case, the Western Australian government has reinforced policies that shaped the colonial frontier. In its severely limited recognition of Aboriginal interests (although it carefully avoided acknowledging 'rights'), the Argyle Agreement broke new ground in this conservative domain. It is hardly surprising that it seems a very poor agreement nearly two decades on, and in the wake of negotiations taking place after the legal and legislative recognition of native title. Even within the terms of debate in the early 1980s, however, CRA's good neighbour approach faced criticism on many grounds, including the following points:

- It undermined Aboriginal self-determination, contrary to contemporary bipartisan federal policy in this area;
- It promoted cultural alienation;

- It created dependency;
- It was paternalistic, lacking Aboriginal control and participation in deci-
  sion making;
- There was little effective communication between the parties involved;
- The programme was poorly managed;
- The mechanisms for determining available funds were unclear to the
  community beneficiaries;
- It was socially divisive;
- It did not include all the communities affected by the social, economic
  and environmental impacts of the mine.

(Christensen 1990; Donovan 1986; Dillon 1991)

Approval of the Argyle Diamond Mine: terms and conditions

Shortcomings of the good neighbour approach seemed to be acknowledged
by the newly-elected Labor government, led by the charismatic Brian Bourke,
when only conditional approval to the Argyle Diamond Mine was given. Cabinet approved the Argyle Mine and accepted an Environmental Review and
Management Plan in May 1983, subject to a number of conditions. At the
same time, amendments to the Diamond (Ashton Joint Venture) Agreement
Act 1981 freed the joint venturers from obligations to construct a new mining
town at the mine site, in return for an additional $50 million royalty payment.
Most of this additional royalty was used by the Bourke government to establish the WA Diamond Trust, which purchased Northern Mining's 5 per cent
interest in the project from Bond Corporation. Some funds were also ear-
marked for distribution to Aboriginal groups affected by the mine, with a view
to overcoming negative aspects of the mine's impacts, and replacing the good
neighbour programme established under the controversial Argyle Agreement.
The conditions for final approval of the mine included requirements for:

- Further discussions with the WA Museum and local Aboriginal groups on
  site protection and management;
- Monitoring social impacts and taking action to control negative impacts;
- Consultation over changes in the administration of the funds contributed
  by Argyle Diamond Mines to its good neighbour programme;
- Modification of the company's Aboriginal Employment Programme;
- Establishment of an Impact Assessment Group, consisting of government, company and Aboriginal community representatives to monitor
  social impacts and oversee social programmes.

Under the new government, then, deals were done to secure the development
of the mine. The close links between aggressive and unconventional Western
Australian entrepreneurs and the Bourke government became a controversial
element of what came to be known as 'WA Inc'. In the case of the Argyle project, a deal done to raise an additional $50 million in advanced royalties, some

of which ended up in the coffers of Alan Bond, who was later jailed for illegal commercial activities. This led to the Western Australian government taking equity in the project, blurring the line between the governments role as a stakeholder, a regulator, a beneficiary, an advocate of Aboriginal rights and a mediator of conflicts. It is also clear, however, that the mine proceeded with several conditions intended to protect Aboriginal interests affected by the mine. The imposition of these conditions as part of the terms under which development would proceed had the effect of defusing and deflecting Aboriginal protests about the project as a *quid pro quo* for the Argyle Agreement's limited recognition of Aboriginal interests in the area, and Aboriginal agreement not to pursue any legal challenges to the mine. These terms and conditions may well have been sincerely framed, consistent and systematic failure to act upon or enforce them raises doubts about the extent to which Aboriginal rights were entrenched in the political framework of the Western Australian Labor Party. The previous government had taken actions whose detrimental effects on Aboriginal people was direct and, in general terms at least, publicly acknowledged by a government which felt Aboriginal concerns *should* be discounted in the face of any alternative proposals. Like CRA, the new government paraded a public commitment to recognition of Aboriginal interests and consultation with Aboriginal communities. Despite establishing a land rights inquiry and criticism of the conservative parties' Aboriginal affairs policies, the Bourke government's legacy for Aboriginal people was at best ambiguous, and at worst, includes responsibility for scuttling the ALP's national policy on uniform land rights legislation throughout Australia, and approval of a number of detrimental resource projects against Aboriginal community concerns.

The Argyle Social Impact Group

Despite approval of the Argyle Diamond Mine being subject to these specific conditions, they appear to never have been properly implemented, monitored or enforced. The Argyle Social Impact Group (ASIG) was established in April 1985, with two areas of responsibility:

- To assess and respond to social impacts; and
- To distribute financial contributions from the WA government and Argyle Diamond Mines.

Public announcement of the new body suggested that the concerns and interests of local Aboriginal groups were to be paramount in its orientation and operation:

> I see this [establishment of ASIG] as the fulfilment of the obligation of the Government and the joint venturers to ensure that Aborigines and other people affected by the mining proposal benefit fairly, and suffer as little distress and inconvenience as possible as the mining proceeds.
> (Brian Bourke, quoted in the *West Australian*, 12 October 1983)

ASIG was planned to run for five years and would then be reviewed. There was no indication that this was intended to be a fixed term of five years, nor that the basic arrangements would not continue after the review. Contrary to Bourke's public announcements, however, the group never undertook any action towards assessing and intervening in negative social impacts of the mine. As discussed above, some independent work on impact assessment was undertaken by the East Kimberley Impact Assessment Project. This work, however, was not integrated into the operations of ASIG. Instead, ASIG simply became a clearing house for funds. Contrary to the stated intention, ASIG also failed to replace the good neighbour programme, which continued to run in parallel with the ASIG funding mechanisms, nor did it address the concerns about the shortcomings of the financial arrangements between Argyle Diamond Mines (ADM) and the affected communities.

Misunderstandings and disagreements between the government and ADM appear to have led to some confusion about the agreed roles of various parties and schemes. As a result, ADM made only a minimal contribution to ASIG, and continued the much criticised good neighbour programme. In the original proposal for ASIG, Aboriginal people were to have an integral role in the structure and operations, including three members on the ASIG Committee. This was reduced, with no consultation with or agreement from the Aboriginal communities affected. Instead of three decision makers on the committee, the communities were left with no representation on the Group's Steering Committee, which exercised control of the funds and the funding criteria and priorities. Aboriginal people on the community-level project committees were relegated to the role of applicants for funds controlled by non-Aboriginal people under pre-determined criteria. Thus ASIG was reduced to a new form of welfare agency, rather than a means to enable people to respond to social upheaval produced by the mine and other regional developments. Between 1985 and 1989 ASIG operated to distribute a total of $6 056 179. ADM contributed a further $1 500 000 through its good neighbour programme in the same period (Table 8.1).

No Social Impact Assessment of the mine's operations was undertaken by ASIG, nor were there any responses to negative social impacts of the mine's operations other than the distribution of funds, although such responses clearly fell within the terms of the Cabinet's conditions for approval of the mine. Not only was ASIG intended to have a monitoring role, but it was also expected to provide an appropriate vehicle, through Aboriginal community representation at board level, for Aboriginal communities developing strategies to respond to and minimise negative consequences of identified impacts. In failing to enforce these provisions, the Western Australian government structurally entrenched the existing marginalisation of Aboriginal groups within the regional economy of the East Kimberley region, and effectively amplified rather than minimised many of the social impact processes set in train by the mine and related development of the mine.

Early in the establishment of ASIG, a serious disagreement of interpretation

Table 8.1 ASIG and GNP funding, 1985–9

West Australian Government		Argyle Diamond Mines Ltd		Interest on ASIG funds	Total (1985–89)
ASIG	*Admin*	*ASIG*	*Admin*		
$3 731 385	$500 000	$1 510 282	$500 000	$314 512	
	Total (WA Govt) $4 231 385		Total (ADM) $1 510 282		$6 056 179
		Plus GNP funding	$1 500 000		$7 556 179

Source: ASIG Annual Reports.

of funding arrangements between the government and ADM led to the company choosing to reduce its contribution and maintain the payments through its good neighbour programme. This was contrary to the original goals of ASIG, but was never pursued and resolved by the government. This arrangement seriously damaged the credibility and performance of ASIG. This problem was further exacerbated among Aboriginal people when their concerns about ASIG's failure to address undertakings, understandings and commitments by the government to the communities most affected by the mine, and a lack of accountability to them, or the Western Australian public more generally, for expenditure and actions affecting the future direction of community development.

As foreshadowed when ASIG was established, a review of the group's operations was established in 1990. The review was required to assess and report on the operations of ASIG[20] including, but not limited to the following aspects:

- ASIG's effectiveness in reducing the negative impact of the mine on the quality of life of local Aboriginal people;
- ASIG's ability to assist Aboriginal people in the area achieve their aspirations;
- The benefits of ASIG to other Aboriginal communities;
- The effectiveness of current structure and administrative procedures of ASIG.

The ASIG review was also required to investigate and report on the relationship between ASIG and the good neighbour programme; to make recommendations on the terms for extension of the life of ASIG and to provide recommendations which would maximise the opportunities from resource developments for Aboriginal communities. The Kimberley Land Council and East Kimberley Aboriginal communities were funded by the Western Australian government to prepare a submission to this review, and led to believe that serious consideration would be given to revising the structure and procedures to overcome serious criticism of its flaws. Despite the extent of this criticism, Western Australian Premier Dr Carmen Lawrence assured Aboriginal people in December 1990 that ASIG would continue operations after the review:

> ASIG has been extremely effective in achieving its original goals and objectives ... In view of the success of the programme over the five year term and the fact that Argyle Diamond Mines are no longer prepared to contribute the decision has been made to continue with the ASIG programme.
>
> (Correspondence, Premier to Chairperson,
> Kimberley Land Council, 14 December 1990)

This statement implies that the government had already reached conclusions based on some evaluation of the performance of ASIG against its stated goals. Although the suggestion that ASIG had been a 'success' appears to contradict the available evidence, the commitment to maintain ASIG in some form was short lived. The ASIG Review was terminated before concluding its investigations and its findings have never been published. The unexpected decision to terminate ASIG seems to have been a response to political rather than policy issues.

By February 1991, Dr Lawrence was backing away from the government's commitment to ASIG. She suggested, for example, that ASIG funds had helped to bring the ASIG communities 'closer to general community standards'. Furthermore, she added, this process of reducing the discrepancies between state and local standards in health, education, housing, nutrition, custody and violence would 'continue under general programmes which [the government] believe are preferable to the arrangement which ASIG represented' (Premier to KLC Chairperson, 13 February 1991). The Premier's position failed to acknowledge the ongoing issues of environmental change, social impact and cultural alienation that had been set in train by the Argyle Mine. It leaves the impression that the ongoing social impact of the mine, the failure of ADM's Aboriginal employment programmes to meet their goals, the outstanding claims for just settlement of land and compensation claims around the Argyle Mine, the effective marginalisation of Aboriginal people and their views from crucial decisions about regional development and adoption of unacceptably paternalistic and poorly managed funding programmes has created a widespread impression that the future of these communities was of much less importance to the government than share deals over Northern Mining with the Bond Corporation and the protection of a massively profitable mining operation.

Aboriginal groups saw the ASIG review as an important opportunity to redress the past inadequacies and to establish a just and fair arrangement between all parties that would ensure a more appropriate future. Concerns about the failure of the government and ASIG to monitor and respond to negative impacts from the mine were also raised. Unfortunately, any criticism or explanation of the perception, widespread in the Aboriginal public, that ASIG had generally failed to meet its goals, or proposals for more effective alternative strategies to pursue those goals, which may have resulted from the review, were suppressed and pre-empted by the government's premature and unilateral termination of both the review and ASIG itself.

From ADM's perspective, the Argyle Agreement and its subsequent role in ASIG were commercial arrangements, intended to enable the company to unimpeded rights to develop their lease areas. From the government's point of view, the purpose was much more ambiguous. Publicly the arrangement was intended to entrench benefits to the affected communities. Privately it appears to have had a much less altruistic motive:

> As with our previous arrangements with Aboriginal Groups in the Argyle area, the assistance will be provided on the basis that the Joint Venture remains free to conduct its mining operations throughout its Argyle tenements.
>
> (WA Premier's Department, file no. 387/83, folio 135)

To some extent the Western Australian government appears to have had a conflict of interest in this matter. As a minority shareholder in the mine, and a direct beneficiary from the failure of Aboriginal efforts to secure greater legal and economic benefits from the mine, the WA government appears to have made decisions about ASIG which have been highly questionable.

Impacts and prospects

The complexity of the social impact processes set in train by the Argyle Diamond Mine have been widely debated, although not as a result of any research undertaken by the Argyle Social Impact Group. In May and June 1980 the Warmun Community wrote to the Australian Institute of Aboriginal Studies, which was at that time involved in a project to review the social impacts of uranium mining in the Alligator Rivers (Kakadu) region of the Northern Territory (see AIAS 1984), requesting an evaluation of the social impacts of the events at Argyle (Dillon 1990: 135). This was rejected by the federal government, which argued that the issues were within the scope of state level legislation. The WA legislation excluded consideration of social and environmental issues from any project evaluation, and although a full Environmental Review and Management Plan (ERMP) was required, and CRA did include a chapter on social and economic issues – against the wishes of the state government and many senior figures within the company – social impacts of the proposed mine were not identified by the government or the company as a priority at the developmental stage of this project (Dillon 1990: 137–40). The WA government's failure to adhere to and enforce the terms and conditions under which the Argyle Diamond Mine was approved has exacerbated the negative impacts, as has the interaction of the mining impacts with the social, economic, environmental and cultural impacts of the government-funded Ord River Irrigation Scheme, for which no compensation or amelioration package was ever considered. Despite the good intentions, the way in which ASIG operated actually contributed to the negative impacts in many ways, as well as providing some specific and identifiable benefits.

The cost to the affected communities is impossible to calculate, but certainly involves a direct and substantial financial, political, economic, cultural and social imposition of WA Inc on one of the poorest and most vulnerable groups of WA taxpayers and citizens. Much of this activity has occurred since the Commonwealth passed the Racial Discrimination Act in 1975. In the light of the High Court's decision in Mabo, at least some aspects of these arrangements are likely to be vulnerable to review.

The East Kimberley Impact Assessment Project

Although the proposed Argyle Diamond Mine was the focus of community concern about rapid structural change in the East Kimberley, it was part of a broader regional context of administrative, economic and social change. The pastoral industry which had underpinned colonial occupation of the region was undergoing rapid structural change. During the 1960s development of the Ord River Irrigation Scheme had introduced commercially marginal intensive agriculture into the region, with Lake Argyle flooding the traditional territory of the Miriwung and Gadjerong people with devastating impacts. Development of the new town of Kununurra to service the Ord Scheme introduced a new public administration structure into the region, with considerable impact on administration of Aboriginal people's lives. An expanding tourism presence in the region has also been an important element of regional restructuring, facilitated by increased attention to conservation values and national parks, particularly in the world-class heritage area of Purnululu (Bungle Bungles National Park). Ross (1991: 3) provides a summary of these structural changes and their impacts on Aboriginal people (Table 8.2). The potentially dramatic impact of the Argyle mining project, then, was not a singular intrusion into a pristine frontier area, but part of a much wider process of regional restructuring threatening to further reinforce the marginalisation of indigenous people. From 1980 to 1984, local Aboriginal groups unsuccessfully sought support for social impact studies of the Argyle project (Williams and Dillon 1985). In late 1984 academic and government support for a project targeting the cumulative impacts of rapid development in the East Kimberleys using a participatory research framework was secured.

The East Kimberley Impact Assessment Project (EKIAP) was co-ordinated through the Australian National University's Centre for Resource and Environmental Studies (CRES) in Canberra. This ambitious three year project aimed to:

- Compile a comprehensive profile of the contemporary social environment in the region from both existing sources and limited fieldwork;
- Develop and utilise appropriate methodologies to social impact assessment;
- Assess the social impact of major public and private development in the region on resident Aboriginal communities;
- Identify problems and issues likely to affect these communities in the future;
- Establish a framework to disseminate research results to Aboriginal communities to allow them to develop strategies to respond to social impact issues;
- Identify areas for further research.

(Ross 1991: 2–4)

Table 8.2 Non-Aboriginal developments and their impacts on Aboriginal people in the East Kimberley region, 1960s–present

Structural change	*Impacts on Aborigines*	*Aboriginal actions and responses*	*Mitigation by government*
Ord River Irrigation Area scheme (1960s) • Dam building • Lake Argyle flooded • Agricultural experiments • Kununurra established	Damage to sacred sites and materials Dispossession and displacement of Miriwung and Gadjerong people	Movement to Kununurra	Creation of Aboriginal Reserve in Kununurra
Equal wages awarded to Aboriginal workers in the pastoral industry (1969)	Eviction of majority of Aboriginal people from stations (many others left voluntarily) Loss of access to land Loss of employment Increased income for the few people left with paid employment	Mass movement to town camps around the region Formation of new settlements Relative political independence New associations between groups	Limited intervention at the local level Welfare assistance Individual government employees help some Aboriginal groups obtain land and services

continued on next page

Table 8.2 (continued)

Structural change	Impacts on Aborigines	Aboriginal actions and responses	Mitigation by government
Removal of barriers to Aboriginal citizenship rights (1971)	Access to alcohol (uncontrolled) Threat to health and social relations		
Provision of social security income (1972)	Financial autonomy	Used for collective purposes to develop Aboriginal communities	Federal government initiatives
Structural change in the pastoral industry (1970s) • Declining economic importance • Rapid turnover of station ownership, management and staff • Closure of the Wyndham Meatworks (1975) • Decline in Wyndham	Further erosion of Aboriginal employment opportunities Fewer opportunities for Aboriginal people to visit land	Attempts to buy pastoral stations to join the industry as owners	Reviews of pastoral industry and land use, although these did not explicitly seek Aboriginal participation nor highlight impacts of structural change on Aboriginal people Followed by Kimberley Regional Planning Study
Exploration and large-scale mining at Argyle (1979 onwards)	Damage to sacred sites Fears of loss of quality of life Conflict and division within Aboriginal population Negotiations with mining company	Increased efforts to return to lands Efforts to establish powers of control over and economic returns from development projects Seeking influence through political means and personal interactions with developers and government agents Initiatives for impact studies	Limited mediation by Commonwealth and WA governments Concentration on physical living standards and social and psychological well-being

Table 8.2 (continued)

Structural change	Impacts on Aborigines	Aboriginal actions and responses	Mitigation by government
Increased tourism activities (1980s) • Development of new destinations including Purnululu	Intrusions into previously isolated areas Damage to sacred sites	Development of outstations in areas of high tourism and conservation value Efforts to establish cultural tourism and co-management arrangements	Regional tourism planning strategies, most of which bypassed Aboriginal concerns
Non-Aboriginal population growth (1960–present)	Further marginalisation of Aboriginal people in towns Political marginalisation in local government and electoral politics Changes in race relations Paternalism of pastoralists replaced by polarisation between 'supporters' and 'opponents' Active promotion of an ideology of exclusion	Institutional development (Kimberley Land Council and a range of local and regional resource agencies, enterprises and community organisations) Political lobbying and organisation	Mainstreaming of services and support Local and regional planning exercises which often exclude Aboriginal concerns

Source: Based on Ross 1991:3.

The EKIAP terms of reference emphasised the provision of information, development of effective participatory methodologies and community consultation (Coombs *et al.* 1989: 139–40). The research team drew on individuals from anthropology, economics, education, environmental studies, geography, history, law, medicine, psychology, social work and sociology. Over the five years of the project, fifteen researchers undertook field-based research within the EKIAP umbrella and the project produced a series of more than thirty working papers, a major report (Coombs *et al.* 1989) and a framework for understanding issues affecting Aboriginal people in the East Kimberleys.

The final report of the EKIAP provided a comprehensive critique of the inability of development to deliver sustainable benefits to local Aboriginal people in its current form:

> If 'development' in the East Kimberley continues in its present pattern and especially if that pattern is encouraged and supported by government, there is a risk that it may result in the denudation of many of the region's natural resources. It is also probable that the bulk of current benefits of the activities associated with development, and the possible capital assets it could finance, would flow to other parts of Australian and international economies.
>
> (Coombs *et al.* 1989: 19)

The recommendations of the EKIAP were often targeted at a general transformation of the processes of development and industrialisation in the region, and urged a rethinking of national and state priorities to better recognise and acknowledge Aboriginal concerns. The project foreshadowed negotiation of a settlement of Aboriginal claims and grievances at a regional level and outlined the areas of concern in a negotiated settlement of this sort (Crough and Christophersen 1993: 4–5). These recommendations were reviewed at a large meeting of regional communities at Crocodile Hole in September 1991, which emphasised the need for strategies that put Aboriginal concerns at the centre of regional development priorities (Kimberley Land Council and Warrangarri Resource Centre 1991: 36–40).

Aborigines and Uranium in the Kakadu Region, a review of the social research

In 1977, following national controversy and much local debate, the Australian government approved development of two uranium mines in the Alligator Rivers region of the Northern Territory (Ranger and Nabarlek). Since the initial impact assessment work undertaken for those projects (Fox *et al.* 1977a, b; see also Australian Institute of Aboriginal Studies 1984),[21] the Kakadu and West Arnhem Land regions have been studied by many people for many reasons. Many studies have looked at general questions about the social impacts of various changes on Aboriginal people in the area. There has also been extensive biophysical scientific research undertaken in the region, much of it under the auspices of either the Office of the Supervising Scientist or the Australian Nature Conservation Agency (previously Australian National Parks and Wildlife Service). In 1996 I was commissioned by the Northern Land Council to undertake a review of the extensive literature on this region as part of the Kakadu Region Social Impact Study (KRSIS).

One of the obvious weaknesses of the material reviewed was that it had not been able to equip Aboriginal people in the region to address many of the problems brought to their country by all the changes set in train by the approval of uranium mining in 1977. To a large extent, the failure of research to effectively empower local Aboriginal people to address and overcome negative social impact processes reflects the circumstances under which mining was approved and consideration of social impacts began. But the process of disempowerment is never simple, nor simply one-sided. As von Stürmer notes:

> If people were truly empowered, it is doubtful whether any social impact assessment or monitoring procedures would be required. People would simply be able to say no.
>
> (pers. comm.)

Few if any of the changes implicated in the Kakadu region have been predicated on Aboriginal people's rights to say no. Aboriginal people themselves have sought to accommodate and benefit from the changes under way in many ways. In von Stürmer' terms, there have been new people–land, people–people, people–institution and institution–institution relationships developing in the region. Too often, however, research has fragmented the totality of the region into elements that make it difficult to simultaneously address the management of resources and the mitigation of negative social, cultural and environmental impacts. In 1997, when Energy Resources Australia, the operating company at the Ranger mine, sought approval to develop a new mine on the traditional territory of the Mirrar people at Jabiluka, the legacy of the shortcomings of the resource management system around the uranium mines over a long period became clearer. In this brief snapshot, I simply want to establish the nature of Aboriginal experience in this region over the twenty year period of closely monitored development.

While there has been concern about negative impacts and disempowerment (AIAS 1984), it is also true that there have been positive outcomes in the region. The activities of the Gagudju Association in managing royalty equivalent and other mining payments, acquiring long-term investments and delivering a range of community services within the region, was a much-admired model for dealing with financial aspects of mining impacts in Aboriginal communities (O'Faircheallaigh 1986). More recent dissatisfaction with the operation of Gagudju and the development of smaller landowner associations in the area should not be taken as a rejection of previous successes. Similarly, the innovations encapsulated in the development of the Kakadu National Park Board of Management have been internationally recognised (for example Yapp 1989; Lawrence 1996).

The Ranger Uranium Environmental Inquiry (Fox *et al.* 1977a, b) was established to provide the Australian government with advice on both the broad issue of the nuclear industry and the specific issues of Aboriginal land claims and environmental and social issues in the uranium province. The Ranger Inquiry played a significant role in evaluating the likely impacts of the Ranger mine (and associated development activities, including the construction of support infrastructure, recommending the establishment of Kakadu National Park, and the development of the new town of Jabiru), and establishing the framework within the impacts of the development on Aboriginal people would be experienced and managed. The inquiry made key decisions that overruled Aboriginal people's ideas about the future of the region.

In a widely quoted passage, which is worth requoting at length because it encapsulates much of the Inquiry's thinking which continues to influence impact processes affecting Aboriginal people in the region, the Inquiry reported that the traditional landowners were clearly opposed to the proposal to mine uranium at Ranger, but their opposition should not be decisive:

> The evidence before us shows that the traditional owners of the Ranger site and the Northern Land Council (as now constituted) are opposed to the mining of uranium on the site … . The reasons for the opposition … would extend to any uranium mining in the Region. Some Aboriginals had at an earlier stage approved, or at least not disapproved, the proposed development, but it seems likely that they were not then as fully informed about the it as they later became. Traditional consultations had not taken place, and there was a general conviction that opposition was futile. The Aboriginals do not have confidence that their own view will prevail; they feel that uranium mining is almost certain to take place at Jabiru, if not elsewhere in the Region as well. They feel that having got so far, the white man is not likely to stop. They have a justifiable complaint that plans for mining have been allowed to develop so far as they have without the Aboriginal people having an adequate opportunity to be heard. Having in mind, in particular, the importance to the Aboriginal people of their right of self-determination, it is not in the circumstances possible for us to say that the development would be beneficial to them … .
>
> There can be no compromise with the Aboriginal position; either it is treated as conclusive, or it is set aside. We are a tribunal of white men and any attempt on our part to state what is a reasonable accommodation of the various claims and interests can be regarded as white men's arrogance, or paternalism. Nevertheless, this is the task we have been set. We hope, and we have reason to believe, that the performance of our task will not be seen by Aboriginal people in a racial light at all. That our values are different is not to be denied, but we have nevertheless striven to understand as well as can be done their values and their viewpoint. We have given careful attention to all that has been put before us by them or on their behalf. In the end, we form the conclusion that their opposition should not be allowed to prevail.
>
> (Fox *et al.* 1977b: 9)

In this remarkable preface to their consideration of the likely impacts of the development on Aboriginal people, the Inquiry identified several key issues that influence the efficacy of the foundations their report lay in addressing the real impacts as they have been experienced since mining commenced. They recognised that there was, in fact, general opposition to uranium mining amongst local Aboriginal people, and that this strengthened as the people involved became better informed. They also recognised that there was widespread pessimism about the ability of white government, and the development interests they represent, to reject the mining proposals because of Aboriginal opposition. They further recognised that this is fundamentally an issue of self-determination, and that the likely impacts of mining would be negative. And most fundamentally, they acknowledged that 'there can be no compromise with the Aboriginal position; either it is treated as conclusive, or it is set aside' (ibid.).

It is hardly surprising to find that in so decisively negating the principles of indigenous self-determination, the Ranger Inquiry set in train processes of marginalisation and alienation that echo through the region to the present day. As one of the conditions of approval of uranium mining, the Australian government established a statutory watchdog to oversee environmental performance of the mines, with substantial funding. In contrast, monitoring social impacts of uranium mining was limited to a five-year project under the guidance of the Australian Institute of Aboriginal Studies, which reported in 1984. With very limited funding, the AIAS Social Impact of Uranium Mining (SIUM) Project provided a damning critique of the impacts of the early development of the industry, and framed recommendations to monitor and mitigate future negative consequences (AIAS 1984).

Despite a bewildering array of research projects and learned reports that have been completed in this region since the SIUM project ceased operations after just five years of study, and despite the investment of millions of dollars in environmental monitoring activities, no coherent approach to social impact assessment or monitoring has emerged. There have been numerous studies, examples of which are cited below, dealing with specific topics such as:

- Payments to Aboriginal groups from mining-related activities (Altman 1983; Altman and Smith 1994; Levitus 1991; O'Faircheallaigh 1988);
- National park and tourism activities (Altman and Allan 1992; Gale 1983; Hill and Press 1993; Lawrence 1996; Press *et al.* 1995; Yapp 1989);
- Alcohol, health and education (d'Abbs and Jones 1996; Langton *et al.* 1990);
- Town planning (Lea 1984, 1987; Lea and Zehner 1985, 1986; O'Faircheallaigh 1987).

In the absence of any concerted government effort to address integrated and cumulative impact assessment monitoring and mitigation, this substantial research effort, however, has failed to produce a practical framework for dealing with cumulative and overlapping impact processes. In the absence of any approach which could be described as 'participatory, empowering and interventionist' (Howitt 1993a: 130), Aboriginal people in the region have simply been pushed aside from the path of development wherever they have not fitted, and criticised for their poor performance wherever they have tried to fit. The joint management of the world heritage listed Kakadu National Park is an exception to this that has been strongly built upon, but even this effort remains entrenched in the non-Aboriginal administrative apparatus of 'conservation' rather than the holistic Aboriginal domain of 'caring for country' (understood as encompassing the biophysical environments and the social, cultural and economic environments with which they interweave).

This fragmented approach has produced a disturbingly high level of incoherence in policy, programme and institutional approaches to important social impact issues and processes. Different spheres appear to be 'owned', usually in a disputed way, by various agencies and organisations, but are rarely genuinely accountable to the Aboriginal people identified as the intended beneficiaries of a particular policy, programme or organisation. Within the Aboriginal domain, many of the social impact processes established by the complex development process (mining, tourism, park development, town development, administrative change and so on), have been played out in different ways along a continuum of contrary tendencies between atomisation (reducing society to autonomous individuals each fighting for what is properly 'theirs', often with built-in assumptions about the ultimate right of individuals to behave without restriction, and as if social groups had no role in shaping individual choices) and collectivisation (reducing individuals to depersonalised units within a larger social unit, often with built-in assumptions about homogeneity, unity and a single, unproblematic identity).

The twenty years since the Ranger Inquiry commenced have seen considerable advancement in the development of Aboriginal institutions in the Kakadu region, but processes of generational change, legislative change and local and regional development itself, suggests that local organisations such as the Gagudju Association are entering a period that might be characterised as a 'second generation' of institution building. There has been surprisingly little work done on this topic in the Kakadu region, although some work (AIAS 1984) has emphasised the need for some critical structural changes. More importantly, there has been no Aboriginal-centred review of the enormously influential non-Aboriginal institutions of the region, including the regulatory bodies such as the Office of the Supervising Scientist, Jabiru Town Development Authority and various NT and Commonwealth government departments and other organisations active in the region. The result is that many practices have become entrenched without due consideration being given to their social, cultural and personal implications for the Aboriginal domain. Given that the Kakadu National Park Board of Management strongly asserts that the region is, in fact, 'an Aboriginal cultural landscape' (Kakadu Board of Management and ANCA 1996: 3), the need for non-Aboriginal institutions to accommodate the Aboriginal domain seems to me to be an issue of absolute centrality to the development of just, equitable and sustainable futures for the Aboriginal people of this region.

Post-Mabo negotiations: reconciliation and decolonisation[22]

In the wake of the judicial decisions and legislative reform arising from Eddie Mabo's historic assertion of the continuity of native title rights on Mer, Australia has confronted legacies of its colonial past. Conservative responses to the recasting of Australian history and Australian identity implicit in the recognition of native title, the reconciliation agenda and the findings of the stolen generations inquiry are often misinterpreted as 'irrational' or 'hysteria', or as a threat to 'conservatives' psychic equilibrium' (Atwood 1996: 105, 106). In fact, conservative responses to the Mabo and Wik decisions and the Native Title Act have been diverse and contradictory, and reflect conflicting rationalities. Vocal opponents of reconciliation and efforts to recognise and respect cultural diversity, difference and legal pluralism in Australia basically draw on two lines of argument:

- Adopting partisan views of history, they assert a universal relevance for traditions rooted in experience on one side of the frontier of the nation's colonial history.
- Adopting positions that advocate primacy of economic issues in social policy and unfettered access to resources, they appeal to poorly defined notions of 'workability','certainty' and 'the national interest' to criticise efforts to legislate, litigate or negotiate for justice of equity.

Resource-based industries, particularly some elements of the mining and pastoral industries and their ideological advocates, have been centrally implicated in both these arguments. In both industries, contrary voices asserting the need to accommodate indigenous rights and implement change have also been apparent in vigorous public debates. Neither of these defensive positions provide suitable foundations for progress towards post-Mabo geographies in which indigenous spaces occupied by resource management systems might be effectively decolonised. Instead, by advocating falsely homogenised visions of Australian society, such arguments risk condemning all Australians to unsustainable futures characterised by division, hatred and violence.[23]

In contrast, pursuit of fair, just and sustainable reconciliation nationally, regionally and locally, built on foundations of recognition of and respect for cultural diversity, offers prospects for a profound and constructive decolonisation of Australian landscapes. In particular, better understanding and accommodation of indigenous rights (including native title) offers avenues for addressing the colonial legacies identified in the preceding examples constructively and equitably. The challenge resource managers face to achieve this is, however, substantial.

Recent discussions of reconciliation and regional agreements provide a window on the difficult and complex task of creating post-Mabo geographies and decolonising resource management systems. There is growing literature on regional agreements, but the geopolitics of local and regional reconciliation inevitably juxtaposes Aboriginal groups' aspirations for just, equitable and sustainable outcomes from regional development activities with the jealous protection of interests privileged by institutionalised ideas of *terra nullius*. There is also some merging experience of negotiation of agreements on the ground with conservative interests.

Terra nullius no more: the emergence of new Australian geographies

The High Court's decision in Mabo, and the passage of the Native Title 1993, despite its shortcomings (see Bartlett 1996), unequivocally overturned the doctrine of *terra nullius* in Australia. The absurdity of the notion that Australian history and geography began when imperial England acquired sovereignty over the continent was never sustainable as an historical fact. And *terra nullius* was supportable as a legal 'fact' only as long as the unambiguously racist foundations of non-recognition of indigenous Australians' humanity were supported legislatively. This legal fiction imposed precisely the sorts of material costs discussed above on indigenous groups throughout Australia. Despite demands to extinguish indigenous rights and re-establish the sort of discrimination rendered illegal by the Commonwealth Racial Discrimination Act 1975, there has been some progress towards new relationships and processes in the wake of recognition of native title.

A celebrated example of such progress is in pursuit of regional agreements such as the Cape York Land Use Agreement, which established a framework of co-operation between Aboriginal people, pastoral industry interests and conservation organisations (Dodson 1996: 140). This voluntary agreement reflects one of the thrusts of Aboriginal responses to the new regulatory landscape – an effort to develop voluntary co-operation between land users through agreements about land use, development and land access. In the specific case of relations between Aboriginal people and the mining industry, there is some experience, at least in the Northern Territory, of negotiating with resource developers. Since the publication of *Exploring for Common Ground*, a report on reconciliation between Aborigines and the mining industry (Council for Aboriginal Reconciliation 1993), there have been many innovative arrangements negotiated with varying degrees of success. Agreements at Cape Flattery and Mt Todd (for a review see O'Faircheallaigh 1995, 1996b),[24] and efforts to negotiate agreements at Century Zinc (see Trigger 1997), western Cape York and elsewhere, all reflect pressures to deal with the new environment in which *terra nullius* no longer exists.

Regional agreements and the emergence of a new, post-Mabo regionalism

Voluntary regional agreements are starting to provide some indication of how the concerns of Aboriginal people to achieve better outcomes in terms of caring for people, caring for country and building sustainable regional Aboriginal economies can be pursued. The Commonwealth government's 'misconceived and regressive' proposals (Dodson 1996: 141) to amend the Native Title Act to achieve 'workability' threaten, however, to undermine this progress. Prime Minister Howard's ten-point plan really seeks to negate the recognition of indigenous Australians as stakeholders in regional economic and social processes. By holding out prospects for a future in which non-Aboriginal interests would not need to negotiate consent from native title holders in order to proceed with developments, it risks reproducing earlier patterns of marginalisation, disadvantage and alienation. Ultimately, it risks setting time bombs of discontent for future relations in such areas (see Filer 1990).

Regional agreements seem to be constructing a new regionalism in non-metropolitan Australia. Already there is wide discussion of questions of Aboriginal governance in the context of regional negotiations (Sullivan 1995, 1997; Richardson *et al*. 1995; Dodson 1996: ch. 6; Finlayson and Dale 1996; Trigger 1997; Yu 1997; Howitt 1997a, b; Ivanitz 1997), and persistent questions of sovereignty (Reynolds 1996) and customary law (Rose 1996a). Similarly, the role of indigenous peoples' ecological knowledge, land management skills and human rights are reshaping debates about so-called 'wilderness' (Langton 1995; Rose 1996b; see also Notzke 1995). The taken-for-granted arrangements of Australia's pre-Mabo regional geography are coming under sustained pressure. Much of the debate, however, risks overlooking key questions about regionalism that have been a focus of vigorous debate within geography for many years. I have previously argued the need for a more critical approach to understanding the 'regions' that are under construction in some regional agreements and the scale at which implementation is envisaged (Howitt 1997a). Getting the scale of such agreements wrong risks further marginalising Aboriginal people from influence over (and ability to benefit from) regional development processes.

Native title and reconciliation

Given the extent to which misunderstanding and misinformation abounds in the media and political debates, it is hardly surprising that many people are unsettled by the prospect of recognising native title interests and pursuing reconciliation within local communities. Pearson (1997) argues convincingly that native title is not a unique sort of property right which consists only of limited use rights. Like all peoples, he argues, indigenous Australians' possession of their territories was complete:

> it [native title] is inherent to the occupation of land and identical to the
> kind of dominion that people of different societies assert over land.
>
> (Pearson 1997: 160)

Aboriginal law and custom allocated between native title holders specific
rights and interests to parts of the whole estate – in ways which are parallel to
the allocation of various rights and interests in the sovereign estates of other
countries by common law and statute law. The rights and interests of indigenous peoples were clearly recognisable to British common law – as exemplified
in dealings with native Americans (Williams 1990; Jaimes 1992), the First
Nations of Canada (Royal Commission on Aboriginal Peoples 1996; Government of Canada 1996), and the Maori in New Zealand (Kaiwharu 1989)
(more generally see also Berger 1991). As Pearson explains, the High Court in
Mabo accepted that native title could be recognised by the common law 'as
against the whole world' in places where there is no other legally created
interest which is inconsistent with the rights and interests established under
indigenous law and custom. Where inconsistencies exist, he argues, the
balance of rights between native title claimants and other claimants 'must be
determined by the common law' (1997:161). The consequence of Pearson's
argument is that rather than placing the burden of proof of continuing native
title onto the native title claimants:

> the task of the common law courts is to assume the existence of a full pro-
> prietary title and to then identify those valid acts of the Crown which have
> qualified that title by regulation or by partial extinguishment of recogni-
> tion by the creation of an inconsistent interest.
>
> (ibid.)

Pearson's view of native title challenges a widespread misconception developing in the native title jurisdiction around the issue of extinguishment. Lawyers
on both sides of the native title debate have been concerned to identify or
avoid 'fatal events' in the title history of particular tracts of land – grants of
inconsistent interests which extinguish native title in a once-and-for-all-time
sort of way. Adopting Pearson's point of view, the critical test for extinguishment of the common law's ability to recognise native title is not to be found in
the title history of the place involved, but in the current circumstances. To
answer the factual question of the continued existence of native title, two basic
questions need to be addressed:

- Does Aboriginal law and tradition currently create recognisable interests
  in country?
- Does non-Aboriginal law create any inconsistent interests?

Framing the questions in this way makes it possible to deal with contemporary
interests and current circumstances in places where indigenous people continue

to claim standing. This approach is preferable to using some imagined set of events in the title history of country which reinscribes emptiness on to indigenous landscapes. It is, surely, unreasonable to argue that a piece of paper issued by colonial governments constitutes extinguishment of native title if it did not lead to dispossession in any practical sense. Clearly, those who seek to advance the cause of extinguishment want to reinscribe the Australian landscape as *terra nullius*.

Contrary to common misconceptions about Australia's imagined frontiers, Aboriginal law and custom continues to exist and create legitimate and recognisable interests in lands, resources and regions. Under the Aboriginal Land Rights (Northern Territory) Act 1976, resource developers have had twenty years to accommodate to a legally-enforced recognition of this reality, and some have made substantial progress towards that end. In other parts of Australia, the legislative recognition of this reality is much more recent. In Rio Tinto's case, movement towards recognition and respect as a basis for reconciliation holds some valuable lessons for all those involved in negotiating agreements about native title and land use.

Changing institutional cultures: the case of Rio Tinto

Following the merger of the Australian-based CRA Ltd and the British-based Rio Tinto Corporation in 1996, the Rio Tinto group became the world's largest diversified resource corporation. As outlined above in the case of Argyle and Weipa, the company's history of dealings with indigenous rights has been chequered. Prior to passage of the Aboriginal Land Rights (Northern Territory) Act 1976, CRA Exploration assessed all Aboriginal Reserves for their mineral potential with a view to acquiring pre-emptory title rights in those areas with geological potential. By the early 1990s, in the wake of a history of great antipathy to indigenous interests within Australia and internationally (see Howitt and Douglas 1983; Moody 1991), CRA held a substantial bank of Exploration Licence Applications in the Northern Territory. The company was unable to secure agreement with traditional Aboriginal owners for exploration (see Howitt 1992). While this situation locked out competitors and provided the company with a substantial future exploration base, it did not advance active exploration.

In the wake of the revolution on Bougainville – which commenced in the CRA-controlled Panguna copper mine – and the Mabo decision, however, a new approach has emerged within RTZ–CRA. From one of the unambiguous leaders of campaigns to minimise recognition of Aboriginal rights in Australia, the company's leadership has emerged in the mid-1990s to lead a proactive shift in company culture which aims to reconstruct RTZ–CRA as indigenous Australians' preferred development partner. This shift in corporate thinking is, it seems to me, important. In the still unresolved Century Zinc case, for example, the company pushed against the previous default solution of special legislation, opting instead to continue negotiations under the provisions of the Native Title Act 1993. At Ngukurr in the NT, the St Vidgeons exploration agreement is widely seen as exemplary. In the Pilbara, the company's iron ore subsidiaries have negotiated compensation arrangements with traditional owners of expansion areas (see CRA Ltd 1997; Davis 1996). On western Cape York Peninsula, the company's bauxite-mining subsidiary Comalco commenced voluntary negotiations in late 1996 with Aboriginal people who have been marginalised by operations of one of the world's largest bauxite mines for more than thirty years. And in 1996–7, the company's major management training thrust dealt with 'Managing Cultural Diversity', including full-day training sessions run under the auspices of the Australian Institute for Aboriginal and Torres Strait Islander Studies (AIATSIS) on 'Understanding Aboriginal Cultures'. Clearly, such changes mark a significant shift from the company's previous hostile position.

These changes have been predicated on policy statements committing the company to giving 'recognition and respect' to Aboriginal people affected by their operations. Chief Executive Officer Leon Davis and Vice-President Aboriginal Affairs Paul Wand, have both given a series of speeches which set up a framework for a cultural change within the company. For example, addressing an audience of Australian executives in Europe in August 1996, Davis suggested that Australia was:

> undergoing radical change. A change of the kind that happened at the beginning of this century, when six British colonies agreed to combine in a single nation. Today [he said], nearly 18 million people are redefining what it means to be Australian.
>
> (Davis 1996: 1)

Having acknowledged his position as a global executive managing substantial assets including major investments in Australia, and that change inevitably frightens some people, Leon Davis provided a perspective on change which challenges many of the dominant caricatures of the transnational executive:

> Nothing demonstrates this process of redefinition and Australia's growing confidence and maturity more than the Mabo debate and subsequent Native Title legislation. The more the nation has looked into the future, the more people have realised the need to come to terms with the past ... . However, just acknowledging the cultural differences that exist will not solve the problem. This will be the task of the Australian people. Just as there must be a deep understanding of Aboriginal needs [in negotiations, for example], there must be an equally deep understanding of the economic imperatives of the system under which we all live.
>
> (Davis 1996: 2)

In March 1995, Davis had expressed CRA's satisfaction with the central tenets of the Native Title Act and committed the company to a new approach in the area of Aboriginal relations, involving:

- Moving away from a litigious framework in dealings with Aboriginal people;
- Opening channels to those not favourably disposed to CRA;
- Developing innovative ways of sharing with and/or compensating indigenous people;
- Developing a genuinely open mind on the key questions and issues.

(paraphrased from Wand 1996: 4)

By February 1996, CRA had released a policy document on Aboriginal and Torres Strait Islander People which shifted the reference point for judging the company's dealings with indigenous interests. While the document was largely produced as a top-down initiative within the company, the management training forums provided a process for its dissemination and discussion within the company. Rio Tinto also set about the process of putting rhetoric into practice – although this is, perhaps inevitably, more fraught in real negotiations than within company forums. It is recognised, however, that implementation of the policy through completed agreements with indigenous people is the only criteria for measuring success:

> There is no benefit in only having a set of headlines and worthy documents. CRA will not be measured by these. The real measure is in application of the settlement and words in arrangements that benefit both Aboriginal people and CRA – deals that have mutual advantage.
>
> (Wand 1996: 7)

By August 1996, in speaking at the conference to celebrate twenty years of the Aboriginal Land Rights (Northern Territory) Act, Wand was willing to go even further. Prior to that speech, RTZ–CRA staff had not publicly criticised the effect of earlier developments on Aboriginal people in any but the most non-specific terms. In contrast, at the Canberra conference, Wand commenced his speech with a public apology on the company's behalf, for:

- The processes involved in negotiating the Argyle Agreement (1980);
- The effects of the company's iron ore developments in the Pilbara in the 1960s and 1970s;
- Standing by and doing nothing when the Queensland government forcibly removed Aboriginal people from the Mapoon Mission on western Cape York and burnt their property.

(paraphrased from Wand 1996).

The Liberal–National Party government announced drastic funding cuts to ATSIC on the day before the conference at which Wand made his apology

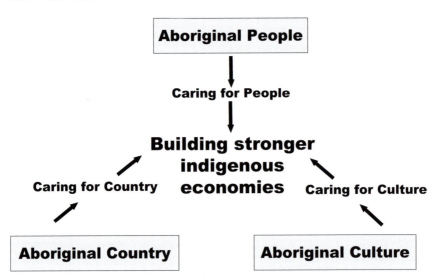

Figure 8.1   The web of relations between Aboriginal economies, Aboriginal people and Aboriginal country

commenced. This was also the period when Prime Minister Howard was characterising critical histories of colonial Australia as a 'black armband approach'. In that context Wand's apology and Davis' contributions since 1994 to the debate over the Native Title Act, stand out as contrary to the emerging mood of intolerance.

As a twenty-year observer of Rio Tinto and its subsidiaries, I find this transformation of rhetoric fascinating and personally challenging. I have had to review many of my own ideological verities in order to put into practice some of my values as an educator in working as part of the AIATSIS team providing management training for Rio Tinto managers. I have invited Rio Tinto staff to speak to students on my undergraduate course about the changing role for geographers within this arena, and sought to engage company staff in discussions about the process of cultural change occurring inside the company (Howitt 1997b). In my assessment, these changes are neither unambiguous nor yet secure within the company. Yet, the fact they have begun at all indicates the extent to which the 'radical change' referred to by Davis, is not a figment of the political imaginations of either the right or left, but a deeply rooted social process within the corporate culture.

Decolonising indigenous Australia

To move toward decolonisation, where indigenous peoples' rights to participate in and benefit from regional economic activity are upheld, it is necessary to build on opportunities such as those provided by the changes occurring, however problematically, within conservative institutions such as Rio Tinto. We might understand the process of such a decolonisation as one of rebuilding Aboriginal autonomy – caring for people, caring for country and building Aboriginal economies in order to strengthen, and in some places re-establish, the web of relations between Aboriginal economies, Aboriginal people and Aboriginal country (see Figure 8.1).[25] We might also think of it as a process of genuinely decolonising these spaces.

In non-Aboriginal domains, such changes require recognition that the Mabo decision and subsequent responses to the recognition of native title unsettle old verities in the landscapes of colonial and neo-colonial Australia. This will, perhaps, enable us to overturn some of the great injustices of colonial and recent history. It will also provide an opportunity to build relations based on neither fear and loathing, nor racialist ignorance (whether sympathetic or hostile) (see Jackson 1996; Craig *et al.* 1996).

For all of us (Aboriginal and non-Aboriginal people, teachers and learners, researchers and scholars, geographers, anthropologists, historians and others), the challenge is how best to contribute to ideas, institutions, relationships, policies and, most importantly, practical outcomes which move towards genuine decolonisation for all of us. The alternative, it seems, is a continuation of a post-colonial frontier in which there is little room for common ground, hope or reconciliation.

9 Dependent nations or sovereign governments?

Treaties, governance and resources in the USA

> For all of the federal government's rhetoric about Indian self-determination, the tribes will not really attain this state until they control their resources.
>
> (Allen 1989: 892)

For many indigenous peoples, struggles for recognition and justice focus on political strategies aimed at achieving legislative changes. Legal issues and legal strategies sometimes appear to dominate indigenous politics. Securing statutory recognition of land rights, constitutional recognition of pre-colonial sovereignty, enforcing constitutional rights and obligations, lobby for anti-discrimination, environmental protection and other laws, and then lobbying for their enforcement have all been a focus of indigenous concerns in many jurisdictions around the world. Debate about resource sovereignty (its nature, extent, legal definition, limitations and implications) is also a central concern of indigenous politics. Negotiating new treaties, enforcing the provisions of existing treaties, balancing tradition and change in relation to subsistence and commercial economic activities, cultural and financial obligations, and securing a capital base in terms of financial, human and cultural capital are all critical issues for indigenous leaders throughout the indigenous world.

In the United States, Indian sovereignty was recognised in the wake of the American Revolution through nation-to-nation treaties and the establishment of an obligation on the US federal government in Indian affairs. Negotiation of 371 nation-to-nation treaties between the United States and Native American nations by the 1860s and the development of the Marshall doctrine recognising indigenous sovereignty, may seem to provide a strong basis for securing just, sustainable and equitable outcomes from resource-based development within Indian jurisdictions in the USA. The legal and practical history, however, tells otherwise.

The genocidal history of US expansionism has been widely documented – although less-widely acknowledged in popular histories of 'how the West was won'. On the one hand, even a superficial study of US Indian legal tradition confirms that acceptance of sovereignty which predates the US Constitution is a fundamental legal reality in dealing with Indian nations:

It must always be remembered that the various Indian tribes were once independent nations, and that their claim to sovereignty long pre-dates that
of our own Government.

(McClanahan *v.* Arizona State Tax Commission,
411 US 164, 172 (1973), quoted in Laurence 1993: 3;
also Canby 1988: chs 3–5)

The Marshall doctrine, which arises from a series of judgements of the US
Chief Justice John Marshall in the 1820s and 1830s, confirms the sovereignty
of indigenous nations, and recognises the treaties signed between the USA
and Indian nations as nation-to-nation treaties. Yet the same judgements, in
what Churchill terms 'a bizarre departure from established principles of international law' (1995: 30) constructed indigenous sovereignty as inferior to US
sovereignty, and reduced Indian nations to the status of 'domestic dependent
nations'. In the same period as Marshall was reaching his enormously influential formulation of Indian status within the US legal system, the territorial
expansionism of the US proceeded with brutal ferocity. The notion of the
United States' 'Manifest Destiny' (a divine duty and obligation to expand
European–American settlement westwards to the Pacific coast) was being formulated. In military engagements in more than forty wars, and civilian atroci-
ties across the continent, tens of thousands of indigenous Americans were
killed. By the 1840s, at precisely the same time as the British were negotiating
the Treaty of Waitangi to protect the fledgling colony of New Zealand from
military defeat, and inscribing the continent of Australia as *terra nullius*, the
Americans were engaged in genocidal clearances of Indian domains. An elabo-
rate system of legislative, administrative, narrative and judicial controls was
well entrenched by the 1890s.

A thorough review of the history and interpretation of the consequences of
US–Indian relations is beyond the scope of this chapter. It is, however,
instructive to consider how resource geopolitics is implicated in this complex
web of treaties, laws, struggles and prospects. In the geopolitics of resources in
indigenous USA, it is possible to read the continuing importance of primitive
accumulation to even the most advanced industrial economies: it is possible to
read the way in which resources are fundamental in constructing wealth and
power; the fundamental connections between economic, environmental and
cultural dimensions of justice; and the limitations and possibilities of legal
strategies in the indigenous geopolitics of resources.

This chapter focuses on the experience of the Navajo Nation in its dealings
with resource politics (coal, water and land in particular) as a case study of the
importance of indigenous sovereignty in contemporary resource management.
The chapter argues that legal acknowledgment of indigenous sovereignty, a
system of indigenous control over resources and the existence of substantial
resources under indigenous control, do not guarantee just, equitable, sustainable and tolerant resource management systems when these things derive from
historical injustice, internal colonialism and systems of governance intended to

undermine sovereignty. It is further argued that the lessons of the Navajo experience, although far from unambiguous, are relevant not just to indigenous peoples, but to all participants in resource management systems.

The development of tribal governments and the acquisition of resource rights

Laurence (1993: 3) identifies three important elements in American Indians' increased influence in resource management systems in the USA. First, Indian tribes were 'in quiet possession of the Americas, governing the land, and conserving its riches, long before Europeans, Africans, or Asians happened upon the place.' Second, many Indian Reservations contain important resources including mineral, energy, water and forest resources, as well as resources significant for conservation and land management practices. Third, Indian tribes have a legal status within American law which carries with it a right and responsibility to participate in resource decisions.

Laurence goes on to explain the situation of tribal governments. His explanation raises important issues for understanding the significance of American examples in providing exemplars and guidance in other jurisdictions:

> These tribal governments did not ratify the Constitution of the United States, nor were they created by it. Indian tribes are *inherently* sovereign, meaning that they do not trace their existence to the United States. As important is the recognition of that sovereignty by the United States is as a practical matter, in the end, the law is clear and tribal sovereignty does not depend on federal recognition.
>
> The legal significance of this recognition of the tribes *as governments* cannot be overstated. The existence of tribal sovereignty makes American Indian law unique. Indians as individuals are treated more or less like every one else in America … . (But) it is only Indians whose groups are recognized by the United States as being *governments*.
>
> (Laurence 1993: 3–4)

Among many Indian groups this idea of sovereignty is highly cherished, and the litany of its treatment by the US government part of their oral cultures. The extent to which this sovereignty was recognised during colonial times is often overlooked by popular commentators in order to avoid dealing with the contemporary grievances of peoples whom they wish to reduce to the status of other minority interest groups faced with the power of the majority. All the colonial powers active in North America – the French, the British, the Spanish, the Americans and the Canadians – have dealt with this issue, and have sought not only treaties, but also strategic alliances with Indian nations against other nations from time to time. In the case of the Iroquois Confederacy in the northeast USA and southeast of Canada, it has been argued that their model of federalism was in fact the source of the key principles in the US Constitution (Johansen 1982, cited by Berger 1991: 58).

Berger's brief and readable account of 500 years of European colonisation in the Americas (Berger 1991; see also Stevenson 1992; Chomsky 1993), documents the consistent efforts of colonial powers to restrain the predations of settlers on Indian lands, and the inability of governments to enforce legal undertakings to sovereign Indian nations. Like others, Berger emphasises the importance of US Chief Justice John Marshall's judgements in the 1820s and 1830s as central in understanding Indian law:

> Marshall's judgements represent the most compelling attempt, in the post-colonial era, to work out the implications of of the occupation by the United States of Indian Land. The United States' example is important, not only because that country is the greatest nation-state to emerge in the New World, seen as an exemplar of democracy and the rule of law, but also because in the United States Supreme Court's formal rationale for European domination in the New World lies the basis for a fair accommodation of the claims of Native people, not only in the New World, but also in other countries.
>
> (Berger 1991: 68)

Marshall's key judgements continue to provide one of the fundamental reference points for legal consideration of questions of native title and sovereignty, including in the Mabo decision in Australia. Marshall:

> accepted the legitimacy of Native sovereignty, Native institutions and Native title to the land and wove them into the American legal system.
>
> (Berger 1991: 73)

The difficulty of constructing practical justice from this weave, however, was demonstrated immediately in President Jackson's reported response to Marshall's judgement in Worcestor *v.* Georgia (31 US (6 Pet) 515, 1832). Jackson allegedly responded to Marshall's approach with defiant rhetoric – 'John Marshall has made his judgement, now let him enforce it' (quoted in Berger 1991: 81). In responding to legal principles, we inevitably find that:

> Events on the ground, the innate prejudices of men, not laws, no matter how carefully crafted, are the determinants of Indian rights.
>
> (Berger 1991: 83)

In the case of the Navajo Nation, whose 25 000 sq m reservation covers lands within Arizona, New Mexico and Utah (see Map 9.1) it has certainly been prejudice and greed as much as the 'rule of law' that has influenced the shape of current relationships. The reservation was created by treaty in 1868. Under

Map 9.1   The Navajo Nation and its mineral and energy resources

the Indian Reorganization Act of 1934, the Navajo tribal government was 'revamped' in the 1920s and 1930s, when:

> in effect, Washington officials 'created' a federally recognized Navajo political institution, the Navajo Tribal Council. Vestiges of traditional political structures remained ... but the United States wielded extraordinary power to grant or withhold both recognition and federal funds.
>
> (Wilkins 1987: xvi)

Diné (the Navajo term for 'the people') have, since 1924, held 'dual citizenship' as members of both the Navajo Nation and the United States (Wilkins 1987: 10). Wilkins provides an overview of the evolution and structure of the tribal council, and the Navajo tribal code, which outlines the structure and functions of the various arms of governance within the Navajo Nation:

[The Navajo Tribal Code] is divided into twenty-three titles. These include US relations, Tribal Administration, Personnel Policies and Procedures, Courts, Domestic Relations, Education, Labor, Land Water, Taxation, etc. Each title contains historical notes showing the organic resolution, cross-references to related matters in the Code, the United States Code, federal Indian law and appropriate state laws. Furthermore, annotations are included which detail how court decisions have interpreted the meaning of certain provisions.

(ibid.: 59)

Resource sovereignty and empowerment: coal, power and the Navajo Nation

The Navajo economy is diversified and distinct. The traditional subsistence sector continues to be dominated by agriculture, particularly sheep husbandry. With a Navajo population of 226 602, of whom in excess 158 149 live on the reservation,[1] construction, manufacturing and services are also significant. Resource industries, however, are particularly important. The mining sector employs nearly 2500 people, or 8.7 per cent of the Navajo workforce (Navajo Nation 1999: 53). In fiscal 1998, mining provided US$66 million revenue, or 65.02 per cent of total internal revenues (ibid.: 46). Oil and gas provided a further US$20.2 million and forestry US$3.7 million. Revenues are generated from mineral royalties, stumpage fees, other charges and taxation of business activities on the reservation (Table 9.1).

The dominance of coal production, particularly from the Navajo mine (BHP 100 per cent) near Farmington, New Mexico and the Kayenta and Black Mesa mines (Peabody 100 per cent) near Kayenta, Arizona, is clear and important. In both mines, Diné are employed at levels which are extremely high by Australian standards. At the Peabody mines, native employment is around 90–95 per cent, including quite high levels within management in the

Table 9.1 Navajo Nation mineral royalties, 1986–93 (US$ million)

Mineral	1986	1987	1988	1989	1990	1991	1992	1993
Coal	27.7	28.2	43.4	48.5	51.3	50.0	52.4	56.0
Uranium	0.004	0.038	0.044	0.044	0.044	0.044	0.011	0.011
Oil	13.6	14.0	13.3	15.6	16.8	22.5	21.3	19.4
Gas	0.8	0.5	0.3	1.0	1.2	1.2	0.8	1.0
LPG	0.4	0.4	0.3	0.2	0.2	0.3	0.2	1.6
Totals	42.5	43.1	57.3	65.3	69.6	73.9	74.8	78.0

Source: Navajo Nation 1994.

office located in Flagstaff AZ. Native employees include not only Diné but also a small number of Hopi, who also receive some royalties from the Peabody operations on an area which was previously designated a joint-use area between the Navajo and Hopi (see below).

BHP's mines in the region include not only the Navajo Mine, but also the San Juan and La Plata Mines, which are located off the reservation. Preferential employment arrangements favouring Diné employment have been extended to the two off-reservation locations, and the company has around 80–85 per cent Diné employees in its operations in the region. Amongst these are included a significant number of management and professional people including lawyers, engineers and environmental scientists.

While there was considerable controversy over the initial negotiation of the mining leases on the reservation (Robbins 1978;Owens 1978), it is interesting to observe that the current arrangements have not only provided the Navajo Nation with a substantial and secure revenue base, and high levels of employment in a range of fields, but that this revenue has in turn enabled the nation to support a wider range of lifestyle choices for Diné on the reservation than would otherwise have been possible. For example, the nation has invested in a decentralised infrastructure for education, communications, transport and power which has enabled people to live much more traditional lifestyles in areas remote from the mining activities in greater comfort and security, and with greater participation from young people, than might otherwise have been the case.

The BHP mining operations are regulated by the innovative US Surface Mining Control and Reclamation Act of 1977, which provides for substantial levels of environmental monitoring and citizen-initiated regulation and review. BHP has recently publicised its high levels of employment and active mine area reclamation programmes (BHP Review 1993, 994). According to Elmer Lincoln, the company's general manager (Tribal and Government Relations), a Navajo lawyer who had previously worked for the Navajo Nation, the mining operations have established benefits for both the company and the Navajo:

> The Tribe is treated as a sovereign government – we treat them the same way as we treat New Mexico and the United States. We are governed by four governments – two tribal; one state and one federal. The big difference with Australia is that here the Tribe owns the resource, and we have to work with them in our best interest.
>
> (Elmer Lincoln, fieldwork interview at
> Farmington NM, 6 July 1994)[2]

Despite the great strengths they identified in the BHP operations from a Navajo perspective, in discussions with Navajo staff at BHP, it was clear that they feel there is a 'glass ceiling' facing Navajo professionals seeking promotion. In large part this reflects a wider tension between the differing aspirations of the company and its indigenous employees. Both Navajo employees and local chapter house officials[3] found it frustrating to have to constantly revise approaches to management every two years or so, as BHP shifts people around. Similar concerns are often raised by Aboriginal people in Australia. From the company perspective, this movement of senior staff has the dual benefit of providing a truly global-company perspective on operations, and also minimising management loyalty to any single locality. In contrast, most Navajo employees are motivated to work for BHP in order to contribute to outcomes on the Navajo lands, or in order to live on the lands. For them, moving to higher levels within the company structure and facing the demands and expectations for senior management will be particularly demanding and challenging. While I am optimistic that some Navajo professionals will break through the 'glass ceiling' within ten years, and that renegotiation of the coal leases early in the new century will facilitate this direction as the Navajo Nation negotiators push, convincingly, for equity in the project, and some form of genuine co-management of the resource, it is also clear that the incorporation of Navajo into BHP's corporate culture will challenge both individual Navajos and BHP.

Certainly the BHP staff consulted during fieldwork saw the company's excellent relations with the Navajo as a positive advantage, not only in dealing with the Navajo mine, but also in expanding BHP's interests in lands owned by native peoples, For example, the company has recently arranged an exchange between Navajo and Dogrib Indians from the company's diamond prospect area near Yellowknife (NWT, Canada). Unlike Australia, where Aboriginal landowners have to deal with whatever mining company has been granted a title on the 'first come, first served' basis estabishing both a *modus operandi* and a good reputation for working effectively, fairly and constructively with native peoples can be the difference between accessing resources and being marginalised in North America.

Similarly, while many positive readings can be given of the Navajos' dealings with resource companies, despite some shaky beginnings, many problem areas remain. Underground uranium mining in the area has left a devastating legacy of cancer and other illnesses among Navajo workers. Disputes over water entitlements have put in doubt water allocations to other users (Back and Taylor 1980). Coping with the impacts of change related to resource development, including roads, money and land degradation, has been neither easy nor unproblematic. Environmental concerns on the reservation have recently prompted the Navajo Nation to establish its own Environmental Protection Authority, with powers related to those of the federal EPA. As yet, however, it remains in a preliminary stage of development, with inadequate data, experience and resources.

Both BHP and Peabody face some environmental criticism over their operations on the Navajo Reservation. BHP has faced some permitting problems related to 'technical breaches' in records of blasting which constituted non-compliance with the demanding requirements of the SMCRA, and the Office of Surface Mining has faced some pressure from Navajo to look more closely at BHP's operations.[4]

Overall, the Navajo Nation seems to be building on its ownership of resources and its status as a sovereign government to secure a constructive and sustainable benefit from mining activities on tribal lands. Its success in pursuing this is reflected in high levels of revenue from mining operations, high levels of employment in the industry, and relatively positive relations between local Navajos and mining operations. This will be tested as leases come up for renegotiation and the Nation seeks to further improve its standing:

> Yes, the Navajo Nation has expectations that the level of Navajo participation in the corporate structure will increase. But this is common ground – so does the company. We also recognise that the new trend will be towards the Nation taking a position as an equity partner. The good relations and understandings are not easy and not an accident. A lot of us have worked very hard at this over a long time. We have tried to educate the Tribal Government about respecting economic opportunity and not strangling it. We recognise that the Tribal Government can act as a very positive partner for example in approaching the federal government to support initiatives that simply would not get a hearing if the approach came from a mining company.
>
> (Elmer Lincoln, interview at Farmington, 6 July 1994)

The legacies of injustice: governmental, tribal and regional politics and the Navajo–Hopi land dispute

One of the most difficult and demanding issues facing decision makers involved in resource development on indigenous lands, in North America and elsewhere, is the contemporary impact of historic (both long past and recent) injustices. American and Australian history is littered with actions by governments and settlers of dubious legality and morality. In some cases, actions clearly in contravention of both statute and natural law were condoned or tolerated by governments unable to enforce laws or exercise sanctions against those who broke them. In Australia, governments persisted in such outrages into the very recent past through policies of removal of 'half-caste' children from Aboriginal families, arbitrary revocation of protected and reserve status for native title lands and Aboriginal reserves, and failure to evaluate, monitor or mitigate negative impacts of social change in Aboriginal settlements. In North America, the landscape is criss-crossed with the Trails of Tears traced by displaced Indian nations, and with legislation with genocidal–integrationist intent, as well as the legacies of the Indian Wars, in which Indian nations were constructed as an external enemy against which to reunify the fragmented and disrupted states following the US Civil War. Many examples could be examined to illustrate the contemporary impacts of such random and arbitrary acts of government, but none perhaps so clearly as the Hopi–Navajo land dispute, which stems from government, bureaucratic, political and judicial decisions stretching from 1868 to the present, and leading to the largest ever peacetime relocation of American civilians as Navajo and Hopi peope who ended up on the 'wrong side' of an arbitrary border were moved in an exercise of ethnic cleansing on a tragic and disturbing scale.

The historic circumstances of the Navajo–Hopi land disputes are outlined in Hasgood (1993) and Benedek (1993). The dispute over an area of Indian reservation, its resources, its use and its inhabitation has produced a legacy of human suffering and confusion. Explanation is to be found in the treatment of Diné and Hopi in the late nineteenth century and in the renewal of interest in the area's coal resources in the 1940s:

> Kit Carson did the job of tracking down the Navajos, burning their fields and crops and chopping down their fruit trees until, in the middle of the winter of 1864, the Navajos, starving, with little clothing, surrendered and marched to [incarceration] at Bosque Redondo at Fort Sumner … . the government had chosen an area for the new reservation that was unfit for farming. The area had no firewood, poor soil, and brackish water that sickened the Indians. After four years, during which time half the population died, the experiment was abandoned. Pestilence had prevented crops from producing, and corrupt suppliers led to shortages that killed many Navajos and Apaches (at Fort Sumner). Finally the government called for settlement talks. When the Navajos were asked where they would like their reservation to be, … one of the Navajo leaders said, 'We do not want to go to the right or left, but straight back to our own country.'
>
> (Benedek 1993: 22)

Thus, in 1868, a reservation straddling the New Mexico–Arizona–Utah border was established for the Navajo:

> Most Navajos had no idea where the boundaries of this piece of land began or ended, and they simply wandered back to their old homes and met up with bands that had evaded capture.
>
> (ibid.: 23)

In the following years, the US government accommodated growth of the Navajo population by adding sections to their reservation until it completely encircled the Hopi reservation, and the Hopi settlements on the tops of the mesas. In 1882, without consulting the Navajos, the US government made a decision which laid the foundations for the long and complex trajectory of the dispute when it:

> withdrew a rectangle of land from the public domain for the use of [the Navajos'] neighbours, the Hopis. Nothing marked the boundaries of their own reservation, and nothing marked the bounds of the Hopi

reservation. The Navajos continued to expand into unused territory. [People's] ancestors settled around the Hopis ... unaware the land had been assigned to the Hopis on a map somewhere far off in Washington.

(Benedek 1993: 23)

In the 1930s, the Bureau of Indian Affairs began moves to clarify ownership of surface and mineral rights in the area. The conclusion, in 1946, was that 'the two tribes held co-extensive rights to minerals in the 1882 area' (Benedek 1993: 134). Following the establishment of an Indian Claims Commission in 1946, lawyers expressed interest in representing the two tribes in resolving the dispute which some felt existed, although the differing relationships each tribe had to the land meant that joint use was largely a reality, and co-existence generally peaceful in practice.

However, the imposition of tribal government in the 1930s through the Indian Reorganization Act of 1934, had produced profound schisms within the Hopi population, with the Hopi Tribal Council facing deep opposition from traditionalists, and support from the US government and a small group of progressives. Despite the failure of the proposal to establish the Council to achieve majority support in plebiscites, technicalities were used to establish it as a representative government. While the traditionalists accepted the realities of co-existence with the grazing activities of the Navajos, the progressives increasingly accepted the argument that the dispute existed and needed to be solved.

In 1944, Navajo living within District 6 (around the Hopi villages of Hotevilla, Bacabi and Oraibi within the 1882 area of the Hopi reservation) were evicted from their homes and relocated, without compensation, outside the boundaries. The first relocations had begun, and the shadow of further resolution loomed over other Navajo families.

In the 1950s, Congress introduced legislation to allow the two tribes to sue each other to clarify entitlements in the disputed area, producing:

> a tangle of legal battles and personal tragedies that ... can be traced not to a conflict between the Navajo and Hopi Indians, but, sadly, to a battle of wills between ... white men.
>
> (Benedek 1993: 33)

The legal argument in the case that was eventually brought to court (Healing *v.* Jones), was constructed by lawyers committed to receiving a fee of 10 per cent of the value of the land in dispute. In 1962, the judgement in this case ruled the Hopi and the Navajo Tribes had:

> joint, undivided and equal rights and interest both as to the surface and the sub-surface, including all resources' of the land of the 1882 area outside of District 6 ... . [a decision which succeeded in] leaving more unanswered questions than had existed before the case was heard.
>
> (ibid.: 37)

The area became known as the Joint Use Area (JUA), but the dispute simmered. The Navajos offered to buy out the Hopis' interest, as the entire area was inhabited by Navajo families. The Hopis refused and their legal counsel insisted on half the area being vacated for the Hopis to use. This quickly led to proposals for partitioning of the land, and the interest of coal companies in getting access to the area's resources added to the push for partition. There is a widespread belief that the coal companies were involved in a deliberate strategy to relocate Navajos from the area in order make it more accessible to open-cut mining (Benedek 1993: 138), but Benedek concludes 'there is no convincing evidence' to support a broad-based conspiracy theory here. As with the case of Mapoon in Australia (see Roberts 1975; Roberts and Maclean 1975; Wharton 1997) it seems more likely that it was a political perception of how best to facilitate 'development', rather than any direct corporate conspiracy, which produced the decisions with which later generations struggle.

The proposal to partition the Joint Use Area was finally endorsed in 1974 with the passage of the Navajo–Hopi Indian Land Settlement Act. After a further failed attempt at mediation, a line was drawn on a map in 1975 and adopted as the partition by a judge in 1977. Expert advice was tabled in this debate by the Navajos, who faced the largest relocation problem. In a warning which was tragically prophetic, Californian academic Thayer Scudder said:

> 'The profound shock of compulsory relocation is much like bereavement caused by the death of a parent, spouse or child.' He also warned that relocation undermines a peoples' faith in themselves, in the family heads who are unable to protect them, and in local leaders. 'Violence, alcohol abuse, and mental and physical illness are all too often intimately associated with forced removal.' Scudder also warned the fate of the Navajo would be worsened by their love of their land as well as the fact that they'd lived under stressfully circumscribed conditions for years before moving.
> (Benedek 1993: 152; see also Clemmer *et al.* 1989)

In 1994, the dispute and relocation process continued to create problems. Further mediation occurred in May–July 1994, and the tragic consequences of incompetent, corrupt and spiteful handling of the relocation process have magnified the impacts predicted by Scudder. Benedek provides a detailed account of the effect of political appointments to the Navajo–Hopi Relocation Commission, of incompetence amongst bureaucrats, of the self-serving actions of politicians and lawyers, of the failures of the media to comprehend the complexity of the dispute and the impacts of this forced resolution, and of competing agendas to address the problems created. The litany of problems described is a tragic warning of the difficulty of using legal strategies to impose resolutions of conflicts constructed in social relationships, particularly when some of the affected parties do not accept the authority of those acting on their behalf. As one lawyer commented:

This case should never have been sent through the court system; the rights to the JUA should have been settled by legislation 'in such a manner as would best serve the public interest' rather than 'requiring the courts to wrestle with a multitude of legalisms and then having to face the possibly unintended consequences'.

(Benedek 1993: 153–4, quoting Schifter 1974)

From an Australian perspective, the Navajo–Hopi dispute holds many lessons relevant to the post-Mabo period in relations between Aboriginal groups and resource industries, and between Aboriginal groups. We have already seen native title claims lodged and disputed by other Aboriginal people who argue from a different historical perspective about the circumstances producing particular relationships to particular sites, particular resources and particular areas. Prior to the recognition of native title, such disputes rarely entered the public domain outside the Aboriginal public. Their entry into the non-Aboriginal domain is often accompanied by assertions of interest which seek to discredit, disempower or silence certain parties. If we look at the lessons of the last fifty years in the Navajo–Hopi lands, this path leads neither to quick resolutions of complex social conflicts, nor just and equitable results for the disputing parties. The continuation of the human consequences of the Navajo–Hopi land dispute, and the continued implication of Peabody Resources' coal interests in the orchestration of the dispute and its resolution remain as a potent symbol of the power of resource systems to turn affected communities' worlds upside down, and the complex interplay of community and resource industry processes in real-world resource management systems.

10 Indigenous rights or states' rights

Hydro-power in Norway and Québec

Indigenous rights and states' rights

Conflicts over resource management strategies between nation states, subnational authorities and indigenous peoples in remote parts of national territories have been common in many jurisdictions over recent decades. Capitalism's long boom in the post-war period facilitated greater penetration of isolated hinterlands previously left to indigenous peoples. This neo-colonial penetration often occurred in periods when economic nationalism was high and tolerance of ethnic or cultural difference was low – periods when the links between nationalism, industrialisation and developmentalism were being reinforced. In many places, resource projects were central to nationalist ideologies and state efforts to forge a stronger and more unified national identity. Such circumstances were hardly conducive to recognition and protection of indigenous rights in these resource-rich areas. Whatever the particular circumstances of the expansion of the resources frontier into new indigenous territories, the specifics of the relationship between the nation state and the affected indigenous peoples was an important element of the development process. Factors influencing this relationship are many and varied:

- The extent to which indigenous rights were previously recognised, protected or respected
- The nature of post-colonial institutions
- The extent to which indigenous identities were possible
- The extent to which peoples' political economy provided a robust or vulnerable foundation for resistance prior to resource-related development intrusions
- The nature and extent of previous colonial and neo-colonial intrusions and their impacts
- The existence and independence or dependence of representative organisations (*inter alia*).

In all circumstances, both indigenous and settler populations and their systems of governance will be complex and dynamic. Neither side is reasonably characterised as unidimensional or static. In each case there are details and nuances to take into account in providing an explanation of circumstances and outcomes, and in interpreting their significance and lessons.

In the late 1960s, hydro-electric developments in northern regions of both Canada and Europe pitted states' rights against indigenous rights. In both cases, the nation involved in proposing the development was a mature and progressive democratic state: Norway was widely recognised as a champion of international human rights, while Canadian federalism was widely admired as an epitome of stability and justice. Examination of these two cases provides important insights into the tension between these two social forces. This chapter argues that the tension between indigenous rights and states' rights is not resolved in any single dispute over resource development. The persistence of indigenous rights, however, goes to the heart of notions of national identity, morality and sovereignty. This means that even where indigenous people fail to achieve specific goals such as non-construction of resource projects such as the Alta Dam and the La Grande project discussed here, their struggles challenge the foundations of states' rights.

Sami rights and the Alta-Kautokeino Dam in Norway[1]

Northern Scandinavia has been the homeland of Sami people (*Saamidaen*) since about 8000BC (Charta 79 1982: 6). In the modern era, its resources (fish, minerals and hydro-power) have been integrated into the national economies of Norway, Sweden and Finland. The hydro-electric potential of Norway's northern wild rivers has long been seen as an important economic resource by the Norwegian government. Like similar state energy agencies in Canada, Tasmania and elsewhere (Crabb 1984), Norway's *Norges vassdrags- og elektrisitetsvesen* (NVE) promoted large-scale regulation of wild rivers as a central feature of their contribution to the national interest. By 1972, sixty of Norway's rivers had been regulated to harness their hydro-electric potential, with significant impacts on Sami people (Kleivan 1978: 61).

The Alta Dam dispute

For energy planners committed to industrialisation and development, the Samis' non-industrial uses of northern resources and the cultural landscapes of which they were an integral component were virtually invisible. Even where they were acknowledged, non-industrial and non-economic uses of the resources were given much lower priority than harnessing them to serve the requirements of industrial development in the national interest. The conflict between Sami interests and the Norwegian government in the period 1979–82 over development of the Alta-Kautokeino Hydro-electric Project demonstrates just how complex place-based resource conflicts really are, and how far-reaching their consequences can be.

The Alta Dam project involved proposals to construct a hydro-electric dam

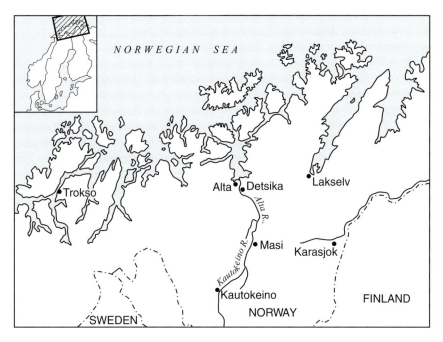

Map 10.1 Northern Scandinavia showing the location of the Alta River

on the Alta River in Finnmark County. Because of the social, cultural and environmental implications of the proposal, it became a focus of major social protest. The dispute mobilised and politicised both Sami and non-Sami Norwegians, and led to the largest post-war mobilisation of Norwegian police in action against protesters at the project site. The dispute also produced two national inquiries into Sami rights (Smith 1987), transformed relations between the Norwegian state and Sami, and contributed to increased legal recognition of Sami rights throughout Scandinavia. Despite the protests, the project was eventually approved by a court decision, and constructed by NVE, signalling a major loss for the Sami in this specific struggle. In the wake of this dispute, however, the Sami 'achieved some unprecedented gains in terms of public recognition, administrative reform, and promises for constitutional, legal and political change' (Brantenberg 1985: 23).

The Alta Dam proposal

In its first permutation in the late 1960s, NVE's proposal to regulate the Alta River envisaged a major project, involving not only a dam on the Alta between the towns of Alta and Kautokeino, but also regulation of two large lakes, and tributaries to the Tana River to the east on the Norwegian–Finnish border (see Map 10.1). In this configuration, the project would have generated 1499 GWh and substantially boosted Norway's power supplies. It would have:

- Flooded the important Sami community of Masi;
- Disrupted reindeer herding over a wide area; and
- Substantially affected fluvial and riparian environments.

There was rapid mobilisation of local opposition to the original plan in the early 1970s, including Sami in Masi. There were also protests from the government of Finland over downstream impacts on the Tana River.

In 1976, NVE submitted a revised, smaller scale proposal. This new configuration involved a relatively small dam on the Alta River intended to generate 625 GWh, principally in the summer months (Kleivan 1978:57). The rationale for the project was the prospect of an imminent crisis in NVE's ability to meet the power demands of the local region, and the need for power to attract development, particularly to the coastal area around Alta. In retrospect, there appears to have been no such crisis (ibid.: 60). The revised proposal required construction of a 100 m high dam on the Alta and a power plant capable of generating 625 GWh, or 0.5 per cent of Norway's total production. In 1976 the city councils of Alta and Kautokeino voted against the revised proposal, but the Finnmark County council approved the construction. In 1977, the resource board supported the Finnmark County decision. In 1979, after cursory investigations, the Storting, Norway's national parliament, considered the project several times. Ultimately the Storting approved construction by a substantial majority. Protests over this decision and its subsequent confirmation by Parliament and the Norwegian Courts mobilised and politicised both Sami and non-Sami in northern Norway, with far-reaching consequences.

The protests at Alta and elsewhere

> The Alta River is the second largest river in Finnmark, the northernmost county in Norway. And it is here that the most important Sami settlement areas are located … . The Alta River is 170 km long and flows through the most important populated districts in the region. The river is thus of unusually great importance – for the local environment, the daily life of the inhabited districts, and as a resort and recreation area. In addition, the wilderness plays a vital part in the Sami's reindeer husbandry, berry-picking, and fishing.
>
> (Stormo and Solem 1981: 2; see also Borring *et al* 1981;
> Simonsen 1985; Hillestad 1992, 1993)

The bare facts of the dispute of the Alta Dam are relatively easy to describe (see IWGIA 1981 for a chronology of principal events; also Charta 79 1982). Local opposition focused on environmental, cultural and economic concerns. Construction of the access road commenced in September 1979, before formal permission had been given. This was met by protests at the site from Sami and the so-called 'river savers', non-Sami Norwegian supporters of conservation and Sami rights. In October 1979, seven Sami established a protest camp outside the Storting in Oslo in a traditional Sami lavvo (tent). This lavvo became a symbol of Sami identity and self-determination, and a focus of international attention. Sami protesters also commenced a widely publicised hunger strike. Police were mobilised to clear the protesters from the square in front of the Storting, and the government agreed to a six-week delay in construction; the hunger strike was then called off. Central to both Sami and conservationist demands for postponement or cancellation of the project was a demand that no further work proceed until Sami rights to land and water were settled. There was prolonged legal and political manoeuvering in 1980. The Storting confirmed its decision to proceed without settling the Sami rights question. In December 1980 the Alta Lower Court ruled that the project should be allowed to proceed (IWGIA 1981: 63). Construction recommenced in January 1981, with large protests at the site. It was at this point that the dispute rapidly escalated into a major national and international matter.

Prior to construction commencing on 14 January 1981, there were protests in Alta, Stilla and Oslo, as well as international support actions. At the site, 900 demonstrators tried to block the road construction. Six hundred officers from all branches of the Norwegian police service were brought to the region in Norway's largest ever peacetime mobilisation of police, at a cost of 1–2 million Norwegian kröner/day (IWGIA 1981: 63). Two thousand people demonstrated against the police presence in Alta. Over the following week, police and river savers and Sami continued to clash in the North. Five Sami commenced a new hunger strike, with one of the 1979 hunger-strikers saying:

> We did not end the [1979] hungerstrike to have the Alta-Kautokeino development postponed one year, but to stop it until the question about the rights of the Sames has been clarified.
> (Nils A. Somby, quoted in Borring *et al.* 1981: 127)

On 3 February a delegation of Sami women met with the new Prime Minister, Gro Harlem Brundtland. At a second meeting three days later they received no clear response and refused to leave the building in protest. They were joined by a further 130 people 'in a spontaneous demonstration' (Borring *et al.* 1981: 130).

In February, it was recognised that the necessary investigations required by the Protection of Ancient and Cultural Monuments Act had not been undertaken. This seemed to provide a way out of the deadlock without requiring government capitulation to the protests. In late February, after construction of nine kilometres of the access road, the construction was halted to allow the required archaeological work to be completed. In response, the protests, including the hunger strikes, were called off.

The Sami legal argument against the Alta Dam

The 1980 decision of the Alta Lower Court was taken as a signal to proceed by the Norwegian government. Despite the decisive response of the government, this judgement was quite ambiguous. While a 4–3 majority confirmed parliamentary approval of the project, the court also confirmed some of the Sami criticism of procedures. Sami had argued strongly that the parliamentary decision was based on incomplete, inaccurate and misleading information. In particular they argued that potential impacts on Sami culture had been misrepresented (IWGIA 1981: 74). The decision was appealed. The government requested that rather than being heard by the Appeal Court, the matter be referred directly to the Supreme Court for decision. This presented a major problem for the Sami; their history within the Norwegian state meant that the rights they were seeking to protect were simply non-existent in Norwegian law. The Supreme Court was, of course, bound to consider the dispute in terms of precisely the law that had been part of the process of Sami dispossession, fragmentation, marginalisation and Norwegianisation. In other words, in dealing with questions of states' rights or indigenous rights, the court was ultimately an instrument of the state.

The Sami argument against the Alta Dam was always principally a political argument. Their opposition to the project, however, had to be advocated in the Norwegian courts in terms of a legal system which steadfastly refused to acknowledge that the collective, indigenous rights of the Sami people had any basis at all in Norwegian law. Their argument has much in common with other legal cases where indigenous peoples have argued claims against unlawful dispossession, usurpation of property and abuse of human rights in legal systems which are predicated on the validity of the acts under challenge. The Sami articulated a *political* argument within the constraints of the *legal* process: the way in which the Sami legal team mounted its case, and the sort of case assembled, reflected this context. The subsequent constitutional and political responses from the Norwegian government and society confirm the 'success' of this strategy in political terms, despite the loss of the legal argument about the Alta Dam:

> As the Sámi wanted to make the most out of this rather unique opportunity they widened the scope of argumentation taking up land rights in principle, ecological analysis, the question of culture viability as well as ... referring to international law.
>
> (Svensson 1984: 163)

The Sami's concerns about the dam's likely impacts were presented to the court through the evidence of expert anthropological witnesses (Björklund and Brantenberg 1981) and in the form of the detailed study by Paine (1982), later published in translation by IWGIA. The Sami team also commissioned a brief from a Canadian expert on aboriginal rights and international law,

Professor Douglas Sanders (1981). While dismissing Sami concerns about the Alta project, the court did recognise shortcomings in the administrative process leading up to the Storting decision, and also found that issues of international law protecting Sami rights had to be considered in situations where 'water regulation caused strong and very damaging encroachment upon Sámi interests with the consequence that the Sámi culture was threatened' (quoted by Svensson 1984: 164).

Svensson's review of the legal case reflects the difficult political choices facing indigenous leaders in such circumstances:

> Going to court ought to bring about more in the way of immediate results than has been accomplished so far (in Sami legal action), considering the vast amount of time and effort expended. But not going to court is no alternative. It is extremely important … to continue being the active party. And in a long term perspective I imagine the pay-off, viewed in general cultural terms, could be quite extensive.
>
> (ibid.)

Concerns about the Alta Dam project: ecological and social impacts

Economic and ecological transformation of Sami society has been a crucial factor in the transformation of intra- and intercultural politics in the region. Understanding the links between the political economy of the Sami and the human-ecological processes underpinning them is central to understanding the nature and implications of the dispute over the Alta Dam. Central to this issue is the geographical and historical context of the Sami group whose section of the reindeer pastoral circuit is most directly affected by the project – the members of the Nuortabealli or East Side *sii'da* (see Paine 1982 for a detailed account).

The protesters were drawn from an alliance of Sami, environmentalists, farmers, fishers and others affected by the project. Their concerns about the impact of the project were that:

- The electricity to be produced by the project could be sourced from other areas;
- The area to be disrupted by construction of the project had high conservation values;
- The project would interfere with salmon fisheries;
- Reindeer migration routes would be disrupted;
- Microclimatic changes resulting from the reservoir would disrupt agricultural systems;
- The failure of NVE and the government to follow legal and regulatory requirements set dangerous precedents for the whole nation (Stormo and Solem 1981: 3–4).

Svensson provides an overview of change over time which puts the changes arising from the Alta project into context. Development of non-renewable resources, particularly iron in the late nineteenth century, and later hydro-electric power and industrial forestry:

> circumscribed the ecological niche of the Sámi; consequently, fewer peo-ple were able to subsist from their traditional means of livelihood. The Sámi communities began to suffer from compelled depopulation. A num-ber of those living in a Sámi community were forced to break out and find other means of making a living.
>
> (Svensson 1988: 80; see also Aikio 1989)

In the region's new industrial communities, the Sami became a minority. The presence of these communities and the access they provided to a range of ser-vices and activities further contributed to local cultural and economic trans-formations. New technologies made it possible for reindeer herding families to develop a semi-nomadic management system, in which fewer family mem-bers were fully engaged. This provided opportunities for some Sami, particu-larly women, to take on full- or part-time jobs in the industrial communities. A move away from reindeer herding as a subsistence industry, providing a full range of materials necessary for nomadic life in the Arctic (meat, milk and cheese, skin, fur and bone), towards a specialised meat production industry also occurred in this period. Increased sedentarism created opportunities for other wildlife harvesting activities to develop, particularly fishing (Björklund 1991).[2] Air transportation increased the value of fish by allowing catches to be transported as fresh fish to urban markets, at much greater value than previ-ously dominant salted and cured products. By the late 1960s, the Sami econ-omy comprised both subsistence and cash-producing activities, both of which were central to the dynamics of Sami culture (Svensson 1988: 80–81; see also Paine 1982: 68–70).

Paine (1982) notes that these transformations of the human ecology of Sami reindeer herding produced a distinctive interdependence between nomadic, semi-nomadic and sedentary Sami in various communities around the region. In his analysis of the potential impacts of the Alta Dam on the Sami community focused on the village of Masi and the Nuortabealli *sii'da*, Paine emphasises the links between cash and subsistence elements of the Sami econ-omy, and the contributions of both reindeer herding and other activities to contemporary Sami culture and identity:

> The tundra has become a 'heartland' for Saami culture, and the sedentary Saami whose settlements are along its river courses are, in so many ways, the 'custodians' of this Saami heartland … . Certainly reindeer pastoralism – *because it is Saami* – makes contributions to the culture as a whole … (and) the pastoralists are standard-bearers to many other … Saami … . (But these two 'arms of Saami culture' basically in Nuortabealli/Masi

comprise) *one* population – *one* culture – whose survival depends upon a *complementarity* of livelihoods and skills among its members.

(Paine 1982: 71–3, emphasis in original)

The ecological requirements of reindeer herding make it a system which is highly vulnerable to interference from development of point-based resources and access to them (for example, mineral deposits, hydro-electric sites and roads). A significant part of the dispute about the potential impact of the Alta Dam project centred upon different interpretations of how the general ecological and management characteristics of reindeer herding, and the specific arrangements of the Nuortabealli *sii'da*, would interact with the specific construction and operational requirements of the revised proposal. To understand some of the details of the dispute over the Alta Dam project, it is therefore necessary to examine the nature of the existing resource management system in place focused upon Sami reindeer management.[3]

The human ecology of the Nuortabealli sii'da and impacts of the Alta Dam project

In Finnmark, reindeer herding in the early 1980s involved about 200 Sami families in the tending of around 100 000 animals in an area the size of Denmark (Paine 1982: 10, see also Ingold). The dialectical relationships between the Sami pastoralists, the reindeer herds and the biophysical environment of which they are part is a complex and dynamic system. Paine points out that the relationship is not one of domestication and dominance by the herders, but of mutual accommodation between herders and their herd, and the places through which they travel:

> The herder learns the behaviour patterns (especially the behavioural imperatives) of his animals: but that is only half the story for the behaviour of a herd is itself influenced by decisions taken by the herder. This is particularly the case regarding dispositions of time and space in the annual cycle. Thus it would not be true to say either that the herder follows his herd or that a herd will follow its herder wherever (and whenever) he may wish. Both parties follow a common schedule or routine; while it is true that this has been worked out by the herder (remembering that the only feasible schedules are those taking account of the animals' needs), the important point to grasp is that as the herd learns the schedule, and adapts to it, it becomes very difficult to make changes in it. To that extent, the herder is held by his herd to the schedule he devised.
>
> (Paine 1982: 11–12)

The fundamental seasonal cycle involves winter pasturing in inland areas where reindeer lichen is available to sustain the herds, and passage along a circuit through spring calving places, summer pastures in coastal and offshore island areas and autumnal rutting places. In northern Norway, these seasonal circuits are fundamental to the reindeer management systems, and are maintained by Sami herding units known as *sii'da*. The *sii'da* comprise a number of closely related families who combine into a single work unit. People will be added and removed from active involvement in the work of the *sii'da* according to the demands of the herd and the season, with some family members being freed to take up other employment or income-generating opportunities (or educational and other cultural options) from time to time. In Kautokeino County, the Sami reindeer management system involved three *sii'da* – Oarjabealli, Guovdajohtin and Nuortabealli (Map 10.2). A report by anthropologists Björklund and Brantenberg concluded that reindeer pastoralism in the area supported 300 people and 30 000 reindeer, and that construction of the Alta Dam would mean that the system could no longer function in its present form (IWGIA 1981: 69); a detailed report was prepared as part of the Sami legal action. Paine provides data on the extent to which kinship relationships underpin this management system: he also emphasises dialectical links between kinship (social) aspects of the *sii'da*, and the ecological relationships situated in both (seasonal) time and (geographic) place. Cows return to known calving places (often the locality of their own birth) and bulls make for their own 'home range' during the rut, while herders develop an intimate (and culturally informed) knowledge of the pastoral terrain in which their herds move (see also Bergman 1991; Fjellstrom 1987; Aikio 1989).

Other management requirements for this system include a need for peace during the critical October rut, when herders must ensure the animals are not disturbed:

> Disruptions to the rut, and poor pastures at that time, are likely to reduce the number of pregnancies and also introduce irregularities in the timing of spring calving.
>
> (Paine 1982: 17)

During the spring, Sami herders' traditional ecological knowledge becomes crucial in successfully managing the herds:

> A calving ground should be snow-free and dry, but *when* ... a hillside or plateau favoured as a calving ground will become snow-free and dry can vary from year to year. Therefore herders must have *alternative places available* ... . [There is also] the pressing matter of timing after calving for those herds which still have to move some distance north to their summer pastures; one has to wait until calves are strong enough to move but one can wait too long so that cows with calves get trapped in the thaw: partially- or newly-opened lakes and rivers become impassable barriers [or many animals drown].
>
> (Paine 1982: 17, emphasis in original)

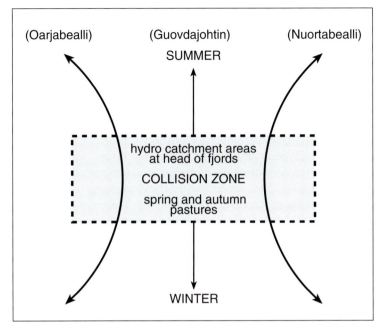

Map 10.2 The three *sii'da*

Source: Paine 1982.

The restricted environmental niches in which suitable conditions are available further complicate the delicate management demands of the spring period. Pregnant and nursing cows in particular have their greatest need for protein at this time, and it is along the river courses and in associated marshes where the snow melts first and the first green plants appear that the reindeer can search for protein-rich feed. These areas are obviously restricted and prized elements of the *sii'da*'s territory. Restrictions on spring pastures, the importance of alternative locations, and links between locations and other parts of the seasonal range (that is, access to the necessary ecological niches in both earlier and later parts of the season), mean that 'the loss of even a small area in the neighbourhood of a river bed can have catastrophic implications for reindeer management' (Paine 1982: 18). At crucial parts of the season, separation of herds from other *sii'da* becomes a major management task as animals are funnelled into a 'collision zone' (ibid.: 19–20) where restricted ecological niches can easily become congested (Map 10.1, p. 281). Similarly, the need for reindeer to rest several times in a 24-hour cycle, and the nature of weather in this region, which might require much longer periods of rest in suitable niches from time to time, creates a further dynamic element to be dealt with in herd management. The needs of both herd and herders in such circumstances must be accommodated to ensure survival. Again, it is principally in the spring and autumn cycles that these demands are greatest. The problem is that wide areas of suitable winter and summer pastures are connected by a relatively narrow range of suitable spring and autumn niches along the migration routes of the herds. In the collision zone, demands on restricted resources are already strained; loss of particular sites within this zone places significant pressure not only on the *sii'da* most directly affected, but it will also have a chain-reaction style flow on to neighbouring *sii'da* as the affected herders adjust their migration to alternative locations, including those used by others. Traditional management patterns involve a high degree of co-ordination and synchronisation (in both time and space) of herd management between *sii'da*s. Disruption to this harmonised arrangement can have disastrous and long-term effects.

Reindeer behaviour, then, was a crucial issue in assessing the potentially disruptive impact of the proposed Alta Dam on the dominant existing resource management system in the Alta-Kautokeino region. The vulnerability of the delicate balance required to sustain this system can be glimpsed in the effect of animals being disturbed in their grazing by a passing vehicle on a road. If this is a persistent disruption, the animals 'either get too little feed or they need longer grazing period' (Villmo cited in Paine 1982: 26).

In the context of the demanding schedules and imperatives for timing and harmonisation with other *sii'da*s, this can have significant consequences for affected animals. Greater alarm can cause energy losses from fright-and-flight responses that can have regressive effects further into the seasonal cycle, by reducing the condition of individual animals and increasing their alarm responses. Disturbances at calving time are seen as particularly significant because cows may abort, desert the newborn calf, or leave the calving ground or post-calving range, perhaps permanently, with significant consequences for the subsequent behaviour of surviving calves. These aspects of individual and herd behaviour mean that even short-term disruptions (single events) can have lasting consequences, and in the case of the hydro-electric proposal, the consequences would certainly outlast the construction phase (ibid.: 25–27).

The combination of the impacts of disruptions on individual and herd behaviour, and the intricate ecological and management demands of the seasonal migration cycle in this resource management system mean that the relatively scarce resources of the 'collision zone' are particularly sensitive to the effects of disruption. It is the nature of the requirements of the hydro-electric industry that it is precisely in this collision zone that the most serious and lasting disruption will occur. In the case of the interaction of the Kvaenangen power station and the Aborassa *sii'da* in Troms (north Norway), where the project occupies only 4 sq km, ten years after construction, an ideal calving ground was rendered inaccessible by the construction and 'is empty to this day despite herders' efforts to get the cows to return to the area' (Paine 1982: 31; the case cited by Paine comes from the report of Björklund and Brantenberg 1981). In this case, the management of the Aborassa herd has been made much more difficult, with reindeers wandering into other herds, moving to autumn pastures too early, and creating tension between these herders and their neighbours. Intervention with 'modern' techniques such as fencing, or longer steps in the migration cycle have reportedly been disasterous where they have been attempted (Paine 1982: 32). In the area affected by the Alta Dam, the collision zone is a:

> large natural 'funnel' bounded on the west by Alta River and in the east by Tverrelv-Stuorajav'ri, and gradually narrowing so that, at Hoal'gir, the animals are in a natural enclosure ... [which is] an ideal [landscape] for working with reindeer.
>
> (ibid.: 34)

On both sides of this collision zone for the Nuortabealli, Guovdajohtin and Oarjabealli *sii'da*s, existing constraints limit the availability and accessibility of alternative routes, and ensure that any disruption in the Nuortabealli *sii'das* would have implications that are serious for all the surrounding *sii'das*. In addition, the predicted environmental effects of the dam, which included a possible localised cold sink related to the reservoir, with an associated delay of possibly two weeks on the commencing of the spring thaw, and an increase in the volatility of local flooding and ice blockages in the river, further add to the management problems imposed on the herders. In his preliminary impact assessment discussion, Paine concluded:

> (i) the chain effects that the hydro project will have on reindeer manage-ment will also flow over to other sectors of the Saami ecology of the area; and (ii) the chain effects thus set in motion in all sectors of Saami ecology will have serious repercussions in the social and cultural life.
>
> (Paine 1982: 58)

Contrary to Paine's conclusions, the Storting felt that the limited area of land directly affected by the hydro project (and excluding the related and necessary additional disruption involved in construction of the site access road of some 34 km) would withdraw only a tiny amount of the reindeer range from use. Further, they concluded that this would only involve a compensable reduction in capacity of the area (Eidheim 1985: 161).

The historical and geographical context of Sami rights

Prior to the sixteenth and seventeenth enturies, Sami economies were domi-nated by fishing, hunting and trapping, and characterised by a seasonal migra-tion pattern ranging from offshore islands and coastal fjords to resource-rich inland areas (Aarseth 1993; see also Terebikhin 1993). During the 1500s, partly as a result of competition from indigenous producers in the new colo-nial territories of North America and Siberia, the fur trade declined. Sami cul-ture began a long period of transformation in which three main trends emerged:

- A coastal culture focused on fishing and livestock herding;
- An inland forest and river culture which emphasised livestock husbandry in combination with fishing, hunting and some reindeer herding;
- A nomadic Saami culture linked through kinship to both coastal and inland groups, but focused on reindeer herding.

(Aarseth 1993: 4)

A second period of trade (Pomor trade) focused on trading Sami fish for Russian flour and other goods developed in the mid-1700s. When it ended in 1914, Sami communities experienced a period of regional social and economic depression (Aarseth 1993: 6). One of the results of this period, which had consequences for the resource dispute examined below, was that it led to:

> a tendency among the Sami [in many coastal villages villages in Finnmark and northern Troms in Norway] to reject all phases of their old life style when an economic boom period began after 1945. Since then the coastal districts' growing involvement in the economic and cultural life of the region has taken on many aspects of the majority [Norwegian] culture.

(Aarseth 1993: 6)

Thus, by the end of the Second World War, reindeer herding was the only distinctively Sami occupational group in Norway. Other Sami were increasingly assimilated into the occupational, income, educational and general social profile of the majority culture, as a direct result of the government's Norwegianisation programmes which had commenced in the 1860s and were most effective in the sedentary coastal Sami communities (Charta 79: 14–15, see also later discussion below).[4]

Social stigmatisation of 'Lapps' (a derogatory term for Sami), suppression of Sami language, culture and pre-Christian religion, and characterising nomadic reindeer herders as 'primitive' and 'inferior' all undermined Sami identity and ethnic solidarity among Sami.[5] This was also further exacerbated by divergent political, legal and social regimes in the four nation states with territorial claims over *Saamidaen* (Samiland) – Norway (Sami population about 40 000), Sweden (about 17 000), Finland (about 4000) and Russia's Kola Peninsula (about 4000) (Cramér 1994: 52). Of these, only about 7000 are reindeer pastoralists, an occupation that is often to characterise Sami tradition. Most obtain their livelihoods from small-scale farming and fishing, along with a range of occupations that are not traditional Sami activities (Svensson 1988: 77–8). Both coastal and inland Sami engaged in trade with their southern neighbours, with fur and pelts, and in the coastal areas also feathers, down and marine products, being the major trading items.

From the Middle Ages colonial intrusions into Samiland involved:

> Territorial, political and socio-economic encroachments by foreign powers and the incorporations of their lands into foreign political entities.

This colonial past was extenuated by economic dependence, cultural and linguistic pressures, assimilation policies, displacement and relocation, loss of resources and continued tutelage and patronage by the states controlling the Sami homelands.

(Müller-Wille 1989: 258)

Until 1751, despite the efforts of its southern neighbours, *Saamidaen* was 'a common territory undivided by national boundaries' (Aikio 1989: 105). In the 1751 Stromstad Treaty, Norway and Sweden/Finland negotiated a territorial border which divided Samiland into national territories. In a codicil to the treaty (the 'Lapp Codicil'), nomadic Sami on either side of the border were guaranteed freedom of movement and the right to use the land in accordance with tradition regardless of national boundaries. In 1826, the boundary between Norway and Russia was negotiated with similar provisions, but this border was closed in 1852, with substantial consequences for the Sami affected (Aarseth 1993: 5). With the dissolution of the Norwegian–Swedish union in 1905, Swedish Sami gradually lost their access to and ownership of grazing lands on Norwegian islands and tundra highlands (Charta 79 1982: 13). Since the 1950s, Sami organisations have been actively engaged in advancing indigenous rights, both within Scandinavia and, since the 1970s, internationally.

Sami rights, Sami identity and Sami politics: political responses to the Alta Dam dispute

The turning point for Sámi politics can be traced to the *lavvu*, the Sámi tent, erected for the demonstration in front of the Parliament in Norway on October 9, 1979. This truly started the discussion in Norway and elsewhere.

(Högman 1989: 38)

The Alta Dam was completed in 1987. Subsequent responses to the conflict over the dam and its implications have produced substantial changes relevant to many of the central demands of Sami activists in the original dispute. In many ways these responses have gone significantly further than many of the key participants anticipated in the early 1980s. These responses have transformed some aspects of relations between Sami and the Norwegian state, and have influenced the tension between indigenous rights and states' rights more widely.

Ten years after the project was completed, Gro Harlem Brundtland, the Norwegian Prime Minister who came to power at the height of this conflict, and whose government persisted in its construction, concluded that the Alta Dam should not have been constructed:

It is now apparent that the development of the Alta riverway was an error of judgement. But this is something one can only say in hindsight.
(Gro Harlem Brundtland, 25 August 1990,
quoted in Brantenberg 1991: 123)

As a result of the Alta Dam dispute, two public commissions were established in 1980, the Sami Rights Commission and the Sami Cultural Commission. The Sami Rights Commission was required to examine:

- Legal rights of Sami in relation to land and water;
- Sami resource rights;
- Prospects for constitutional recognition of Sami, including the possibility of establishing a Sami assembly.

(Torp 1992: 86)

When it reported in 1984, the Sami Rights Commission recommended:

- A codicil to the Norwegian constitution safeguarding the legal status of the Sami;
- A special Sami Act on their legal situation;
- Creation of an elected Sami Parliament.

(ibid.: 87)

These recommendations were accepted by the Storting in 1987. The Committee's recommendations aimed at establishing recognition that protection of Sami language, culture and society is a responsibility of the Norwegian nation and parliament.

The Sami Cultural Commission was commissioned to examine issues related to culture, education and language. It reported in 1985, recommending that Sami and Norwegian be recognised as equal languages with the need to have equal recognition as official languages (Steinlien 1989 7–10; see also Smith 1987).

Relationships between Sami, the Norwegian state and non-Sami Norwegians have long been ambiguous and multifaceted. For example, just before the Alta dispute at the 1978 UN Conference to Combat Racism and Racial Discrimination in Geneva, Norway 'stood out as one of the most vehement champions of the rights of indigenous peoples (including Sami)' (IWGIA 1981: 67), while pursuing a policy at home that IWGIA characterised as 'internal imperialism supported by the judiciary' (1981: 68). Despite the implicit recognition of Sami territorial interests, and limited rights in the 1751 'Lapp Codicil', Sami land, water, resource and property rights were not well protected by Norwegian law. In fact in 1902, to encourage Norwegianisation, the Storting passed legislation that 'established that only those who could read and write Norwegian, and who used this language daily, could become land-owners in North Norway' (Charta 79 1982: 14).

Although not enforced after the Second World War, this statute remained until it was revoked in 1959 in response to a thinly veiled rebuke from South Africa when Norway criticised that country's laws under its racially based apartheid system (Högman 1989).

To a great extent the development of Norwegian democracy and the welfare state has been predicated upon an assumption of a single Norwegian ethnicity as the basis for citizenship:

> In the set of institutions constituting the Norwegian system of state there is no distinction between Norwegians and Saami. They are supposed to be equals with respect to rights and duties. In Norwegian political discourse it is said that individuals of either category are equals ... as citizens. Formulations like these easily conceal the fact that the Saami must regularly bend to political, judicial and bureaucratic decisions that take no account of their basic collective rights as a people.
>
> (Eidheim 1985: 158)

Only in the case of reindeer herding was there any legal recognition of a special Sami identity. This occupation was also stereotypically 'Sami' in the eyes of many Norwegians, and its legal protection had the effect of entrenching negative stereotypes. For many coastal Sami, this characterisation of reindeer herding as the archetypal 'Sami' lifestyle emphasised the social and cultural distance between their own lives and those of the archetypal Sami.

In many ways, the construction of state-endorsed caricatures of multifaceted Sami culture further marginalised Sami from both mainstream Norwegian culture and various aspects of Sami identity. Nystö points out that the material conditions of economic activity in northern Norway undermine the political advances secured by Sami:

> As we enter the 1990s, living standards in most sectors of Nord-Troms and Finnmark are on a par with other regional districts of Norway. The critical factor is employment and earning opportunities which are far from satisfactory. The employment market is suffering from a structural imbalance and it is difficult to recruit people for various jobs requiring a high level of training. At the same time, unemployment levels are also high in the fields of agriculture, fishing, industry and construction work ... . The Samis' everyday life is very closely linked to the opportunities such communities provide in terms of employment, public and private services, cultural activities and other aspects of social life.
>
> (Nystö 1991: 37–8)

Structural legacies from previous eras of government antagonism or paternalistic forced assimilation have meant that the Sami population is at times ill equipped to grasp the limited opportunities that exist:

Many of the external features of Sami culture were linked to old trades and occupations which disappeared as a result of the adaptations to the economic and social changes. For generations, Sami culture had been regarded as an unimportant and unwanted element of Norwegian society and therefore found itself the victim of a harsh assimilation policy. This made it difficult for the Samis to attach any positive associations to the concept of a Sami identity in their dealings with the country's other inhabitants. This made it tempting for the Samis to shed their cultural heritage.

(ibid.: 37)

The prolonged political and economic attack on Sami cultural identity was also reinforced by a lack of any recognition of collective rights to ownership of land, water and other resources in Norway, or indeed anywhere in Scandinavia. Northern Norway's natural resources, particularly mineral, marine and water resources, have been central to the construction of economic opportunities, and an important element in Norway's strategic interest in the region. Failure to secure Sami resource rights in combination with the assimilationist policies meant that for many Sami, emphasising Norwegian rather than Sami identity produced greater economic opportunities. These complex social, economic, political and cultural processes produced deep divisions within the ethnic Sami population. These divisions were further exacerbated by evolving differences in all these realms among Sami in other nation states. Development of pan-Sami political links in the 1960s provided some basis for comparison of national approaches to questions of culture, legal rights, language and autonomy. Institutions and organisations such as the Nordic Sami Council (founded in 1956) and the Nordic Sami Institute (founded in 1973), along with national and regional organisations and the national Sami parliaments or assemblies in each of the Nordic nations (Finland 1975, Norway 1989, Sweden 1992) have provided a strong political framework for discussion of Nordic co-operation on recognition of Sami and nurturing Sami rights and culture.

As Brantenberg points out, this historical and geographical context meant that the Alta Dam dispute did not simply polarise Norway into pro- and anti-Sami rights responses. As he puts it 'the dispute went beyond the ethnic border, dividing not only Norwegians, but Saami as well' (1985: 23). Brantenberg's carefully nuanced reading of the dilemmas created for Sami from various backgrounds by the Alta Dam dispute and its consequences, concludes that simplistic readings relying on militaristic metaphors of battles, campaigns, victories and losses risk reducing complex identity politics to a zero sum game. He suggests that the dispute and the subsequent commissions exposed 'dilemmas and ambiguity in Saami ethnicity and Saami–Norwegian relations', which 'contributed to a process of ethnopolitical change and mobilisation', with Sami from diverse backgrounds taking the opportunities created by these debates and openings to pursue 'different interpretations of Saami ethnicity' (Brantenberg 1985: 44).

There can be no doubt that these Norwegian initiatives provided a catalyst for change in Sweden (Torp 1994; Lasko 1987) and Finland (Aikio 1987). The moves to institutionalise Sami rights and self-determination are in marked contrast to earlier hostility and paternalism. While no 'model' of self-determination is ever simply transferable to other geographical, cultural and historical circumstance, the Sami experience demonstrates that indigenous self-determination or autonomy:

> does not indicate isolation or withdrawal from existing political and economic structures. Instead, the newly adopted status ... is important – on own cultural and social terms [sic].
>
> (Dahl 1992:187)

Central to the reautonomisation of Sami communities and protection of the cultural viability is recognition of their interests in terms of rights (including an enforceable right to veto and exclude) rather than statutory concessions:

> Because of demands of industrial development from the dominant society, community development in the north is paramount. Ecological niches to which the various ethnic minorities are adapted [throughout the Arctic] must be protected. Such protection can be realized only if the locally defined ethnic community has sufficient political strength to act as an autonomous group in conflicts of interest. Such a minium degree of local autonomy, which to a large extent is based on protection of land and water, appears as the most crucial requirement in maintaining cultural viability ... [and] without viable ethnic communities, native peoples will find it very difficult to avoid assimilation pressures.
>
> (Svensson 1986: 214–15)

The fragmentation of the Sami population into distinct communities of interest, reindeer pastoralist Sami and non-reindeer pastoralist Sami, with divisive consequences in struggles such as the Alta Dam case, resulted from 'deautonomisation' of the reindeer herding *sii'da* through the Reindeer Pasture Law. Under this law, only reindeer-herding Sami have been:

> entitled to rights *qua* Sámi ... . By means of this de-autonomization, the Sámi came directly under non-Sámi jurisdiction, although a special group of people with a particular occupation, were granted certain exclusive rights, such as monopoly rights to herding reindeer and favorable rights to hunting and fishing on their reindeer pasture lands.
>
> (Svensson 1988: 78)

Revised regulation of the reindeer-herding industry under the new Reindeer Husbandry Law 1971 was a limited step towards 'reautonomisation', introducing a degree of localised decision making within the reindeer-herding

Sami communities. This did not, however, address the longstanding Sami demand for legal recognition of their *rights*, particularly to land and water (see also Svensson 1984, 1986; Torp 1992; Aikio 1993) and left them without legislative or veto power over those territories under claim. While the struggle continues, one of the legacies of the Alta Dam dispute is a diverse and dynamic politics of identity and sustainability in Scandinavia.

Chernobyl and its impact on the Sami (April 1986)

The development of sophisticated ethnopolitics in *Saamidaen* reflects diverse influences, including the Alta Dam dispute. These influences include the demand for electricity, growing markets for Sami products, growing political co-operation between Sami in the Scandinavian countries, changing geopolitical relations within the Arctic in the wake of the Cold War, and the Samis' geopolitical links to the international indigenous peoples movement: the meltdown of the Soviet nuclear power plant at Chernobyl in the Ukraine in April 1986, however, had dramatically negative consequences for Sami reindeer herders. This episode provides a salutary reminder of the interconnectedness of localities, communities and apparently autonomous resource management systems. It also reinforces the idea that in terms of resource management failure, we all live at Ground Zero. The complex and unpredictable nature of the consequences of breakdowns in industrial resource management systems is demonstrated in the effects Chernobyl had on Sami reindeer pastoralism; even relatively empowered groups such as the Sami, are unable to exercise control, or even to obtain adequate information, compensation or protection from global-scale catastrophic consequences beyond the control of specific ecosystem-based cultures such as theirs. The Chernobyl incident demonstrates the limits of bioregional systems, single nation states or even international alliances of nation states (such as Norway, Scandinavia and the European alliance) in addressing the consequences of resource management failure:

> The nuclear fallout from Chernobyl is considered the single most demolishing event ever to come upon the Sámi: by means of the most advanced modern technology a catastrophe from real life has made the [traditional and successful Sami approach to] ... readaptation entirely non-operative.
>
> (Svensson 1988: 84)

In Sweden, too, destruction of Sami reindeer herds after Chernobyl had far-reaching consequences:

> These circumstances have had a profound effect on the well-being of the Saami and Saami identity. One man interviewed ... expressed great worry as to the future. He seriously doubted that his sons could find much motivation in continuing as reindeer herders.
>
> (Broadbent 1989: 135)

The reindeer lichens of the southern Sami lands, where winter pastures are located, were badly affected by the fallout from Chernobyl. They received substantial fallout within two days of the disaster. The lichens absorbed and retained high levels of Cesium 137, a carcinogen: this contaminated reindeer meat, which was withdrawn from sale for human consumption. Meat from some areas showed extremely high readings (Svensson 1988: 84–5; Broadbent 1989: 132–4), and thousands of contaminated reindeer were slaughtered as a result. Even fish from mountain lakes and rivers showed high levels of contamination. The incident has had far-reaching effects on Sami health, economy, demography and emotions. Svensson identifies increased risk of cancer along with depression, anxiety and emotional stress (sorrow, rage and dispiritedness) as key health issues. He suggests that the loss of income, much of which may be permanent as consumers refuse to buy reindeer products, regardless of place of origin, because of the radiation issue, will have long-term impacts on the Sami communities; other damage will result from the loss of important cultural knowledge and the cancellation of cultural activities such as the community occasions and events that used to be involved in the slaughter of a reindeer. These complex and interacting impacts of a disaster completely outside the orbit of the existing ethno- and geopolitics will, Svensson argues, create longer-term demographic changes (1988: 85–6).

Svensson makes the point that in the previous conflicts over Sami cultural survival, such as the Alta Dam, the 'enemy' was to some degree easily identifiable. However difficult it might have been, dialogue about the grievances raised in the conflict was possible:

> The Chernobyl disaster represents a completely different situation. In this case the opponent is basically invisible. Even though the party responsible can be identified, it remains beyond the reach of any direct Sámi counter-action. This new state of affairs makes any struggle for justice extremely difficult to bring about.
>
> (ibid.: 86)

As Svensson notes, one response of the Sami to the Chernobyl disaster has been to guard even more jealously their remaining pasture resources and to emphasise even more starkly the 'undeniable urgency for a revised political form in which the Sami minority will be able to act from a relevant basis of power' (ibid.: 87).

Hydro-Québec's impacts on the Cree and Inuit

Hydro-Québec and resource sovereignty

Harnessing the energy of northern Québec's wild rivers has fundamentally shaped the relationship of the region and its First Nations with Québec, Canada, the USA and the wider global economy. In the early part of the twentieth century, it helped forge the corporate empires of the international aluminium industry. More recently, the institutional, political, legal and economic development of Hydro-Québec has been closely linked to the reassertion of French-Canadian identity and Québec separatism. Ironically, the struggle of the Cree and Inuit to assert their rights in the region has also been fundamental to the revitalisation of aboriginal nationalism, both within Canada and internationally, and construction of powerful political and economic institutions advocating aboriginal self-determination (see McCutcheon 1991; Jhappan 1992).

The relationship between Québec's energy-centred development trajectory and the concerns of the indigenous peoples, whose traditional territories have been the object of development, has been enormously important in shaping not only the future of particular communities, but also in internal debates about Québécois identity and sovereignty, Canadian federalism and Canada's international political, legal and economic standing. The juxtaposition of the frontier and the homeland in northern Québec has continued to be an important element of social and political debate over many decades. The parallel between conquering the wilderness with energy-harnessing technology and conquering the peoples whose homelands comprise these wilderness territories has been a recurrent theme in the political rhetoric and national development sentiment of Québec.

In the late 1960s, the provincial government developed a plan to harness the 'wasted' potential of the rivers draining into James Bay, Hudson Bay and Ungava Bay from the so-called wilderness areas of northern Québec. Many words have been written about this proposal. Focusing on the development and implementation of the La Grande Project (James Bay phase I), I wish to argue that Québec's assertion of unconstrained sovereignty over Cree and Inuit resources and territories in northern Québec, like the actions of the Norwegian government in relation to the Alta project, set in train political, judicial, social, economic and environmental processes that transcended the immediate context of Hydro-Québec's development proposals. In a later discussion about the development of project assessment guidelines, I will take up specific concerns with James Bay phase II (the Great Whale River proposal), but in this section, the La Grande Project is a valuable focus for consideration of the complexity arising from competing claims over resources, and the interplay of history, geography and society.

The James Bay Project: alternative visions of northern futures

The northern part of Québec, targeted for development of the three-phase James Bay Project (Map 10.3), was originally 'acquired' by the British Crown in the seventeenth century, without treaty or occupation, and granted to the Hudson Bay Company in 1668. In 1912 the Canadian federal government transferred title of the northern part of the Québec–Labrador Peninsula to the province of Québec, conditional upon them making treaties with the area's

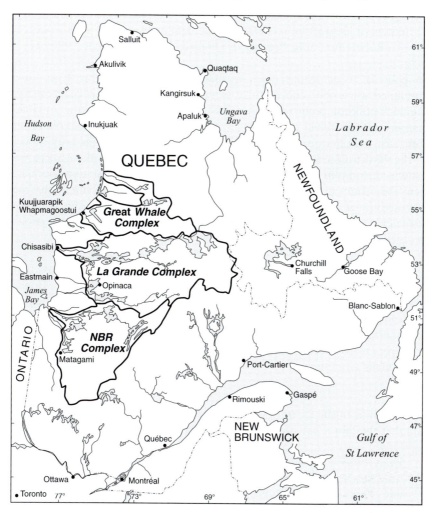

Map 10.3   The James Bay Project Area

indigenous peoples. As Charest (1982: 415–6) notes, the industrialisation of Québec has long been linked to the production of hydro-electricity. Alcan and Hydro-Québec, which was nationalised by the provincial government in 1963, were the principal proponents of this hydraulic industrialisation. In 1971 Hydro-Québec announced a proposal involving construction of four power stations, four major dams, eighteen spillways and control structures, 80 miles of dykes, the creation of a number of large reservoirs and the flooding of over 3000 square miles of territory (O'Reilly 1988: 33). Divergent views of the place of mega-projects in the future of northern Québec emerged in response to announcement of the James Bay Project. The government position was unequivocal and positive:

> Québec is a vast hydro-electric plant in the bud, and every day, millions of potential kilowatt hours flow downhill and out to sea. What a waste!
>
> (Robert Bourassa, Premier of Québec, quoted in Gorrie 1990: 21–2)

In contrast, Cree leaders such as Billy Diamond and Max Gros-Loius announced their opposition. But, as Feit points out, the absence of an existent regional political body for James Bay Cree delayed any formal regional response (1985: 39–41). At a July 1971 meeting initiated by Philip Awashish, thirty-five people from seven Cree communities considered the implications of the Hydro-Québec proposals, unanimous opposition to the project was expressed, and 'the starting point for a new regional representation process, a new leadership and eventually a new organization among the James Bay Cree' was established (ibid.: 40). Billy Diamond, then Cree Chief at Rupert House and later lead negotiator in the James Bay and Northern Québec negotiations, described events:

> For the first time in history, the Cree sat down together to discuss their common problem – the James Bay Hydro-electric Project. But we found out much more than that – we found out that we all survive on the land and we all have respect for the land. Our Cree Chiefs also found out that our rights to land, our rights to hunt, fish and trap and our right to remain Crees were considered as privileges [not rights] by the governments of Canada and Québec.
>
> (quoted in Feit 1985: 40)

For the Inuit of the far north, it also seemed that the stakes were very high, and that the James Bay proposal had enormous implications for the future:

> Everything was at stake. Our region had always been neglected. We needed schools, safe airstrips, health care, and other basic community services. Some people wanted to relocate away from the enormous hydro-electric complex with all its harmful environmental impacts. The territory of our ancestors, on which we hunted, fished, and trapped for subsistence – our land – was at stake. Our sovereignty, the right to govern ourselves and make decisions concerning our territory and its use – our authority was at stake. The very survival of our people as a distinct and proud people was at stake.
>
> … our rivers were at stake, and so were our fish, our livelihood from game hunting, our right to survival. Our future and that of our children were at stake. But it was not easy to convince the courts and the governments that we had any rights at all.
>
> (Watt 1988: 54)

The Canadian government was approached by the Cree to take action against the provincial proposal based on its trust responsibility for native peoples. The federal government, however, was already concerned about emerging separatist sentiments in Québec and was reluctant to intervene, adopting only a position of 'alert neutrality' (Feit 1986: 195).

Like many development proponents before and since, the Québec government felt it held an unassailably strong position, with wide public support, and no legal barriers to implementing its development mandate:

> The government of Québec refused to negotiate, stating that the plans for the hydro-electric project were not negotiable and that Indian people had no special rights.
>
> (Awashish 1988: 43)

The Cree and Inuit disagreed and initiated court proceedings to halt construction. Although the decision was later reversed, they succeeded in obtaining an injunction to halt development work. In a dramatic judgement, Justice Malouf 'decided that the Crees and Inuit had apparent rights to the territory' and granted an injunction stopping further work on the project. Although this judgement was later over-ruled by a higher court, O'Reilly argues that it was 'a turning point in the attitude of the government of Québec' (1988: 35). A provincial negotiator, John Ciacca, was appointed and the prospect of a negotiated settlement emerged. The assumed balance of power favouring the proponents of hydro-electric development in the region shifted, and a complex and influential period of negotiation, institutional development and political activity ensued, focused on the Inuit, Cree and provincial interests. By the end of this process, the first comprehensive land-claim settlement of the modern era, the first of Canada's 'modern treaties', had been concluded in the shape of the James Bay and Northern Québec Agreement 1975 (JBNQA). The fledgling Cree leadership had formed the Grand Council of the Crees and the Inuit had established the Northern Québec Inuit Association and Inuit Tungavingat Nunamini, which continued to oppose the JBNQA. The JBNQA established further institutional structures including environmental protection, community development and regional governance authorities. In return for a substantial compensation package, alteration of the project and guarantees of environmental protection, the indigenous negotiators extinguished any continuing rights and approved development of the La Grande Project – James Bay phase I.

Negotiation and treaty making: the James Bay and Northern Québec Agreement 1975

The stakes in the negotiations were high for all parties. The James Bay and Northern Québec Agreement negotiations were the first to occur under the auspices of new federal guidelines for comprehensive regional claims settlements. Given the conditional transfer to Québec of title to the northern region by the federal government, the very existence of native rights and the mechanism for their extinguishment were at stake. Likewise, the establishment of a mechanism to address injustices suffered by native peoples, and to balance these against the need for, impacts of, benefits from and alternatives to the James Bay project (see Ciacca 1988; Diamond 1988; Watts 1988; Sivuaq 1988 for discussion of the 'stakes' from a variety of standpoints).

For the Cree negotiators, the recognition that even legal success could be overturned by legislative action to extinguish aboriginal rights and impose an unacceptable settlement also pushed them towards negotiating a settlement. While their initial actions focused on modifying the James Bay project, the circumstances led to a much wider negotiating agenda:

> The negotiations were a rare opportunity for us to demand recognition of our rights and to demand remedies for our serious claims concerning our distinct society and way of life. The Cree nation of Québec sought a satisfactory agreement on the following:
>
> 1 Modifications to Complexe La Grande, with remedial works to assist Cree hunters, fishermen, and trappers;
> 2 Recognition of Cree hunting, fishing, and trapping rights;
> 3 Land and territorial rights;
> 4 Community development with sufficient lands;
> 5 Programmes and assistance for Cree hunters, trappers, and fishermen;
> 6 Police and justice;
> 7 Cree local and regional self-determination and self-government;
> 8 Protection of the environment, and future development;
> 9 Economic and social development;
> 10 Health and education; and
> 11 Monetary compensation.
>
> (Awashish 1988: 44)

From the perspective of the Québec government, it was not only the right and power to determine the province's development trajectory that was at stake, but even the province's territorial integrity: 'In Québec ... the territory is felt to be threatened every time native peoples raise the issue of their title to it' (Vincent 1988: 245). The negotiations took two years to complete:

> We thought that with the Agreement we had secured the means to adapt to the damages caused by the La Grande Project and to the changes to Cree society that would surely result from increased contact with the larger society. The Agreement index reads like the constitution of a new country and, in many ways, that is what it was meant to be.
>
> (Diamond 1990: 27)

Nearly two decades on, and faced with the impending impact of the Great Whale Project in the area to the north of the La Grande Project, reflection on the adequacy of the agreement, the effectiveness of its implementation, the meaning of its terms and the role of Cree and Inuit in the development of northern Québec continue to fuel social and political debates in Canada:

> We have had fifteen years of constant struggle to try to force Quebec and Canada to respect their commitments under the ... Agreement. If I had known in 1975 what I know now about the way solemn commitments become twisted and interpreted, I would have refused to sign the Agreement.
>
> (Diamond 1990: 28)

Hydro-Québec and megaprojects as a preferred development path

At Hydro-Québec, there continues to be a technological imperative embedded deeply within the institutional culture – if it's possible, it should be done. Referring to the proponents as a group, including not only Hydro-Québec but also their engineering consultants, McCutcheon characterised them in the following way:

> They are the kind of people who would make real the metaphor Spaceship Earth, who would engineer all natural ecosystems so as to manage them for human gain. They propose continental scale improvements to the planet's energy and water systems ... . (The former head of one of Canada's major engineering consultancy firms) once summed up their ideas and values for me. 'In my view,' he explained, 'nature is awful, and what we do is cure it'.
>
> (McCutcheon 1991: 148)

Hydro-Québec has also consistently argued that the negotiations in the 1970s effectively overrule Cree protests about the social and political implications of further development of the James Bay Project:

> Even if it is accepted, for the purpose of discussion, that local populations have a right to veto major projects, the fact remains that the James Bay and Northern Québec Agreement has already settled their case of the Grande-Baleine (Great Whale) project and the Crees and Inuit of Québec.
>
> (Hydro-Québec 1993: 41)

Successful construction of phase I of the La Grande Project was completed by 1985 at an estimated cost of C\$13.7 billion (Varley 1995: 477). Maxwell *et al.* (1997) argue that Hydro-Québec's success in the project has reinforced a commitment to megaprojects as a preferred development course, despite the problems that emerged during its negotiation, construction and operation. They identify its success with large-scale projects; the organisation's engineering culture and the intimate link between its operational objectives and nationalist goals for economic and cultural autonomy have all reinforced this commitment; and also that a public policy engagement with serious analysis of alternative development paths, commitment to principles of social and environmental justice, acceptance of public participation by provincial authorities and strong and independent public regulatory oversight of development decisions is needed to challenge this orientation and its public interest implications.

Impacts and mitigation at the La Grande Project

While the La Grande Project positioned Hydro-Québec as one of North America's lowest cost power producers, and positioned the company as a major competitor in US electricity markets, a number of serious unanticipated implications emerged during construction and operation of the project (McCutcheon 1994). Anticipated negative impacts were addressed through design modifications and remedial or corrective works. The JBNQA established compensation regimes for unavoidable negative impacts and institutional arrangements for strengthening community development processes and responses. As McCutcheon notes (1994: 3), the scale of the mitigation efforts in the La Grande Project was unprecedented. Despite the enormous public interest and the wider significance of this effort, he asserts that a comprehensive retrospective evaluation of the effectiveness of impact monitoring and mitigation programmes is not possible because of incomplete records, limited access to Hydro-Québec's in-house documentation and the partisan nature of much of the research undertaken by or on behalf of interested parties to the continuing conflicts over the geopolitics of water and territory in northern Québec.

McCutcheon summarises the predicted negative impacts affecting Cree and Inuit as follows:

- Loss of traplines and fishing zones;
- Difficulties related to increased debris in the reservoirs, including for example damage to fishing craft and nets;
- Difficulties relating to access using float planes in areas where river flow was reduced;
- Loss of fish habitat;
- Problems with drinking water supplies in villages at the mouth of regulated rivers;
- Difficulties in using and crossing the lower section of the La Grande River because of increased water flow and unnaturally early break-up of ice;
- Problems of erosion.

(McCutcheon 1994: 10–11)

Mitigation and management of all these matters was addressed in the JBNQA. There was, however, a major unanticipated impact which remained unrecognised until 1979: impounding the reservoir waters produced significant elevation of mercury levels in the reservoirs and bioaccumulation posed a threat to human health amongst those relying on reservoir-caught fish. Ironically, one of the mitigation strategies aimed at reducing the impact of disruption to traditional fisheries was to support and encourage exploitation of fish from the reservoirs. The mercury contamination arises from the decomposition of organic matter submerged by the reservoirs. The elevation of organic mercury or methylmercury in hydro-electric reservoirs is reported by Hydro-Québec as a temporary phenomenon, 'with conditions returning to normal after 20 to 30 years' (Hydro-Québec 1993, Highlights 17 – Mercury). The health effects of long-term exposure to low doses of methylmercury remain uncertain. While episodes of mercury poisoning from industrial sources, such as waste disposal in Minimata Bay in Japan, have well-documented physiological effects, monitoring of people by the Cree Board of Health and Social Services since 1982 has produced valuable data but no unequivocal conclusions. The cultural value of fishing, its mythic and social implications, the importance of fishing in native mobility and the economic value and nutriotional importance of subsistence fishing activities all make mercury contamination a complex and difficult impact to address; even Hydro-Québec acknowledges that 'it is probable that abandonment of fishing could create worse problems than mercury itself' (1993 Overview 17 – Mercury: 2).

In developing responses to the problems of mercury contamination, the risks of exposure and the need to manage use of reservoir resources, Cree authorities have faced several challenges. Contaminated fish appear healthy and communication of statistical risks arising from exposures proved difficult. The concept of contamination and bioaccumulation was translated into Cree as *nemas aksun* (fish disease), which 'connotes a contagious disease of fish that can spread to humans' (McCutcheon 1994: 59). This linguistic problem led to a perception that bush foods such as fish carried a risk of cantagion and 'undermined faith in the integrity of nature, a faith central to the Cree's conception of themselves' (ibid.). Limiting consumption of fish has wide-reaching implications for health, identity, culture, land use and the bush economy. Long-term effects of exposure on human neurological systems are monitored by the Cree Board of Health and Social Services, and the Cree themselves also 'worry that the health of otters, ospreys, herons, sandhill cranes, and other fish-eating creatures may be adversely affected by mercury' (ibid.: 60). Since 1986–7 the James Bay Mercury Committee has reported on monitoring and mitigation activities, with annual reports published in Cree, English and French. While stabilisation of or decreases in mercury levels in water and accumulation of mercury in hair samples of Crees in high-risk groups has been reported (James Bay Mercury Committee 1992: 9–10), this unanticipated effect of hydro-electric development illustrates the complexity and wide-reaching implications of resource-based development for indigenous groups and their territories.

Institutional development in Northern Québec

There is no doubt that conclusion of the James Bay and Northern Québec Agreement and construction of the La Grande Project heralded dramatic changes in the affected areas. The cost of the negotiations for the Cree was $2.2 million, which was reimbursed by the provincial government to the Grand Council of the Crees (Diamond 1988: 115). The costs of implementation were to be met by funds generated by the JBNQA. New institutions delivering service delivery in health, education, housing, environmental protection, employment and enterprise development as well as co-ordinating bodies approximating regional self-government authorities all emerged from the negotiations in both Cree and Inuit territories.

Political development through these organisations has led to a more sophisticated and perhaps better prepared leadership amongst First Nations in Northern Québec than existed when the James Bay Project was first announced in 1971. With hindsight, the breakthrough achievements of the JBNQA negotiations seem to have been achieved at great cost. In 1993 the Grand Council of the Crees reported to the Royal Commission on Aboriginal Peoples that they felt the negotiations had forced them to extinguish fundamental rights in exchange for things to which they should have been entitled (Varley 1995: 484). Ted Moses, one of the negotiators, reflected that during the negotiations and for nine of the ten subsequent years, he had been all but unable to enjoy the rights the agreement had secured (1988). Billy Diamond, the principal Cree signatory to the 1975 agreement reported in 1990 that the Cree had by then negotiated three major revisions to the La Grande Project, producing with each new negotiation compensation for damage to Cree lifestyles and Cree rights. The La Grande Project has expanded in terms of output, in terms of coverage and, since the decision not to proceed with the Great Whale Project, in terms of the scope of its role within the overall 'grand vision' for hydraulic development in the region:

> We only accepted changes that we could live with and that would not destroy our way of live. We have had fifteen years of constant struggle to try to force Québec and Canada to respect their commitments under the overall James Bay Agreement. If I had known in 1975 what I know now about the way solemn commitments would become twisted and interpreted, I would have refused to sign the Agreement.
>
> (Diamond 1990: 28)

As Alan Penn points out, since completion of the agreement in 1975, substantial changes have been implemented in government policies in relation to public land administration, land access, forest tenure, mineral exploration, environmental protection and hydro-electric development. 'These policy developments have taken place independently of and largely without reference to advisory structures in the agreement, raising questions about the relevance or utility of these advisory structures' (1995: iii). Penn also makes the point that the intentions of the Cree negotiators in shaping the 1975 agreement were 'directed towards the preservation of a hunting society' and that the agreement was not, therefore, targeted at the 'needs of an expanding and diversifying native society' (1995: 18). The Cree and Inuit institutions established under the 1975 Agreement have met many of the anticipated and unanticipated challenges arising since then, including the need to deal with the Great Whale proposal (Cohen 1994), issues on Québec separatism (Grand Council of the Crees 1975) and Cree sovereignty (Cree Eeyou Eschee Commission 1995). Not surprisingly, differences in political, economic and strategic directions have emerged within and between the Inuit and Cree since conclusion of the JBNQA (Puddicombe 1991; Sviuaq 1988). The extent to which the institutions that arose from the JBNQA are able to evolve as adaptive institutions of the sort advocated by Jacobs and Mulvihill (1995; see also Jacobs and Chatagnier 1985), remains to be seen. In terms of the current argument, however, their experience dramatically illustrates the nature of the complexity arising from resource-based intrusions into indigenous territories and its wide-reaching implications for everybody affected by such intrusions – whether as proponents, beneficiaries or 'victims' (if such a simplistic rendering can be allowed for a moment). As Sylvie Vincent concluded in her ten-years-on reflection on the JBNQA:

> The more complex a problem is, the more tempting it is to reduce it to a single variable, quantifiable if possible. But in this case it seems better to take a broader perspective. The relationship between native and non-native people in James Bay and Northern Québec can be considered in the Canadian context of the debate around patriation of the constitution, but also in the context of Québec's reconquest of a national identity and therefore of its territory …
>
> Hydro-electricity – which is at the crossroads of development, nationalism, and territorial concerns – is a good example of the way in which these three issues can fuse and form a field of tension between native and non-native people that is evident in many other areas.
>
> (Vincent 1988: 245)

Conclusion: self-determination as an outcome of struggles over resource management

Like all natural resources, the potential of sites within a national territory for producing electrical power represents a part of the resources power of the state whose territory is involved. Realising (and managing) that potential, however, is far from simple for the states who seek to assert exclusive control. In many modern states, cultural, economic and biophysical environmental consequences of major resource decisions inevitably draw other stakeholders into the decision making process. Resource claims contrary to those of the claimant state (for example the claim of the Sami in the Norwegian case discusssed here and the claims of Canada, Cree and Inuit in the Québec case) further complicate issues.

In terms of the dominant ideologies that focus on development and industrialisation, proposals such as the Alta-Kautokeino and Great Whale dams seem unproblematic. The threat of non-state intrusions into the sacred realm of state sovereignty by federal or tribal organisations is commonly raised in such circumstances. Compromising state sovereignty intertwines such issues with the complex processes producing nationalism and national and regional identity. In most jurisdictions the transfer of sovereignty through various lease arrangements which abdicate state control of resources in favour of a form of corporate sovereignty which empowers transnational resource corporations or autonomous empires in the guise of ostensibly accountable statutory authorities such as Hydro-Québec, NVE or in Australia SMEC, Tasmania's HEC, State Forests in NSW, CALM in WA and so on, does not raise the same sort of concerns – even though indigenous peoples and environmental activists are often citizens meant to be constituents of (and represented by) the state in ways not reflected in the relationships between governments and resource corporations.

Part V

Ways of doing

11 Diversity and world order

Professional practice and resource managers

Yet another 'New World'? Resource management for the twenty-first century

The case studies discussed in the previous section provide a window on the geopolitics of resources and their place at the heart of the modern world and contemporary life. Resource management systems are core institutions of modern industrial production and contemporary world order. The ability and desire to manage the means of survival – to augment, enhance and to some extent control them – is one of the distinguishing characteristics of humanity. And yet, for many people – individuals, families, communities and cultures – the dominant resource management systems are failing. These systems are ostensibly aimed at fostering not just human survival but also human development, but they fail to do this for vast numbers of people.

Earlier sections of this book argued the need to cultivate new ways of seeing in order to make better sense of what is happening in contemporary resource management systems. Drawing on the words of a seventeenth-century English radical, it was suggested that resource management professionals need not only to turn upside down their own worldviews, but also to recognise the extent to which their decisions turn upside down the everyday worlds of others. For this reason, the need to put in place new ways of thinking about these issues was also advocated. In particular, it was argued that resource geopolitics needs to be dealt with in context. Resource managers need to become more literate in the geopolitics of resources, and less dependent on inadequate metaphors of marketplaces and level playing fields; this has led to the use of ambivalent and messy analogies concerning complex geographies and social processes. In constructing this argument, the book has drawn on insights from indigenous peoples' experience of the effects of industrial resource management systems. In several case studies, the nature and implications of various aspects of complexity and current practice were explored. In these chapters, the analysis of resource management practice drew on indigenous experience and, at least implicitly, on at least three other sources of critique of the modern era – feminism, environmentalism and post-modernism. In part the intention was to emphasise the significance of dissenting voices in society, and to suggest that thoughtful responses to them provide a basis for improved outcomes on the ground.

One of the common themes to emerge from these dissenting voices is the significance of diversity in constructing better outcomes in resource management. The framework so far has demonstrated that resource management is clearly 'part of the problem' which produces injustice, unsustainability, inequality and intolerance of diversity. The framework has provided a toolkit for identifying, analysing and responding to these problems. This was the task of deconstruction and critique identified in the first section. The task now is to use the toolkit as a means of moving resource management towards being 'part of the solution' – to apply the conceptual tools and the empirical insights arising from the case studies to produce more just, equitable and sustainable outcomes that are more tolerant of human diversity. Using these core values of sustainability, equity, justice and diversity as a reference point, this section commences the *re*constructive task; the task of rebuilding resource management practice to provide a vision of constructive possibilities for change. In this chapter, it is argued that the dominance of exogenous resource management systems – systems in which the key imperatives driving the system are outside the local area – and the related marginalisation and disempowerment of locally indigenous resource management systems, is clearly 'part of the problem' because:

- Institutionalisation of the crises at a global scale facilitates reductionist homogenisation of complex, dynamic and diverse, but interrelated, problems as beyond the intervention of the people affected by them.
- The imperatives of exogenous, externally oriented resource management systems, virtually by definition, exclude local perspectives on both costs and benefits and impose barriers to incorporating local judgements into systemic decision making.
- Historically developed structures of injustice at global, international, national and sub-national scales simultaneously support and are supported by the dominant resource management systems, which tend to reproduce and refine these power structures as part of their own operation.
- Markets for many of the outputs from traditional resource management systems (including traditional ecological knowledge) have been subject to a variety of forms of commodification which undermine both their autonomy and their sustainability.

It is further argued here, particularly in relation to the place of indigenous peoples in resource management systems, that resource management can be constituted as 'part of the solution' if:

- Structures for accommodating cultural diversity within equitable political institutions can be linked to sustainable resource use.

- Management of non-renewable resource systems is evaluated against social, cultural, regional and intergenerational criteria of justice and equity rather than solely against market-linked criteria.
- The rights and responsibilities of indigenous peoples are entrenched as the basis for developing overlapping and interacting indigenous systems of resource management at relevant scales, with accommodation of the rights of non-indigenous populations protected in various political and economic institutions which are not solely accountable on economic criteria.
- The professional practice, values and ethics of resource managers treat cultural diversity, indigenous values and a holistic land ethic as central goals in reconstructing flawed resource management systems.

It is these overlapping and interacting propositions – a critique and an alternative reconstruction[1] – that provide the conceptual focus of the vision of human geography and humane resource management developed in the remainder of this book. It is also in the dialectical tensions between critique and reconstruction that social science in general and human geography in particular has something of substantial value to offer in educating resource managers to dedicate their professional effort to the creation of a more just, equitable and sustainable future in which diversity and difference are valued and supported.

Diversity has been a critical issue in feminist, environmentalist and post-modernist critiques of the dominant models and theory of social life. In each case, the response to essentialised, totalising models of human experience has been to assert diversity and difference as an important element – to value diversity and difference, rather than treat it as pathological, destablising or menacing. Feminist critiques of male dominance in metaphor, in power and in thought counter not with a proposal for a female alternative, but for a humanising alternative that includes and celebrates difference, and values and nurtures solidarity and support. For environmentalists, ecological diversity has become a metaphor with substantial political implications, and for many people has been extended to include metaphors in which distinctions between the human realm and the non-human world are treated as arbitrary and often counter-productive. And amongst post-modernists, the notion of the 'other' has become an important, if somewhat abused and often confused image of the cultural and human diversity that is smothered by modernist totalising visions.

The sort of critical reflection in which I have been engaged in relation to resource management is mirrored in much wider theological, intellectual, geopolitical and social debates. At the end of the second millennium since Christ, some have reflected on the current period as one of millennial change – a sea change in ways of seeing, ways of thinking, ways of doing – even ways of being (Maybury-Lewis 1992). Others refer to a crisis of identity, or representation, or understanding, which has sown seeds for new forms of representation of human experience and ambition (Clifford and Marcus 1986). Still others have seen and seized opportunities to create new worlds from the legacies of the old (Le Guin 1989; see also Howitt 1993b). People who are privileged, empowered and enriched by the operation of industrial resource management systems appear to be able to delude themselves that they have a unique monopoly on knowledge, wisdom and truth. This disparate collective of competitors and antagonists, this patchwork of theological divisions, national cultures and passionate individualists, this often ethnocentric, often racist, often terribly divided collectivity is embedded in structures in which *their* needs, values, history, knowledge and futures have become central. Meanwhile, all else is at risk of being marginalised or reduced to a caricature of both itself and its western parallel. Even within the imagined 'West', class, gender, ethnicity, creed and so on have reduced the reference point further. A wealthy, young, western, male, heterosexual, urban élite becomes the focus for defining 'success' in the contemporary post-modern world (see also West 1990). Sometimes reified and abstracted into dehumanised structures such as 'market forces', the 'national interest' or 'the international community' or simply the 'West', this imaginary centre subsumes resource management systems, along with virtually every other aspect of human experience. This imaginary centre has displaced localised meanings of success and value and imposes ostensibly objective measures as the unquestioned (and unquestionable) foundation of value and success for specific resources, specific resource projects, and specific resource management systems. And at the core of this imaginary centre lies the rationalist proposition that all 'value' is ultimately reducible to money values; that all value is ultimately tradeable.

Indigenous or exogenous? Models for local resource management systems

By and large, there are two approaches to the use and management of resources in a particular locality, whether that locality is constructed as a small ecologically defined area, a bioregion, or a national political jurisdiction. One is internally oriented, and one is externally oriented.

Exogenous resource management systems

In the externally oriented systems, market forces provide the most common measurable system of exchange between overlapping, interacting, competing systems of production, exchange, distribution and consumption of resources. Almost inevitably, certainly within existing visions of capitalist and socialist market forces, productivity and competitiveness have had higher priority than sustainability and equity in most resource systems. The result has been tragic for many of the participants in the production of wealth in the modern world. Not only dispossession, alienation and marginalisation, but also pauperisation, poor health, inequity and uncertainty have been produced by these resource management systems, along with massive monetary wealth, political power and extraordinarily refined technology. In these systems, there is a constant juxtaposition of wealth and poverty, of power and misery. This can be seen in many examples:

- Women in Australian and Canadian mining towns forced to rely on a social welfare net when families break down under the pressure of labour processes required to produce the world's most competitive and productive coal mines.
- The tragic legacy on migrant and Aboriginal workers of a working life spent producing asbestos in Wittenoom and Baryugil.
- The Aboriginal victims of violence, alcohol and alienation stranded outside communities in remote Australia where every non-Aboriginal family's yard seems to be graced with a four-wheel drive and a boat.
- The mineworkers' families left without fathers, husbands and sons as a result of underground deaths in the mines of Australia, Poland, South Africa, the USA, Russia, and so on, as safety margins are sacrificed to profit margins.
- Tribal communities whose forests are laid flat and empty in order to provide chopsticks and newsprint: whether it is relative pauperisation in a rich country like Australia or the absolute immiseration of civil war and genocide in Kurdestan, Chiapas, Georgia, Amazonia, Yakutia and Somalia.

In all these cases, the twin products of these resource systems (commodities and power) sit in awkward juxtaposition. In such contexts, the notion of sustainability can *only* be considered in relation to justice. What is the underlying human value of economic sustainability if it is attainable only at the cost of a complete loss of cultural sustainability? What is the underlying human value of ecological sustainability, if its cost is the imposition and institutionalisation of repression and poverty? At what scale does the generation of wealth justify the annihilation of biodiversity? At what scale might it justify the annihilation of human diversity? At what point does distribution of wealth as well as its generation become the responsibility of participants in and beneficiaries of resource management systems?

Such questions abound in real-world resource geopolitics. Yet the production of resources in the modern world is dominated by such exogenous systems. The booms and busts of commodity markets hold regions and their human (and non-human) populations to ransom. Economic rationalism puts economic accountability way ahead of human responsibilities. Resource management in this context becomes a matter of crisis management – constantly seeking to balance costs and profits; to maximise productivity; to pursue 'world's best' practices (in the selected fields of industrial relations, technology and financing rather than environmental assessment and social responsibility!), without risking access, markets or profits.

In these exogenous resource management systems, regions are reduced to interchangeable components and competitive alternative sources for global commodities for the inexhaustible demands of industrial production systems. This is the geographical reality of metaphors of level playing fields and free markets. The major players in overlapping commodity systems – iron and steel, aluminium, energy and so on – pursue strategies of 'global sourcing', and the complex geographies of these places are reduced to and subsumed by the homogenisation of the market. Geography is no longer a domain of cultural landscapes, of lived meanings and living cultures; it is reduced to a strategic tool to protect market share, to expand sales, to level the playing field, to enhance profits. For global resource companies, the global landscape has no geographical dimensions other than costs. It becomes a constantly changing *surface* of prices and wages and taxes and regulations and markets and commodities. To a great extent, the everyday lives and needs of the regional communities that populate *places* in this global landscape disappear.

Dealing with this version of the global landscape requires neither cultural nor environmental sensitivity. It just requires market analysts and computer programmes. To protect one's place in this econo-system, it is the institutional framework of nation states, world banks, transnational finance systems and massive wealth that is needed. Within these structures and institutions individuals become interchangeable components in the larger system, and so do places. They become commodified and reduced to a reified monetary value in exactly the same way as the components of what Aboriginal Australians refer to as country – the soils, the minerals, the trees and the creatures of the earth – have been commodified. Personal knowledge, personal relationships with the sources of value – the sorts of knowledge and relationships that underpin the cultures of many indigenous cultures – are generally treated as irrelevant to the efficient (and profitable) operation of these systems.

Clearly, the dominant resource management systems *must* be understood as constituting 'part of the problem' which produces and supports resource management systems that produce unjust, inequitable, unsustainable and homogenised outcomes. Efforts to change these systems are essential to any transformation of the existing problems. Yet we are told that the institutions involved are extraordinarily resistant to reform (Shiva 1991) and self-interested participants in the problems (Anderson 1983; 1992; Ekins 1992: 206). Scientism, developmentalism and statism, the three forces that Ekins identifies as central to the 'global problematic' (Ekins 1992: 207), are deeply implicated in modern industrial resource management systems. It is these that lead to proclamations of 'solutions' based on faith in technological approaches to resource management; absolute commitment to resource exploitation as the basis of national economic development; and the key role of resources in defining national sovereignty that have contributed to the kinds of global crises considered earlier (see Chapter 1).

The extent to which a practically achievable alternative vision of resource management can be forged in opposition to this dominant model, however, is a critically important question. Do the diverse cultural traditions of different peoples and different places provide foundations for an approach to resource management where local resources are no longer subsumed to the service of cleptocratic neo-colonialism? As I see it, the elements of such a foundation are already available, and the task of reconstruction is already under way. There is nothing inevitable or irresistable about this reconstruction, however. The task pits its proponents against the most powerfully entrenched vested interests in the contemporary world.

Indigenous models of resource management

In contrast to these externally-oriented approaches to managing local resources, indigenous systems provide an entirely different set of reference points or criteria against which success might be measured. While they may be limited in their scope, flexibility and transportability, they do offer an appropriate starting point for thinking about reconstruction of current industrial resource management systems. It is not that these systems require resources to be unconnected to other systems, other places, other ideas. As Dahl noted in relation to the Sami (see Chapter 10), regional self-determination:

> does not indicate isolation or withdrawal from existing political and economic structures. Instead, the newly adopted status ... is important – on own cultural and social terms [sic].
>
> (Dahl 1992:187)

In the current context some important issues *must* be constituted as global scale problems: climate change, global commons, atmospheric and oceanic pollution and the breakdown of international relations; but an insistence on only local criteria for 'success' is also regressive. Part of the necessary arena for action has to be global: action cannot be restricted to limited, isolated and autonomous 'locals', as if local places around the world did not interact with or affect each other. For indigenous people, however, local systems of meaning provide the foundations for relating to the wider world. This is a point that is well made by Knudtson and Suzuki (1992). In making judgement on any resource project or activity, locally oriented indigenous decision making would measure value and balance costs and benefits, first in relation to local questions of rights and responsibilities, local visions of sustainability and quality, and local structures of accountability for performance. These are the very things that the dominant exogenous industrial systems sacrifice and replace with an homogenised system of value and exchange in which the priceless and invaluable too often becomes the unpriced and valueless.

In thinking about the development of indigenous resource management systems, it is not just the traditional resource management systems of indigenous people that need examination and critical support; it is a whole range of locally constructed and locally or regionally responsible resource management systems, initiatives and innovations amongst local governments, community organisations, responsible resource companies and so on that can be characterised as endogenous and internally referenced. This notion of an indigenous orientation and internal referencing should not, however, be misinterpreted as inward-looking or self-referential. Indeed, it is often the dominant paradigm of Western scientism that disguises deeply solipsistic deception and self-reference as 'objectivity'. Within that paradigm:

> The world is formed around dualities: man/woman, culture/nature, mind/body, active/passive, civilization/savagery, and so on and on in the most familiar and oppressive fashion ... Stripped of much cultural elaboration, this structure of self–other articulates power such that 'self' is constituted as the pole of activity and presence, while 'other' is the pole of passivity and absence
> ... the 'other' never gets to talk back on its own terms. The communication is all one way, and the pole of power refuses to receive feedback that would cause it to change itself, or open itself to dialogue ... The pole of power depends on the subordinated other, and denies this dependence.
> The image of bi-polarity thus masks what is, in effect, only the pole of self. The self sets itself within a hall of mirrors; it mistakes its reflection for the world, sees its own reflections endlessly, talks endlessly to itself, and, not surprisingly, finds continual verification of itself and its world view. This is monologue masquerading as conversation, masturbation posing as productive interaction; it is a narcissism so profound that it purports to provide a universal knowledge when in fact its practices of erasure are universalising its own singular and powerful isolation.
>
> (Rose 1999: 176–7)

Parochialism is as unlikely as naïve globalism to produce better outcomes for the many 'victims' of industrialisation and development. Non-parochial localism, multiscalar activism and pluralist regionalism are the underlying orientations that need to be practised and cultivated in these endeavours. While no simple 'model' or 'universal solution' is awaiting discovery, many examples at various scales can be explored as a basis for further specific action.

'Footsteps along the road'[2]

The four main sources of critique consulted in this examination of resource management issues (the international indigenous peoples' movement; the international women's movement; the international environmental movement; and post-modernist movements in social theory) have each contributed elements toward a reconstruction of resource management systems in which social justice, environmental sustainability, economic equity and cultural diversity can be seen as products rather than victims of the way a system operates: that is, systems that produce autonomy and commodities rather than power and commodities.[3]

In practice, there are no unproblematic examples of exogenous resource systems being transformed by local, indigenous pressures, or of indigenous systems adapting to a wider set of industrial connections. Despite this lack of any unproblematic models to use, there are many concrete examples of movement in both these directions in action. In the practical work of managing minerals, fisheries, cultural resources, natural heritage and forests, the interplay between the indigenous and exogenous models has begun to produce new forms of resource management.

Perhaps the most telling examples come from Australia, where the traditional owners of some of Australia's most important cultural icons – Uluru and Kakadu – have negotiated joint management agreements to produce quite new approaches to managing the ecological, economic and cultural resources of those areas. These agreements represent a major advance in national and international terms; many other indigenous peoples whose lands are covered by (or of interest to) conservation zones are seeking ways of applying 'the Uluru Model' outside the specifics of the Aboriginal Land Rights (Northern Territory) Act 1976.[4] The limitations of this approach to co-management (see also Chapter 14 below) must also be recognised, as traditional owners have faced both scientific and political responses to their efforts to pursue self-determination on their own lands.

In New Zealand Maori groups claim that the Treaty of Waitangi recognises Maori sovereign control over resources for which the crown has claimed sovereignty. In that jurisdiction, issues as wide as regional land-use planning, gold mining, coal mines, forestry, inland and marine fisheries, geothermal energy, the sale of state-owned land and resources and tourism have been directly affected by the transformation of both indigenous and exogenous resource management models. In the New Zealand case, the exercise of Maori sovereignty has been tempered by the accommodation of existing *pakeha* interests, while the operation of even basic *pakeha* resource management systems has simultaneously been transformed by accommodating Maori interests. Again, let me emphasise, this has been problematic – it does not provide a universal model for application elsewhere. One can see both optimistic and pessimistic trajectories for the future in such settings.

Likewise, our focus on the international experience of indigenous peoples, contains grounds for both optimistic and pessimistic trajectories. On the one hand, there is evidence of increasing recognition of indigenous rights and interests in some jurisdictions. In Canada and Australia, the momentum produced by some limited legislative reform has been compounded by major legal recognition of indigenous rights (for example, Calder and Mabo respectively, and more recently in Delgamuukw and Wik). This movement may signal a shift towards the decolonisation of indigenous territories, yet there is simultaneously a recolonisation of indigenous territories in many countries, with new resources, new technologies and new ideologies directed against 'parochial indigenous minorities' whose backward attitudes are seen by intolerant and impatient national majorities as standing in the way of national self-realisation through resource projects on and in their lands. These 'new resource wars' (Gedicks 1982, 1993) challenge both sides. On the one hand, they threaten the integrity, continuity and survival of land-based cultures across the planet; but conversely, they also challenge the dominant values systems and institutions with a tradition of integrity, continuity and survival – a resistance based on people's relationships with places in their heart rather than pesos in their own (or more typically, someone else's) pocket. While the balance of power in conventional terms undoubtedly vastly disadvantages the tribal peoples, there remains an extraordinary depth to their resistance and struggle. In global systems indigenous peoples are always out of place, always transgressive. Their case is increasingly supported by dissident voices from within the exogenous systems: voices which heed the message of people such as Schumacher (1974), Ekins (1992) and Max-Neef (1992).

From a wide range of resistant, transgressive and dissident voices, one can envisage new approaches to the management of resource systems. At the margins of the dominant paradigm one can find new procedures for assessing (and responding to) the impacts of all sorts of actions; new regulatory mechanisms for popular participation in setting social goals; new notions of compensation; new expectations for resource projects to leave behind more than dust, and contribute more than high wages for an élite group of male workers, corporate profits to be expatriated out of the producing regions and government taxes to be used to suppress resistant, transgressive and dissident voices. The interplay and debate between the elements we have drawn on for our particular reconstruction are not to be discounted or overlooked. There is no natural alliance between environmentalists and feminists; nor between environmentalists and indigenous peoples. As Gray (1991) has pointed out, in the biodiversity debates, there are several conservation strategies which effectively sacrifice the interests and rights of indigenous peoples to the goals of conservation of biodiversity, although this antagonism is not inherent to either position.

It is possible that we might also be able to target progress towards a practical 'joint management' model in that most conflict-ridden field of mining on indigenous peoples' lands. In Papua New Guinea in the last five years local traditional landowners have taken strong equity interests in companies developing mining projects on their land. In Weipa, industrial location decisions concerning proposed alumina refinery infrastructure were revised in consultation with local traditional landowners. In Greenland, 'resource trusts' have been investigated by the local self-governing authorities as a mechanism for the pursuit of common goals. Harnessing externally oriented commercial opportunities to locally defined development aspirations, in ways that parallel equitable and sustainable joint management of resources where genuine self-determination is resisted, seem well worth exploring in coming years.

In effect, we have taken what Wagner (1991) appropriately calls a few tentative 'footsteps along the road'. We have not reached the goal of constructing sustainable, just and equitable resource management systems in which subsistence and cash economies are complementary rather than competitive. We have begun to see the shape of alternative and preferred futures to that being put in place by the dominant institutions of the planet. We have made new connections, constructed new ways of thinking about the issues and laid the basis for a new practice of the trade of resource management. Central to this reconstruction is a reassessment of the geographical scale at which resource management systems are held accountable. On the one hand, there is an imperative to reorient these systems to a more local and regional account- ability in terms of production, investment, distribution of benefits, costs and direct and indirect affects. On the other, there is also a need to recognise the interdependence of all these overlapping systems in a global ecological con- text. This changed view of accountability inevitably affects notions of ethical, acceptable and responsible practice in resource management. In the following chapters, then, we consider professional practice in a range of applied resource management settings.

12 Social impact assessment

This chapter focuses on social impact assessment (SIA) in the context of the professional practice of resource managers. It is beyond the scope of this volume to provide readers with detailed information about each relevant piece of legislation affecting SIA in various jurisdictions or detailed evaluation of all the current methods and theories utilised in this field. Rather, the chapter offers an overview, some examples, and an opportunity to think about the contribution SIA can make to improved resource management practices.

Impact assessment is commonly undertaken as an interdisciplinary exercise, using diverse skills assembled into multidiscplinary teams. Human geographers have particular contributions to make that are discussed here. The particular relevance of SIA to indigenous peoples and cross-cultural issues in resource management are also considered. The chapter argues that social impact assessment offers one way of pursuing 'applied peoples geography' (Harvey 1984: 9, see also Howitt 1993a). The focus on cross-cultural factors addresses many issues, methodological and conceptual, which are important but remain largely hidden in many impact assessment settings where the absence of a clear 'cross-cultural' dimension makes some issues less clear.

Impact assessment as public policy

In many resource management systems, some form of impact assessment has become standard. It has become legislatively entrenched in many jurisdictions. It has also become an important arena for public participation in environmental and resource allocation decisions; and often an important mechanism for securing accountability and evaluating public responses to development proposals. In the case of large-scale resource projects, impact assessment has clearly become part of governments' project planning and management system – often being seen by developers as a significant cost burden. In some jurisdictions, impact assessment has been entrenched as a mechanism for better linking development to public interests. Growth in the impact assessment industry has contributed to rapid employment growth for professionals. The field boasts an international association, international journals, many national and sub-national professional groups and regular conferences and professional exchanges.

The terminology of impact assessment has been a regular feature of public policy discussions and media reporting of environmental and development issues since the mid-1980s. Impact studies are prepared for a wide range of proposals. In Australia, it is typical for an 'EIS' (Environmental Impact Statement) to be prepared by professional consultants, often engineers or project managers, under instruction from a project proponent, with terms of reference provided by a government authority. With variations this arrangement is common around the world, although in some jurisdictions, there is a greater separation of the proponent and the assessment task. The EIS report often consists of several volumes, with weighty 'scientific' appendices with detailed data and analysis. It is typically reviewed by an environmental agency before final revision and evaluation. To many observers, these documents appear independent and authoritative. It is easy to mistake authoritative style for substance; and to mistake a commercial consultancy contract for independent research. As with so many elements in resource decision making and management systems, the impact assessment process itself needs to be examined critically to enable our analysis to move beyond simplistic and superficial understanding. It needs to be read for absences and consequences, analysed in terms of power rather than just read for information and considered in terms of the values and ways of thinking that are embedded in its methodologies and procedures.

Despite its widespread acceptance as an important tool in resource and environmental management, dramatic, and often systemic problems with IA generally, and SIA in particular, persist. Many projects that have been subjected to impact assessment have experienced systemic failure – the *Exxon Valdez* oil spill in Alaska (Palinkas *et al.* 1993; Shaw 1992), large-scale destruction of rainforests in the Amazon, Southeast Asia and Melanesia (Cummings 1990; Smith *et al.* 1991), species and habitat loss, cyanide leaks from gold mines, overflows of contaminated water in radioactive tailing dams at uranium mines, dam failures, climate change, damage to the health and wellbeing of workers and communities affected by even 'modern' resource systems and so on. As one critic recently observed:

> These problems do not arise out of ignorance. They have not occurred because developments were unplanned, nor their impacts unforeseen. Rather, they are a result of a flawed conceptualization of impact assessment and its role in environmental planning and resource management.
> (Smith 1993:1)

Smith's observation recasts the challenge of impact assessment not as a technical challenge, but in terms of its inherently political and human dimension.

The politics of impact assessment

The first element of this challenge arises in the legislative setting in which impact assessment is authorised. Impact assessment is conventionally seen as arising from the requirements of the US National Environmental Policy Act of 1969 (Finsterbusch 1995: 14). This enormously influential legislation has now operated for more than thirty years, and requires any US federal agency to prepare an Environmental Impact Statement (EIS) prior to taking 'actions significantly affecting the quality of the human environment' (Interorganizational Committee 1994: 1).

This relatively short history in terms of scientific and policy endeavour has meant that impact assessment is often seen as a new idea. The focus of impact assessment is clearly on investigating interaction and change in relations between human activities and environmental circumstances, which is, of course, a central theme of the social sciences, and the object of much recent theoretical endeavour in social theory. It is also a traditional focus for geographers.

In Australia, the key piece of legislation until the late 1990s was the Commonwealth Environmental Protection (Impact of Proposals) Act of 1974, which deals with proposals where the Commonwealth has jurisdiction, or by consent of state authorities. This legislation, framed by the Whitlam Labor government, was directly responsive to the NEPA in the USA. In addition to this Commonwealth act, there are over 330 individual acts and ordinances involved in this field (Australian Environment Council 1984: 27–38), plus a wide variety of special legislation covering individual projects, along with many other procedures, regulations and administrative arrangements (Formby 1977; 1987; Thomas, 1987: 34–55; Harvey 1998). In 1992, the Australian government's One Nation Statement put in place procedures for 'fast-tracking' major development proposals through this unwieldy complexity of acts and regulations. While this is now established procedurally, the highly politicised nature of the approval process has not been diminished. In 1999, passage of the Commonwealth Biodiversity Conservation Act substantially changed environmental management arrangements in Australia, re-empowering sub-national state governments in environmental protection and land use, and reducing the scope for Commonwealth intervention. Accompanied by policy changes that have privileged privatisation of public assets (including some environmental assets), reduced attention to sustainable environmental and social outcomes (including highly publicised interventions against increased greenhouse gas controls at the 1998 Kyoto convention) and reduced public investment in environmental and social justice, the new act has reduced the pressure on resource projects to enhance environmental and social performance, although many resource corporations have continued to pursue new policies of sustainability, participation and transparency.

The intensely political (and politicised) nature of impact assessment and the project approval processes built upon it is sometimes explicitly acknowledged, although the rhetoric of value-free and objective environmental science is often harnessed to disguise their power implications. As Canada's Justice Berger observed:

> If you are going to assess impact properly ... there is the ineluctable necessity of bringing human judgement to bear on the main issues. Indeed, where the main issue cuts across a range of questions, spanning the physical and social sciences, the only way to come to grips with it ... is by the exercise of human judgement.
>
> (Berger 1981: 393)

In practice, the institutions in which such judgements are exercised are often highly politicised. In Australia, the Commonwealth government tried to depoliticise environmental decision making by establishing a Resource Assessment Commission (RAC) as a forum for debate and neutral consideration of all relevant viewpoints (see Galligan and Lynch 1992; Lane *et al.* 1997). The government referred issues to the RAC that were particularly difficult and conflictual – the proposal to mine gold, platinum and palladium at Coronation Hill in the Kakadu Conservation Zone, management of Australia's coastal resources, management and use of old growth and other forest resources (for example Resource Assessment Commission 1991, 1992, 1993). The RAC inquiries produced high quality research and credible reports, but its recommendations inevitably required political judgement and decision. The Commonwealth disbanded the RAC in 1993, largely in frustration at its inability to achieve the desired depoliticisation and deliver detached technical decisions about resource conflict.

Several commentators provide broad frameworks for considering the basic function of social impact assessment: Craig (1990) distinguishes between technical and political approaches (Table 12.1); Howitt (1989a) considers the power relations SIA reflects (Figure 12.1); Gagnon *et al.* (1993) highlight differences between 'formal' and 'informal' SIA research (Figure 12.2). While such accounts provide valuable insights into SIA's role in resource geopolitics, the distinctions they draw are not cut and dried. The way one considers impact studies depends on one's purpose and vantage point.

SIA cannot be divorced from the practicalities of power politics at several scales. With some significant exceptions, impact assessment has, like social science more generally, followed a tradition whereby it is 'done *on* the relatively powerless *for* the relatively powerful' (Bell 1978: 25, emphasis in original).

Issues of ethics, equity and power exist in every step of the impact assessment process, and impact assessment itself is an arena of struggle over alternative futures. For most developers, an impact study should provide a *post facto* justification for decisions that have already been made within their own decision making domain. They seek to use impact assessment as a means of protecting and enhancing their interests. For them, therefore, it is generally a partisan tool, as it is for many groups who feel their preferred futures are

Table 12.1 Characterising impact assessment research

Formal	Informal
Required by statute or regulation	**Independent of formal proceedings**
• generally required prior to major decisions by governments or investors	• often linked to formal procedures
• generally undertaken in accordance with specially prepared guidelines	• generally aimed at influencing decision making processes, although not formally part of them
• product oriented (special report to inform decisions)	• specific guidelines, if they exist, generally *ad hoc* or negotiated by participants
• project proposer and government agencies have major role, with public generally limited to responding to draft publications	• less product oriented (report less important than processes involved such as lobbying, community mobilisation and direct intervention)
• generally funded by proposer	• project proposer and government agencies have little if any role
Public inquiry	• often funded from community resources with small budgets, or from independent sources
• usually established in major conflict situations	
• usually funded by government, often with very large budgets	
• not only government and proposer have standing but also community and other interests	
• generally limited to advisory rather than decision making role	
• published reports often major policy reference points for other decisions	

Technical	Political (participatory)
• product oriented	• community development oriented
• claims to be objective and scientific	• aims to be democratic and educational
• relies on technical experts	• emphasises community understanding and importance of value choices in alternative futures
• negative impacts generally assumed to be able to be dealt with by technological solutions	• negative impacts seen as requiring community intervention, monitoring and mitigation
• recommended changes to projects generally limited to technical or 'cosmetic' changes to original proposition	• recommended changes to projects generally include possibility of veto or radical transformation of original proposition

Table 12.1 (cont.)

Short timeframe	Long timeframe
• research period constrained by investment and political horizons	• research period extends to monitoring construction, operation and decommissioning periods
• period for public participation limited and often restricted to written responses	• many options for public input, including non-written contributions
• importance of research products emphasised	• research processes likely to be valued as highly as research products, which will be much broader than just an impact statement
• past history and locally preferred futures given little importance	• local history, diversity and aspirations explored

Source: Developed from Craig 1990.

Note: Craig (1990) distinguishes between technical and political approaches to SIA (e.g. Table 1: 44). In an unpublished paper in preparation, Gagnon *et al.* point out that it is clear 'that the technical approach, which Craig distinguishes from an overtly political approach, also has significant political content and consequences' (1993). In that paper we prefer to distinguish between managerial–technical approaches and participatory–emancipatory approaches. Many of the comparisons Craig makes in her table are, however, valuable ones.

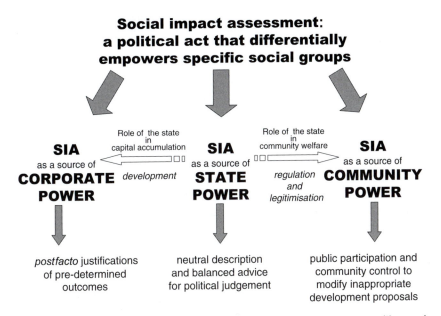

Figure 12.1 Social impact assessment and power. Social impact assessment, like much social research, is conventionally done for the powerful and about the powerless. However it is done, SIA has consequences for the construction and distribution of power in society. It can reinforce corporate or state power, or be a vehicle for community empowerment

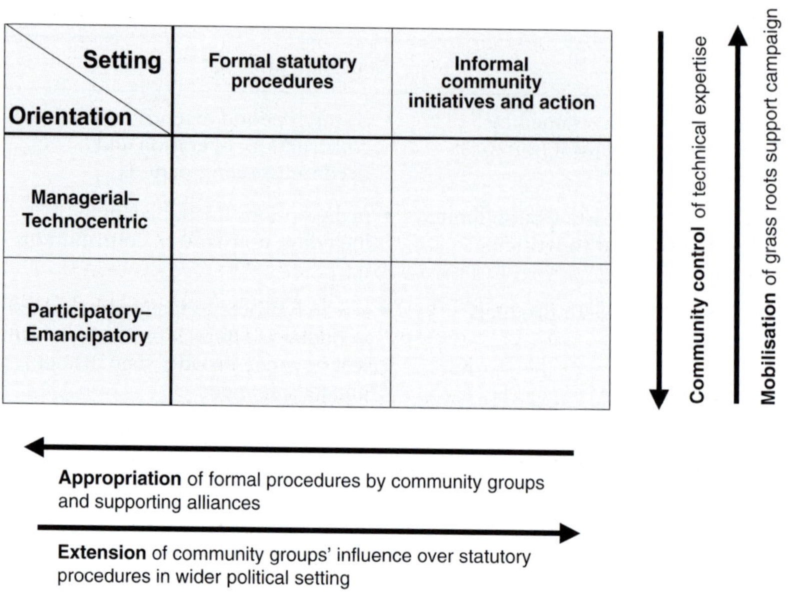

Setting / Orientation	Formal statutory procedures	Informal community initiatives and action
Managerial–Technocentric		
Participatory–Emancipatory		

↕ Community control of technical expertise

↑ Mobilisation of grass roots support campaign

← **Appropriation** of formal procedures by community groups and supporting alliances

→ **Extension** of community groups' influence over statutory procedures in wider political setting

Figure 12.2   Social impact assessment and community empowerment. Community interests, including local indigenous groups, can excercise influence and increase autonomy through SIA in many settings. Gagnon *et al.* (1993) argue that taking over formal procedures, extending influence in statutory processes, controlling their own experts, commissioning their own research, and political mobilisation all provide avenues for community empowerment through SIA

threatened by proposed resource projects. For example, American Indians commenting on social impact assessment in relation to their own needs observed that 'SIA is important if – and only if – it is a source of power. SIA is a tool or a weapon – or an arsenal – in a war for our survival' (quoted in Wolf 1980: 3–5).

Yet many people continue to hope that the state can play the role of neutral arbiter, despite the existence of both clear and fuzzy lines of conflicts of interests that such circumstances reveal. Our examination so far of resource management systems should have already signalled that the state is rarely, if ever, a simplistically neutral arbiter of resource and environmental conflict. In many situations, where state control of natural resources is central to state claims of sovereignty, its role(s) can be highly politicised.

Bringing together the typologies developed by Craig (1990) and Gagnon *et al.* (1993), it can be seen that, although most SIA research is likely to be formal and technical, and to operate within relatively short time horizons, formal SIA research does not have to be short term or technical (Table 12.1). In fact, SIA offers an opportunity to consider how to apply the resource manager's toolkit advocated in this analysis of resource geopolitics.

Both social impacts and impact assessment are viewed and constructed differently from different vantage points and by different participants (including researchers). It is also the case that impact processes related to a single project might appear differently and operate in different ways at various geographical scales, and that these matters raise many practical, ethical and professional issues.

Despite the predominance of manuals advocating a particular methodology as 'best practice' in SIA, what constitutes 'best practice' changes rapidly. Rather than offering its own version of 'best practice' for application in all circumstances and times, this discussion critically examines SIA as one important way of 'doing' resource management using the visionary and conceptual tools in the resource manager's toolkit. Fragmentation and diversity in resource management systems have already been identified as central in understanding the geopolitics of resources. They also offer a substantial methodological challenge to both resource management in general and impact assessment in particular. What appears to be a positive impact (benefit) to one party (or at one scale), is better understood as a negative impact (cost) to another party (or at another scale). Any methodology that assumes simplicity, stability and stasis will have dangerous and inequitable implications for grappling with the complexities and contingencies of resource geopolitics.

So it is again necessary to emphasise that, in framing SIA research, it is necessary to consider what the goals and purposes of various stakeholders in a resource management system might be. This is as true for resource managers working within resource enterprises as it is for people working as volunteers for community organisations affected by development proposals. As a minimum, this probably involves considering what role there is for SIA (or any other management tool) once we recognise that the target of resource management (like restructuring and social change in general) is actually a moving target. We are not aiming at a fixed, stable, centred goal that we can confidently measure when we reach (see Figure 12.3). The context of SIA research is always affected by a complex array of institutional, cultural, environmental, social and political arrangements. Whether one participates as an entrepreneur, a marginalised local, a government regulator, an environmental advocate or a grassroots activist, the constancy of interaction and change, the dialectical relations within and between various elements, changing expectations and aspirations, contrary processes of fragmentation and integration, and interpenetration of resource management systems with other aspects of social life (*inter alia*) will constantly shape the possible avenues for action, and the value of and appropriate approaches to work such as SIA.

What are social impacts? What is SIA?

All of this discussion does not, however, define the nature of social impacts and SIA. Official US guidelines define social impacts as:

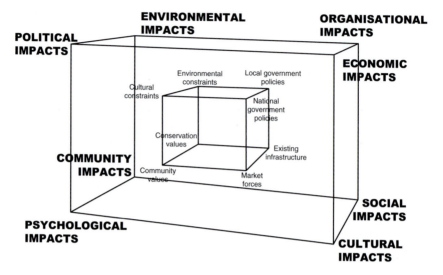

Figure 12.3 Visualising the complex context of social impact assessment. Like a set of Chinese boxes, the ecological, economic, political, cultural and infrastructural context of any set of social (or other) impacts will both reflect and influence circumstances, perceptions, constraints and possibilities at several scales

The consequences to human populations of any public or private actions that alter the ways in which people live, work, play, relate to each other, organize to meet their needs and generally cope as members of society. The term also includes cultural impacts involving changes to the norms, values, and beliefs that guide and rationalize their cognition of themselves and their society.

(Interorganizational Committee 1994: 1)

More simply, Armour (1992) suggests that social impacts are changes that occur in:

- *People's way of life* (how they live, work, play and interact with each other on a day-to-day basis);
- *Their culture* (shared beliefs, customs and values); and
- *Their community* (its cohesion, stability, character, services and facilities).

The US guidelines define social impact assessment as:

Efforts to assess or estimate, *in advance*, the social consequences that are likely to follow from specific policy actions (including programmes and the adoption of new policies [or the approval of action by third parties]), and specific government actions (including buildings, large projects and leasing large tracts of land for resource extraction), particularly in the context of the US National Environmental Policy Act of 1969.

> (Interorganizational Committee 1994: 1, emphasis in original)

A NSW government document defines SIA more simply:

> Social impact assessments are ... [studies] that ... attempt to answer the question: What are the impacts of a project or development on *people*.
>
> (Cox 1994: 5, emphasis in original)

SIA addresses many of the core issues of social science research. It uses the basic methodologies of social science. It is often characterised by quite sophisticated research activities. So why isn't SIA just 'social research'? While it shares many characteristics with much social research, which might be strongly applied in nature and policy relevant in orientation, SIA research is always action oriented. As Cox puts it, SIA:

> is focused on outcomes: that is, not only the significant impacts of a proposal, but what can be done about them ... . Decisions will have to be made, ultimately, whether [a] development will proceed or not. Impacts will have to be weight-up. Local interests will need to be balanced with regional and national interests, or the wider social good. Social impact assessment precedes social impact management.
>
> (ibid.: 6–7)

In other words, unlike general applied social research, SIA is:

- Action focused;
- Practically oriented;
- Usually undertaken within strict (and often short) time frames;
- Often participatory in approach;
- Linked to management outcomes.

Armour (1992) identifies four key factors affecting the nature and significance of social impacts:

- The nature of the proposal;
- The characteristics of the affected people and places;
- The perceptions of the affected people;
- Impact management strategies put in place.

The following issues should also be added to this list:

- The nature of the project's proposer and management;
- The policy (regulatory) and market setting (both broadly and specifically).

In practice, however, these last two matters are rarely addressed in impact studies.

The US guidelines use five major headings to characterise social variables to which attention might be given in SIA research, and at various times in a project's lifecycle (see Figure 1.3, page 18):

- Population characteristics;
- Community and institutional structures;
- Political and social resources;
- Individual and family changes;
- Community resources.

(Interorganizational Committee 1994: 8–9)

A widely used Canadian reference, the Blishen–Lockhart model (Blishen *et al.* 1979) emphasises the importance of:

- Economic conditions (community economic viability);
- Social behaviour (community social vitality); and
- Links between economic and social wellbeing and participation in political empowerment (political efficacy).

A general model of social impact assessment processes

Specific legal requirements for SIA vary between jurisdictions. Similarly, the details will vary with circumstances such as the nature of the proposal under consideration, the circumstances of the impact study, and the responses of various community sectors. But in general terms, the tasks of SIA can be divided into six stages (Table 12.2). As Taylor *et al.* point out (1990: 82), while it might make sense in some frameworks to separate the monitoring phase from the actual assessment phase, effective management of (that is, intervention in) impact processes demands that monitoring and intervention is seen as integral to the SIA process, rather than separate from it. Let us consider each of these stages in turn.

Scoping establishes the scope of the research required and where the notional boundaries will be drawn. It involves the setting of terms of reference and research guidelines, and:

> begins with a determination of the *decision-making* context of the assessment. This is followed by preliminary identification of the baseline and direct project inputs … , the main factors that could cause social change, and estimation of the variation in these alternatives across alternatives. The scoping activities include collection of initial information about the existing social environment and determination of the appropriate geographic and topical focus of the assessment effort. The formulation of possible alternatives based on these initial data may also be included as part of the scoping activities.
>
> (Branch *et al.* 1984: 17)

Table 12.2 Steps in the SIA process

Scoping	Identification of issues, specification of key variables to be described and/or measured, identification of populations and groups likely to be affected, setting of temporal and spatial boundaries of the study, setting the terms of reference, securing resources for the study.
Profiling	Overview and analysis of the current social context and relevant historical trends, preliminary interpretation of descriptive statistics, review existing literature relevant to the study area and issues, refinement of proposed study methods, data sources and study plan.
Formulating alternatives	Identifying alternative project configurations (and/or alternatives to the project as proposed) including 'no development' option, reviewing the proposal in terms of local, regional and national development goals, comparing relative merits of alternatives.
Predicting effects	Estimating the possible and probable effects (positive and negative) of one or more options against specific significant criteria, comparing predicted outcomes to baseline studies and projected growth/change without the proposal, estimating scale, intensity, duration, distribution and significance of predicted effects.
Monitoring and mitigating	Collecting information about actual effects and applying this information to mitigating negative impacts and enhancing positive impacts.
Evaluating	Reviewing both the social effects of the change and the SIA process used systematically after the event.

Source: After Taylor *et al.* 1990: 84, see also Branch *et al.* 1984: 17

Scoping also involves designing the way in which the research and other activities involved in the assessment will be undertaken (in other words, the research proposal stage). In some jurisdictions (for instance Queensland and the Australian Commonwealth), this phase may involve formal public partic-ipation, through the formulation and circulation of impact assessment guidelines for public discussion. In other cases, this phase may involve consultation with affected communities to inform them about a proposal and undertake a preliminary investigation of issues of concern. It is at this stage that issues of communities' vulnerability to (or preparedness for) certain sorts of impact because of their previous experience can be reviewed. Similarly, it is also at this stage that cultural and social values affecting the

SIA process should be identified (for example conceptualisation of 'valued ecosystem components'; Evaluating Committee *et al.* 1992). During the profiling phase of a SIA, it is also appropriate to identify the existing research relevant to the topic and the area, including baseline studies, comparable studies, accessible data (including census and other statistical data) and other relevant literature for review. In many cases this will be an ongoing activity throughout the SIA process. Where it is appropriate, the scoping phase of an SIA will also formulate mechanisms for public participation and control. In many recent intercultural studies involving projects on Aboriginal land in Australia (O'Faircheallaigh 1996a, b, c), community-based steering committees to oversee SIA research have been established at this phase and given training in relevant areas.

Profiling generally involves construction of a baseline study prior to the commencement of impact processes, and the compilation, analysis and interpretation of existing descriptive statistics and other data. It generally involves developing a clear profile of the affected community/communities and important current and historical issues. In the case of multicultural environments (that is, cases where there is more than one cultural group involved in use of an affected landscape), profiling should also include consideration of the available information about cultural values and priorities, including ethnographic and anthropological studies, ethnobotanical and ethno-ecological research, cultural histories and other materials. This phase will identify all the affected stakeholders in a resource management system. The analytical and interpretative work undertaken in this phase of the SIA will refine the definition of key criteria which will address significant impacts. It will also be at this point that research strategies are formulated, ethics and research protocols finalised, interviewees and focus groups for qualitative or survey work identified and prioritised, key information sources identified, questionnaires and survey instruments designed and so on. In some cases, where public participation is a high priority, this phase will report in some detail to community-based SIA steering committees, establishing a more-or-less detailed dialogue about existing research, historical experience, current issues and strategic aspirations etc. It is in this phase that some differences between statute-based SIA and community-based SIA may emerge – although the focus on making and debating decisions remains common to both sorts of SIA study (Figure 12.2, page 330).

Formulation of alternatives is often severely constrained in formal, project-based research, as serious consideration of a 'no project' alternative is commonly avoided by project proponents in anything other than a perfunctory sense, and other alternatives are often framed in a way that favours the desired alternative. Where public participation is supported (for example the Berger Inquiry), a wide range of alternatives to the original proposal may emerge through consultation and citizen review of proposals.

Projection and estimation of effects should not be limited to a detailed consideration of the preferred option, but a careful consideration of all realistic options, including the 'no-action' option. This phase will generally involve both quantitative and qualitative research, focusing on the way(s) that the proposed development activity will affect critical criteria and social, cultural or ecological elements previously identified as significant. Areas to be considered may include, for example, population characteristics, community and institutional structures, individual and family changes, community resources, valued ecosystem components, sacred or significant sites and locations, intergenerational or gender relations, distributional issues and political and social resources within and between communities and groups. Work may be undertaken through community-based focus groups to assist communities to identify needs, formulate priorities and develop negotiating strategies (for instance, Lane *et al.* 1990 utilised a strategic perspectives analysis; Burdge 1994 advocates community needs assessment techniques; O'Faircheallaigh 1996a advocates a focus on negotiation). In any case, it is necessary in this phase to formulate some meaningful, reliable and rigorous analysis of at least the three elements prioritised by the Blishen–Lockhart model (Blishen *et al.* 1979) advocated by Craig (1988) – economic conditions (community economic viability), social behaviour (community social vitality) and links between economic and social wellbeing and participation in political empowerment (political efficacy).

It is at this point in the process that key decisions are typically made – although, of course, proponents, governments and community groups are making crucial decisions throughout the process. For example, in the wake of completion of (draft) impact assessment findings, companies often have internal mechanisms that will trigger the next phase of project finance and design work. Governments will review material at this point (that is, Draft and Final EIS stage) to formally provide a Yes/No decision for specific statutory approvals required by the developer. Similarly, community groups will use the EIS report as a focus for community decision making about responses to proposals and alternatives.

Geography is complexly involved in many ways. In the projection and assessment phases, geography needs to be considered in all possible configurations of regional futures – futures with the proposed changes, futures with all possible changes, and futures with no additional changes (Figure 12.4). It is not, however, only local geographies which matter, because both the impact processes and the points of intervention through which we might affect them are simultaneously constructed within the locality, and through the locality's interactions with other places and other scales. To successfully analyse, understand and intervene in these processes, we need to develop what Massey refers to as a 'non-parochial view of place' (Massey 1993a: 144).

As Boothroyd *et al.* (1995) point out in their Canadian study, even large-scale projects with very substantial predicted impacts on communities and environments often stop at this point. Having paid a project consultant to draw up the reports required to achieve statutory approval, it is often the case that proponents proceed to build the project as approved, without reviewing

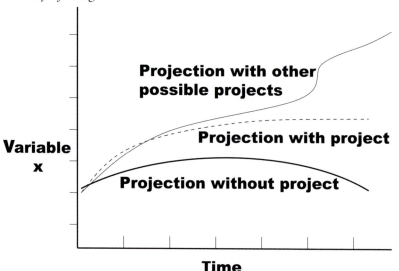

Time

Figure 12.4 Baseline projections: with and without proposed changes. In undertaking social assessment of a resource project, it is necessary not only to assess implications of the project for any particular variable, but to contextualise this against projections based on a 'no development' option, and the cumulative impacts of other possible projects in the area under consideration

or responding to impacts as they arise. Even in cases where there has been substantial community mobilisation to deal with a project at the initial notification and assessment stages, communities may face a lapse back into apathetic acceptance and passive resistance of changes once decisions are made. Ivanitz, for example, points out the implementation challenges facing community groups who negotiated for up to twenty years over comprehensive land-claims settlements in Canada (Ivanitz 1997). Yet the purpose of impact assessment research is fundamentally to equip someone – governments, developers, communities – to understand, identify, and intervene in changes *as they happen*.

Monitoring, mitigation and management, as pointed out earlier, are integral to the overall SIA process, and should be integrated into the operational management of projects together with the programmes to manage and mitigate negative impacts (and enhance positive ones). It has become relatively commonplace to put in place quite sophisticated (and expensive) monitoring provisions for biophysical environmental impacts. In the case of Australian uranium mining, for example, the Commonwealth government established the Office of the Supervising Scientist with its own legislation (Environmental Protection (Alligator Rivers Region) Act 1978) with a multimillion dollar funding base over the past twenty years. In contrast, dealing with social environmental changes is often treated as an issue for government welfare agencies and marginalised in post-project approval decision making. For example, in the case of the uranium decisions of twenty years ago, an extremely modest (in other words, severely under-resourced) social impact monitoring project was eventually put in place with a limited time horizon of five years (the Social Impact of Uranium Mining Project: see AIAS 1984; Howitt 1997d). In many cases, the participation of affected groups will be central to the success of these programmes, but it is often not supported by successful developers until the risk of local alienation and unrest become obvious (see Filer 1990a, b; Gerritsen and MacIntyre 1991; Banks 1996).

Finally, evaluation, in terms of the original goals, and in comparison to similar or related projects elsewhere, should be seen as an important way of strengthening both ongoing programmes related to a particular project, and the development of SIA more generally.

SIA in cross-cultural settings

In terms of resource management systems, formal project-oriented SIA research has often trivialised the concerns of marginalised and minority groups affected by a project as merely parochial. In the case of Aboriginal groups in Australia, formal SIA has often disempowered indigenous efforts to intervene in local development decisions. In Queensland, for example, treatment of issues of direct concern to Aboriginal groups has been far from reasonable in several cases. In assessment of one sand-mining proposal on Cape York Peninsula, Chase concluded that:

> [The] state government, developer and consultant all ignored the Aboriginal perspective [in ways which denied the affected Aboriginal groups] the most fundamental right – the right to exist.
>
> (Chase 1990: 15, 17; see also Lane and Chase 1996)

In other cases, little credence has been given to legitimate Aboriginal concerns. In CRA's initial consideration of its Century base-metals mine (Hollingsworth *et al.* 1991; see also Trigger 1997b; Harwood 1997), Aboriginal concerns were not mentioned at all! It is hardly surprising, therefore, that in Queensland Aboriginal groups have not seen formal impact studies as a vehicle for their empowerment. On the contrary, for many Aboriginal people, formal impact assessment has become just another of the many structural impediments to Aboriginal participation in regional development planning – another item on someone else's development agenda to which they must respond.[1] Lane and Yarrow (1998) suggest integration of indigenous people's right to negotiate under the Native Title Act with impact assessment processes, but it is also clear that some negotiations have proceeded amicably because developers were not forced to negotiate but did so voluntarily (O'Faircheallaigh 1996b).

Although integration of impact assessment and indigenous negotiations may increase certainty in the planning process, even the limited opportunities indigenous people have to participate in impact assessment are resented by some commentators. Such resentment reflects a position that any restriction on development, whether it is related to indigenous or environmental review, contributes to unreasonable and expanding interference from red, green and black tape in their investment decisions. Their responses have called for streamlined environmental reviews, fast-tracked governmental approval procedures for major development projects, and simplified industry–bureaucracy interfaces with so-called 'one-stop shops' where development applications are expedited by a single government entity. Similar issues have been raised in relation to the participation of other affected communities in impact studies in other areas (on Thailand and Québec, see Gagnon *et al.* 1993; on American Indians, including some examples from South and Central America, see Geisler *et al.* 1982; for other international examples see Smith 1993; Omara-Ojongu 1991).

When undertaking SIA research in cross-cultural settings, concerns arise which are less significant in settings with less cultural diversity. In particular, questions of power, intercultural communication and equity, cultural sustainability, the extent to which culturally specific impacts are understood and evaluated appropriately, the extent to which there is diversity of opinion and impact in affected groups are issues which need to be considered in any effort to deal with cross-cultural impact assessment. In the case of indigenous peoples, the specific history, culture and aspirations of indigenous groups has both conceptual and methodological implications for doing impact assessment research. Different cultural values, different ontological and cosmological approaches, definitions and valuations of resources and landscapes, unequal power relations in development processes, historical marginalisation and alienation of Aboriginal groups and so on all limit the extent to which mainstream methodologies can be transferred to Aboriginal SIAs.

Several Australian and Canadian studies have faced both cross-cultural concerns and indigenous interests. Many methodologies build on the work of the Canadian Berger Inquiry (Berger 1977), which recognised the value of indigenous peoples' knowledge of ecological processes, and the extent to which local indigenous peoples' communities and lives would bear the brunt of social and environmental impacts from a proposed trans-Canadian energy corridor.

In Australia, specific consideration of Aboriginal concerns in impact assessment work can be traced to the monitoring programme established following the approval of the Ranger uranium mine, an an area excised from Kakadu National Park (AIAS 1981; Tatz 1982; Kesteven 1986; for a review of this material see Howitt, 1997d). This was followed in the mid-1980s by an innovative project to examine social impact processes in the East Kimberley region of WA, following the approval of CRA's Argyle Diamond Mine against Aboriginal wishes, but also in the context of a general marginalisation of Aboriginal people from regional development processes in this region (Ross

1990a, b; Coombs *et al.* 1989c; Howitt 1989). The Resource Assessment Commission Inquiry into the Coronation Hill project in the NT included an impressive SIA component (Lane *et al.* 1990; see also Howitt 1994b, 1997d). Development of participatory, negotiations-focused impact assessment driven by indigenous people through native title claims has gained currency in Australia (O'Faircheallaigh 1996a; Ross 1999a). In Canada, the social and environmental assessment of the Great Whale hydro-electric project (see Chapter 10, above) added new dimensions to the way in which cross-cultural and indigenous social impact assessment can be thought of (Evaluating Committee *et al.* 1992). In this case, there were several innovations which might be considered international 'best practice' in the field at the moment, and which provide enormously important reference points for future.

All these cases confirm that SIA research needs to be participatory, empowering and interventionist (see Howitt 1993a). As Lane *et al.* (1990) observe in relation to the Kakadu Conservation Zone SIA (see below), this takes us a long way from the principles of reductionist, objectivist, positivist science that characterise the dominant paradigm of resource management.

Changing methods and principles in SIA

In practice, of course, many impact studies are prepared in ways that do not reflect international 'best practice' in the field. Many, for example, continue to be sponsored by project proponents, and to reflect many longstanding problems with SIA research in indigenous areas (see Wilson 1984; also Chase 1990). Recognition that SIA requires a degree of specialised expertise in social and inter-cultural research, rather than simply being a sub-set of environmental impact assessment which could safely be handled by an environmental scientist or engineer as part of a wider impact study brief, has been central to the changing practice of SIA. There is also wider governmental acceptance of public participation in SIA.

To provide an effective foundation for 'better' resource management, social impact assessment should be what Berger referred to as 'full and fair' (Berger 1977, 1988). While such descriptions of SIA are appealing and make SIA sound somehow routine and relatively easy, the practice, of course, is neither. In 'real-world' circumstances, a wide range of complicating factors comes into play. To appreciate the challenges facing real-world SIA, it is necessary to consider:

- The 'positionality' of the knowledge created in the SIA process;
- The multiple vantage points from which to understand impact processes;
- The ways that scale affects impacts and impact processes;
- The diverse imperatives affecting the behaviour and decision making of various stakeholders implicated in a specific resource management system.

It is all too easy, under the pressure that comes from real and imagined dead-lines imposed when doing social impact studies, to limit the temporal or geographic scope of impact assessment work to those issues conceived as important from a single vantage point – to privilege a particular version of reality above others. And, of course, while we may think we are carefully following an appropriate and justified management agenda, it is easy to find one-self caught up in other people's hidden agendas – with unexpected consequences.[2] Burdge and Vanclay (1996: 82–3) in a 'state of the art review' of SIA, identify three areas in which recent SIA practice has added value to resource management processes:

- Raising awareness of how projects, policies and political change alter the cultures of indigenous populations;
- Providing increasingly realistic appraisal of likely outcomes of particular policies or actions;
- Integrating impact mitigation into planning processes more effectively.

They also recognise, however, that although it should ideally be an integral part of the planning and development process, SIA faces a significant scale problem in bridging the gap 'between project level research findings and the larger scale assessments needed for regional and national policy decisions'.

It is this complexity, the existence of overlapping, contested, differentially empowered domains influencing resource management systems, that provides the practical imperatives for refining our approaches to environmental and social evaluation, monitoring and oversight of resource management systems. Clearly, SIA is no panacea for resource management problems. Even the most comprehensive (the most full and fair) impact studies cannot hope to deal with the totality of influences affecting any particular situation (see Figure 12.3, p. 332). SIA does, however, have an important role to play in under-standing and managing issues in these hotly contested domains. SIA is no magic solution for resolving resource-based social conflict; no matter how, or by whom it is done, SIA is a politically significant process in a politically charged setting. In SIA there can be no simplistic appeal to naïve objectivity, because SIA is all about trying to intervene in (usually carefully and thought-fully), to change the processes being studied. Some useful lessons can be drawn from recent practice. The following discussion refers to studies from Australia, Canada and China to identify some important issues of general relevance.

The Kakadu Conservation Zone SIA, Australia

One of the most important SIAs conducted in Australia in recent years was the Resource Assessment Commission inquiry into social impacts of various land-use options in the Kakadu Conservation Zone (Lane *et al.* 1990). The RAC Inquiry was established to resolve a conflict between proposals to extend the boundaries of Kakadu National Park into what was known as 'Stage III', extending conservation-oriented joint management arrangements developed in the park areas to the north, and a proposal to exploit a gold–palladium–vanadium deposit in Coronation Hill, within the declared conservation zone. Anthropological evidence revealed the mine would damage a sacred site complex, and the environmental and Aboriginal protests about the mining proposal deeply divided governments. The social impact study was one of a large number of consultancy reports commissioned as material for the RAC Inquiry. It provides a good example of a wide-ranging impact study aimed at informing decision making on a complex and divisive resource conflict. While most of the public controversy focused on mining issues at Coronation Hill, the SIA report was required to cover a much wider scope. Its terms of reference were to:

1   Compile a comprehensive profile of the socio-economic environment of the region using mainly existing information sources. The key social variables ... being:
   - lifestyle – the way people behave;
   - attitudes, beliefs, values – the way people think;
   - social organisation – the way people meet these needs, including services, facilities and infrastructure;
   - populations – the way people are distributed on the land;
   - land use and tenure – the way people use the land;
   - economic and employment profile.
2   Explore potential socio-economic and cultural impacts of development in a social, historical and regional context by examining how the community works, the cumulative effects of past developments, and the combined effects of new developments and identify alternative development options.
3   Explore how people themselves perceive events and their impacts using appropriate participatory techniques.
4   Enable the SIA process to become a part of the peoples' means of defining their goals and aspirations.
5   Relate peoples' aspirations to the context of development in the region, including potential new development and new populations.

(Lane *et al.* 1990: 1)

Using an approach which combined the so-called 'technical' and 'participatory' aspects of SIA, Lane *et al.* aimed to provide a research process which produced mitigation strategies for various potential impacts. The Kakadu Conservation Zone SIA adapted the Blishen–Lockhart model to research dimensions of community economic viability, social vitality and political efficacy. It also utilised a participatory research approach known as Strategic Perspectives Analysis which was 'used to assist actors to consider their objectives' (ibid.: 5). In conducting their strategic perspectives analysis, Lane *et al.* note that:

the focal point of land use contention in the study area is Coronation Hill. The legal status of the Conservation Zone means that a variety of activities, including mineral exploration are permitted ... until such time as the Governor-General proclaims that the area becomes available as either national park or for the recovery of minerals.

(ibid.: 61)

They identify three basic proposed land uses for the Conservation Zone: mining, conservation and tourism, as well as Aboriginal land use. As part of their strategic perspectives analysis, they identify the numerous actors (or stakeholders) in the resource management system. The detail and internal diversity identified in groups that have often been lumped together not only in media reports on Coronation Hill, but also by many professional observers, provides a useful reminder of the practical challenge facing SIA research. Their list includes:

- Jawoyn custodians – Aboriginal traditional custodians of the land involved;
- Jawoyn Association – the community representative organisation of the Jawoyn Aboriginal Nation;
- Pro-custodian Jawoyn – Jawoyn people who supported the position of the senior custodians who opposed mining at Coronation Hill because of the damage threatened to sacred sites;
- Pro-custodian non-Jawoyn;
- Dissenting traditional Jawoyn women;
- Senior Jawoyn men employed by CHJV;
- CHJV Aboriginal workers;
- Pro-CHJV Jawoyn;
- Pro-CHJV non-Jawoyn;
- Coronation Hill Joint Venture;
- The Northern Territory government (including Department of Mines and Energy, Department of Lands and Housing, Department of Industry and Development, Department of the Chief Minister, NT Tourist Commission and NT Conservation Commission);
- Australian National Parks and Wildlife Service, the Commonwealth Department most directly involved in management of Kakadu National Park;
- Northern Land Council, the statutory body representing Aboriginal interests throughout the northern part of the Northern Territory;
- Aboriginal Areas Protection Authority, the Northern Territory government department involved in protection of sacred sites;
- Northern Territory Environment Centre, a community-based environmental non-governmental organisation; and
- Northern Territory Museum.

(Lane *et al.* 1990: 64–77)

The report identifies two groups of compatibilities among these groups – one among groups with conservation objectives and one among groups with pro-mining objectives (ibid.: 83–4).

The report goes on to assess likely impacts of various land-use options, and to consider strategies for monitoring the impacts:

> Monitoring is a fundamental management technique which is designed to provide maximum benefit at minimum social cost … . Monitoring, there-fore, is a natural extension of the methodological approach of this study in that it views SIA as a continual, cumulative process, rather than a product. Monitoring can contribute to knowledge about induced socio-cultural change over time, as well as facilitating continual feedback to refine and improve mitigation strategies.
>
> (Lane *et al.* 1990: 127)

The Western Cape York Economic and Social Impact Study 1996

In late 1995, following overtures to the Aboriginal communities of western Cape York from Comalco, Cape York Land Council commissioned a team led by Professor Ciaran O'Faircheallaigh, a political scientist from Griffith Uni-versity, to undertake a major economic and social impact study of the effects of more than thirty years of bauxite mining and related activities. The ESIA report, which remains confidential, was intended to assist the communities to develop a negotiating position, and to provide Comalco with a benchmark from which to better understand the impacts of its operations. The research was undertaken between December 1995 and July 1996, with the final report being provided in October 1996.

The ESIA methodology developed by O'Faircheallaigh and Holden in nego-tiations with Mitsubishi over the Cape Flattery silica mine on eastern Cape York (O'Faircheallaigh and Holden 1995a; O'Faircheallaigh 1996a) and the Skardon River kaolin project near Mapoon (O'Faircheallaigh and Holden 1995b) was refined to encompass consultations and detailed interviews with hundreds of people in four major communities (Napranum, Aurukun, Mapoon and New Mapoon). The scope of the project was in many ways greater than anything undertaken since the innovative East Kimberley Impact Assessment Project in the early 1980s. Negotiations between Comalco and Aboriginal groups in the region are continuing. Holden and O'Faircheallaigh also com-pleted an ESIA for Alspac's new bauxite mine just north of Weipa in late 1996 (Holden and O'Faircheallaigh 1996). This process produced an innovative agreement over compensation and impact management between Alcan and the area's traditional owners prior to transfer of the Alcan project area to Comalco.[3]

The Kakadu Region Social Impact Study 1996–7

Following the defeat of the Keating Labor government, the Three Mines uranium policy, which had restricted development of uranium projects in

Australia since 1983, was lifted. In mid-1996, Energy Resources Australia announced their intention to seek approval for a new mine to exploit the Jabiluka deposit north of its Ranger mine within the boundaries (but excised from) Kakadu National Park. ERA quickly moved to complete a conventional EIS, which faced considerable criticism from environmental, indigenous and Northern Territory government sources when it was tabled in early 1997. Following the expression of great concerns among local landowners of the Jabiluka site, the Northern Land Council and traditional owners initiated action to secure a regional social impact study which would address the complex issues of overlapping impacts of uranium mining, national park development, tourist, town development, royalty revenues, road development and other matters within the Kakadu region (Howitt 1997d provides an overview of the Kakadu literature). The terms of reference for the KRSIS required a high level of control resting with a community-based Aboriginal Project Committee. Oversight of this SIA was undertaken by a stakeholder advisory committee, assisted by an independent expert and chaired by Council for Aboriginal Reconciliation Chairperson Patrick Dodson. The KRSIS reported in mid-1997 (see Dodson *et al.* 1997; Levitus *et al.* 1997). While the final KRSIS reports provide only limited guidance on impact management in the region, the process and priority given to SIA as a tool in shaping the decision making environment reflects the extent to which it is entrenched in the Australian geopolitics of resources.

Towards integration of SIA and negotiation of indigenous rights in Australia

Drawing on his experience in Cape York, O'Faircheallaigh discusses this negotiations-based approach to SIA in some detail (1996b). He emphasises the importance of integrating consideration of economic and social issues, and of equipping Aboriginal negotiators with information, advice and skills that integrate development of a community-based analysis of likely impact issues and development of a well-grounded negotiating position within the community interests involved. O'Faircheallaigh acknowledges significant challenges in producing 'better' outcomes from the negotiations-based approach he advocates (1996b: 26–8). Most significant is access to financial, time and human resources required, which may limit the scope of both impact assessment and negotiations. The legislative framework may also constrain the process, limiting the extent to which indigenous rights to negotiate are recognised as enforceable (see also Lane and Yarrow 1998). He also recognises that the link between the consultative processes related to impact assessment ('hearing people speak directly about the impact of "development" on their lives' 1996b: 27) and implementation of negotiated agreements is fragile and vulnerable to disruption. Finally, he identifies the pressures that arise from the scheduling of negotiations as threatening less than optimal impact assessment. Ross (1999a) offers a preliminary framework for linking an evaluation of the context of impact assessment to a judgement about prospects for successful negotiations. She suggests that the 'social impact-negotiation potential' reflects the balance between characteristics of the location, the affected communities and the development and the proposed development project, and that these need to be considered in terms of broader history, economic climate and legal, political and administrative frameworks. In practice, this requires precisely the resources and cautions that O'Faircheallaigh (1996b), Lane and Yarrow (1998) and Lane *et al.* (1997) identify. International experience of negotiations between developers and indigenous peoples (see for example Ivanitz 1997; Ross 1999b) reinforce these cautions.

The Three Gorges Dam in China

Hydro-Québec (Chapter 10, above) returns to attention in the case of the enormous project to build a dam on the Yangtze River. This project will displace nearly three-quarters of a million people and inundate significant areas of the Yangtze floodplain. Technical problems with siltation have also raised concerns about future flood impacts.

In 1988 a feasibility study endorsed the project: this study was funded by the Canadian International Development Agency (CIDA) and undertaken by a consortium led by Canadian International Project Managers Ltd (CIPM) with substantial participation from Hydro-Québec. The feasibility study – ostensibly an 'impartial technical review ... to assist in reaching a decision and to provide the basis for securing funding from international institutions' (quoted in Fearnside 1994: 27) – comprised 13 volumes. In an excellent review of the Canadian involvement in this work, Fearnside (ibid.:28) notes:

> The CYJV [CIPM Yangtze Joint Venture] report is remarkable in the way it strains to emphasize positive aspects of the scheme. Most incredible is its listing of resettlement as a benefit: 'resettlement construction and development would spur growth in the area bordering the reservoir'. Among the benefits ascribed to the Three Gorges Project is to 'encourage development of the region with resettlement fund'. On the contrary, resettlement is a major negative impact of the project.

This project proposes displacement of more people than any other project ever undertaken. Fearnside argues that displaced persons are more likely to be moved to western desert provinces for resettlement rather than areas adjacent to the reservoir. He also notes that the figure of three-quarters of a million displacees is an arbitrary one, produced by dubious means, and in part a result of a lowering of the reservoir depth (though not the dam wall). He notes that the Chinese government's previously announced reservoir depth would displace 1.1 million people. He notes that having funded the 'lower' level inundation, donor agencies would have no leverage against a Chinese government which opted to increase the reservoir:

China can promise to do anything that the Canadian government or the World Bank might want to hear. Promises can be made, for example, to operate the dam at [the lower level] and to handsomely compensate all displaced persons … . Once the dam is completed … nothing prevents China from changing its mind. No additional construction would be required; the stroke of a pen and the turn of a valve could raise the [level]. The additional 465 000 displaced persons might receive little or no assistance.

(Fearnside 1994: 35)

Like much of the grey literature in impact assessment, the feasibility study was not easily accessible. It was eventually obtained in Canada under that country's freedom of information legislation – but with more than 2000 pages deleted because of commercial confidentiality provisions. Again, Fearnside notes that the Canadian study was far from independent:

These companies stood to gain a great deal if the report were to result in a Chinese decision to build the dam and in approval of international financing for it. CIPM estimated that Canadian firms could potentially take $300–400 million in engineering and managerial sales and $1.0–1.5 billion in equipment sales. One should not be surprised that the feasibility study found no problems that would impede international approval of financing for the dam.

(Fearnside 1994: 27–8)

The Canadian study faced enormous logistical difficulties in compiling a comprehensive report within the two-month timeframe provided for field visits in the affected area. The lack of individual accountability for specific methods used in the research and conclusions drawn, the lack of overall transparency, the enormous potential for conflict of interests between the companies engaged in providing 'impartial' assessment of the project and also likely to be involved as tenderers for phases of design, construction, supply and operation all represent a major failing of the impact assessment approach used in this case. Fearnside's conclusion is that organisations such as the International Association for Impact Assessment and professional engineering organisations should have policies that address the sort of secrecy and conflicts illustrated in this case.

The independent review of the Sadar Sarovar Projects in India 1991–2

The Sadar Sarovar dam on the Narmada River in India is one of the most controversial dam projects on the planet. It commenced in 1987, and was reviewed for the World Bank in 1991–2 (see Berger 1994). Following considerable controversy about this project, widespread criticism of the World

Bank's involvement in development projects which generated substantial negative social impacts, and wider World Bank review of resettlement and indigenous rights issues, the independent review revealed that despite policy changes intended to ensure that development projects such as Narmada would not be used to 'justify the nullification of ... basic human rights' (ibid.: 11), the Bank had not required compliance with reporting and other conditions of the loans it had made available for the project. Berger concluded:

> We found it impossible to separate our assessment of resettlement and rehabilitation and environmental protection from a consideration of the Sardar Sarovar Projects as a whole. The issues of human and environmental impact bear on virtually every aspect of large-scale development projects. We concluded that unless a project can be carried out in accordance with existing norms of human rights and environmental protection – norms espoused and endorsed by both the Bank and many borrower countries – the project ought not to proceed.
>
> (Berger 1994: 19)

Towards a relational view of SIA: the contested terrain of SIA research

Contemporary social experience is characterised by high levels of interaction and change, and resource management systems are embedded in wider processes of interaction and change than just those linked directly to any particular resource project. For the resource manager, this insight can be both debilitating and liberating. Operationalising a relational view of resource management systems and SIA presents some particular practical hurdles.

On the other hand, establishing exact boundaries in the field of resource management becomes impossible; professionals in this field often have to read extraordinarily widely. Where it might have been acceptable to deal with a timber concession or mining lease as if it were disconnected to either its immediately surrounding geography, or the wider world, SIA practitioners need to address questions as diverse as social protest, international treaties, global climate change and local employment plans. In my own recent work, for example, I have needed to research topics as diverse as community psychiatry, corporate strategy, the economics of pastoral production, the management of ballast water, details of funding programmes for apprenticeship training and many other topics. Such links and relationships to wider social processes provide a way of making sense of the impact of resource decisions.

Without a framework to render linkages accessible to analysis, many social impacts previously could only really be seen as *externalities* – uncomfortable or inconvenient intrusions from 'the outside' into the comfortable certainties of the resource management system. The dynamics of the Bougainville rebellion, for example, become more comprehensible when the 'causes' of the problem no longer have to be found within the resource management system alone. And the field of action and response is also broadened, so that the resource manager no longer needs to feel obliged to accept the artificial boundaries of their management system as the limits of their professional interest, responsibility, obligation or action.

As a methodological tool for resource managers and part of the 'resource manager's toolkit', social impact assessment provides a valuable set of ways into the complexity of social change which characterise participation in resource management systems. Although it would be easy to limit a review of SIA research to formal preparatory work prior to an investment decision, a lot of the best SIA work has been done outside the constraints of formal project-oriented studies. Impact assessment fits into a much wider context of research into social change:

> Large-scale industrial and resource development projects, infrastructure development, social policy changes, the introduction of new technologies, and alteration of economic structure all can produce change in how people in the affected areas live. During the last decade, increased attention has been given to the effects of large-scale projects and rapid technological change on the natural and human environment. Efforts to anticipate and evaluate these effects have expanded with them.
>
> (Branch *et al.* 1984: xiii)

While some manuals (Branch *et al.* 1984; Taylor *et al.* 1990; Smith 1993) emphasise applications related to 'large scale projects and rapid technological change' (Branch *et al.* 1984: xiii), and the need to provide information for 'more informed decision making' (ibid.: xiv), impact assessment work is not exclusively aimed at large-scale projects. It can usefully be directed at a range of other sources of social change at a variety of social, economic and geographical scales (policy changes, technological changes, small-scale projects, cumulative regional change).

This then is the contested terrain in which social impact research of all sorts is undertaken. In this arena, what general issues affect the practice of SIA? Wildman identified seven key aspects of the SIA process in the Australian setting:

- Purpose of the social impact analysis;
- Social impacts may affect different functional systems of society;
- The role of the impacted community: partner or patient?
- Co-ordination;
- Direct and indirect impacts;
- Impacts have a time-specific lifespan;
- Not all impacts are predictable.

(Wildman, 1985: 138–41)

In reflecting on the East Kimberley Impact Assessment Project, Ross (1990) argued for an approach to SIA research which entrenches community control, community values and community history in the framework for developing impact research (see also Coombs *et al.* 1989). Another researcher in the EKIAP team, anthropologist Nancy Williams, has pointed out that whatever framework is adopted ethical issues and values questions exist at every step of the SIA process, from definition to completion of the task (1986).

SIA and the research agenda for human geography

Some years ago, I suggested a tentative research agenda to link SIA and human geography (Howitt 1989a: 160–1). While SIA cannot 'belong' to any single discipline, geographers' contributions build on the application of insights and understandings of the linkages between local scale impacts and wider scale elements of change and interaction, and the contextualisation of general processes of social change in space, time and political settings. Three elements emerged as shaping future geographical research in this area:

> First, single project proposals need to be understood in their regional and corporate/institutional context … .Second, there is a need to address 'issues' as well as 'tasks' – to provide (strategically relevant) policy advice, both to governments and community interests … Finally, our effort … needs to be directed at tackling … the mechanisms which translate structural constraints into real events in processes of social change.
>
> (ibid.)

Drawing on traditions in economic, social, political, cultural and regional geography, geographers have much to offer SIA research. Theoretically informed research will greatly assist practical understanding of impact processes, not only by regulatory authorities, but also by those affected by negative social (and cultural and economic) impacts; and also by those responsible for the operation and management of resource projects producing these impacts. However, it is worth pointing out that, without a commitment to long-term monitoring, the costs of resistance to negative social, cultural and economic impacts are often constructed as 'timebombs' which can become extraordinarily disruptive to the productive operation of resource projects long after SIA research for project approval has been forgotten (see Figure 12.5). The term 'timebombs' in this setting comes from Filer's analysis of the Bougainville rebellion (1990a), which provides a timely reminder of the extent to which such resistances can disrupt conventional management practices.

But no single discipline can perform *all* the research required in a complex SIA project. Specialist skills in anthropology, planning, economics, politics and other disciplines often meet in multidisciplinary SIA teams. But just as no single discipline suffices, technical expertise alone is insufficient. Appropriate SIA research also requires appropriate non-technical experts from the affected communities to bring insider understanding and insights to bear on the issues

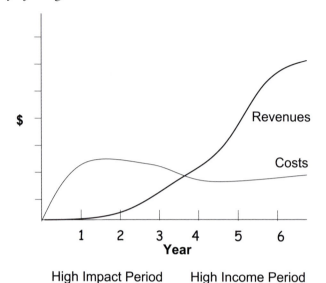

High Impact Period   High Income Period

Figure 12.5 A model of financial flows from the Misima Gold Mine, PNG. This
graph of Gerritsen and MacIntyre's 'capital logic' model identifies the
juxtaposition of the early period of a major resource project as a high
impact – high cost – low income period, with later medium-to-low
impact – operating cost – high income periods. The social tensions and
environmental consequences that arise from this juxtaposition have been
seen by Filer (1991) as creating 'timebombs' for resource managers

Source: Based on Gerritsen and MacIntyre 1991:38.

revealed in research. Being a geographer is no guarantee that a researcher will
bring an adequate framework to capture the scope of SIA research required
for any particular project. The geography journals are littered with unfortunate reminders of the myopia, ethnocentricism, gender bias, ahistoricism, disciplinary jealousies and atheoretical weaknesses of many geographers. Even in
the promising literatures synthesising post-modernist, post-positivist and
post-structuralist critiques of earlier paradigms in geography, there is often a
glaring inability to engage with many of the issues raised here as central to the
practical challenges for participation in resource management systems. The
extent to which many geographers in SIA have implemented the goals of an
applied peoples' geography is certainly questionable (see Howitt 1993a).

   In the end, SIA research is not an arena for a single discipline. It requires
transcendence of disciplinary boundaries to synthesise diverse analytical and
interpretive strengths in understanding and intervening in diverse human
experience of social impact processes. However, geography in all its complexity (both as a real-world phenomenon and as a discipline) is implicated in all of
them.

SIA, reconciliation and change in Australian resource management systems

Tuesday 26 May 1998 marked the first National Sorry Day in Australia, an effort by many Australians to come to terms with and respond to the persistent legacies of genocidal polices towards indigenous Australians by Australian governments over many generations (Australia 1997). Twenty-one years earlier, when hundreds of Aboriginal children my own age were still being forcibly removed from their families, an historic referendum symbolically extended citizenship to Aboriginal and Torres Strait Islander peoples within the Australian nation. It was confirmed by 90.7 per cent of Australian voters that the Australian constitution should be changed in order to accommodate the special interests of indigenous Australians (Bandler 1989). Despite this overwhelming commitment to change, and despite the commitment of substantial resources to solving various aspects of a situation that many have labelled as 'the Aboriginal problem', injustice, inequity, intolerance, alienation and division persist. Nowhere is this clearer than in the field of resource management. The Mabo and Wik decisions confirmed that Australia has inherited a system of land and resource management which privileges corporate developers against indigenous and environmental interests (Bartlett 1993b). Yet changes have occurred, and the pressure for change continues (Howitt 1997a, b). Incorporation of environmental values, principles of gender equity, intergenerational justice, sustainability and anti-racism has produced some remarkable changes in Australian and regional geopolitics since the late 1960s. As the debate over a national apology for the lasting consequences of the tragically flawed but well-intentioned genocide policies of the past confirms, the shift towards humane sustainability is strongly contested and resisted.

Indigenous peoples' struggle for recognition and respect is an important element in the continuing struggle for better outcomes in resource management. Better outcomes, however, cannot be restricted to the symbolic domain; they must also have an effect in the economic and environmental domains. In Australia, despite the resources allocated to targeted programmes, and despite the technical sophistication of experts engaged in service delivery efforts, concrete improvements in some key areas of Aboriginal health, education, employment, self-determination, incarceration, housing and rights have remained elusive. The depth of mistrust and the gulf in understanding revealed in recent resource industry disputes such as Century Zinc (Trigger 1997a, b; Harwood 1997) and Jabiluka[4] serve to remind us of the gulf that continues to exist over resource-based regional development. They also serve to demonstrate that while there may be substantial technical challenges in such situations, there is a simultaneous conceptual challenge – how to think about needs, values and aspirations that refuse to treat things given ontological priority in our own taken-for-granted, common-sense view of things as obvious or desirable. Drawing on recent Australian work on SIA, native title, indigenous identities and resource-based regional development, some practical lessons for the future can be identified.

Seeing the country: towards social, economic and environmental justice

Aboriginal Australians use the word 'country' in a way which divorces it from the conventional discourses of post-colonial politics (Rose 1996b: 7–15; see also Widders and Noble 1993). Aboriginal people's efforts to look after their country encapsulate many of the key issues of the struggles against colonialism and its legacies. To the extent that these struggles unsettle the assumptions of privilege, power and identity in post-colonial Australia, they involve a shift towards genuine decolonisation. In meeting and mounting these challenges, however, the indigenous rights movement faces a massive backlash of privilege, prejudice, poverty and myopia. With limited economic resources, negotiation and persuasion have been the preferred modes for tackling this work rather than direct confrontation.

One of the genuinely urgent practical challenges facing resource geopolitics is to develop ways of addressing the apparently contradictory forces of economic, cultural and environmental processes. In terms of just and sustainable outcomes, this means evolving both visions and skills which practically assist in developing strategies which simultaneously engage with:

- The identity politics of cultural diversity;
- The distributive politics of economic survival;
- The environmental politics of sustainability (see Chapter 2).

As already glimpsed in the cases of SIAs on Coronation Hill (Lane *et al.* 1990), western Cape York (O'Faircheallaigh 1996), Century Zinc (Trigger 1997b) and Kakadu (Dodson *et al.* 1997; Levitus *et al.* 1997, Howitt 1997d), the way in which the privileged position of resource projects shapes new geographies is changing (see also Howitt 1995). In Australian federalism, relatively autonomous states maintain largely unfettered power over allocation and management of land and resources. The federal native title legislation and the High Court's 'interventionist' orientation in Mabo and Wik (Bartlett 1993a; Hiley 1996) faced strong criticism from state governments that saw this as erosion of states' rights (see also Chapter 10). Similar criticism has persisted in some quarters over the constitutional power of the federal government to intervene on environmental matters to maintain certain international standards and obligations (for instance the Franklin Dam case, Toyne 1994; Crabb 1984). The fragmentation of environmental decision making and social policy inevitably makes it difficult to 'get the scale right' in many resource management situations. Setting the spatial, temporal and political scope of what we try to do (for example in SIA, in post-development monitoring, in teaching) presents problems of power, influence and relevance (Howitt 1997c).

Recognition of native title has shifted the ground of much EIA and SIA work. It is no longer possible to pretend that Aboriginal people are not stakeholders in particular regional development processes, when there is a property right involved. Whatever problems are faced, there is increasing expectation that informed decision making requires information about likely social consequences of alternative decisions. Using the mining provisions of the Aboriginal Land Rights (NT) Act 1976, and the right to negotiate provisions of the Native Title Act 1993, more indigenous stakeholders are requiring developers to facilitate indigenous communities doing their own SIAs and then negotiating arrangements to secure the best possible outcomes for all parties. O'Faircheallaigh emphasises the importance of community participation and education in this approach to SIA (1996a, b).

Towards multicultural definitions of environment

The metaphor of co-existence that emerges from the High Court's Wik decision reminds us that most Australian landscapes are genuinely multicultural in character. Different groups value elements of the landscape and the environment in quite different ways. The idea of co-existence reminds us that displacement and marginalisation are not the same as extinguishment or extinction, and that dismissing the value that one group of citizens places on various ecosystem components is fundamentally discriminatory and unacceptable in a society committed to intercultural tolerance and reconciliation.

To tackle these complex issues, it is necessary to provide some conceptual clarity and clear tools that allow Aboriginal community members to construct their own way of seeing them. Drawing on a range of anthropological and other research, the following can be identified as core goals of many Aboriginal groups:

- Caring for country;
- Caring for people;
- Caring for culture.

In many Aboriginal organisations, there is a clear statement of goals that reflect these ideas. Many groups would, for example, identify the following as important strategic aims for their community organisations:

- Moving back to country;
- Looking after the law;
- Getting recognition;
- Security for the community's future;
- Getting good services where people want to live.

Such goals provide pointers to a view of development as those activities and interventions that help to bring together Aboriginal people, Aboriginal

Figure 12.6   Integrating economic, cultural and environmental justice

country and Aboriginal culture in ways that produce better practical outcomes (recognition, health, opportunities, environmental care, housing, mobility, family relationships and so on). This is illustrated in Figure 12.6, in which appropriate language terms and/or graphic icons can substitute for the English words, and in which the dynamic of doing something to bring the key elements of country, people and culture back together can easily be demonstrated.

13 Policy arenas
Reform, regulation and monitoring

In thinking about ways of 'doing' resource management, it is very easy to limit our horizon to what resource managers 'do' in their professional work (in some form or another), and to neglect the broader social and political arenas in which decisions crucial to resource management are made. For example, it would be easy to overemphasise the importance of 'doing' impact assessment, and ignore actions in the wider political arena that shape what impact assessment is required. In this chapter the focus shifts away from the specifics of impact assessment to consider the broader policy arena as an arena of practice. The discussion also considers how public policy processes interact with corporate and community actions in shaping reform processes and practical outcomes in resource management systems.

Taking Leftwich's definition of politics as a starting point (see Chapter 1, above), both politics and resources are to be found everywhere there are people – not just in governments and resource operations. Clearly, examination of the interaction of the policies of various players in resource management systems will shape our understanding of both the systems as a whole, and their effects on people and environments. Restricting our interest to some dictionary definition of policy is unlikely to be very helpful (see for example New Zealand Ministry for Environment 1994: 1), and there is a need to consider how corporate strategies respond to, influence and sometimes even control government policies – and vice versa.

The field of policy making that is relevant to our consideration, therefore, is very wide. We could usefully consider matters as diverse as the ways in the World Bank funds development projects, to the ways in which traditional landowners in Papua New Guinea cope with change. This discussion, however, will focus particularly on the processes involved in reforming resource management, planning and environment protection programmes by governments in several jurisdictions. This allows us to consider how resource companies and citizens have had to respond to policy changes, new knowledge, and even entirely new resource management regimes, and to consider the interaction of public policy with corporate strategies and community responses.

Policy reform as a focus for action by resource managers

Academic researchers are often urged to undertake policy-relevant work – research that contributes directly to policy making, or teaching that produces good policy makers. Similarly, much political activity in the environmental movement is taken up with efforts to influence policy. The definition of policy adopted in such efforts, however, is often limited to decision making activities of governments. Many groups other than governments, however, frame policies aimed at achieving specific social goals, and policy-relevant academic or activist work can, and perhaps should, target constructive policy contributions not only in the public sector, but also in the private and community sectors. Understanding the links between public policies and the decisions and priorities of other groups is crucial in contributing to improved resource management practices.

Public policy making about resource management has, always and everywhere, needed to balance interests that are in tension, if not open conflict. The axes along which balances need to be found include:

- Public versus private interests;
- Short-term versus long-term goals;
- Opportunism versus sustainability;
- Sectional versus general public ('national') interest;
- Economic versus social interests.

Political philosophies and ideologies clearly influence the ways in which individuals and institutions (political parties, parliaments, public service departments, companies and so on) judge the balancing acts involved. There are no objectively correct solutions to the challenges of just and sustainable resource management. What appears ethical and in the national interest to one group, seems to pander to sectional interests and be reprehensible to another (see for instance conflicts over Coronation Hill: Jawoyn Association 1989; Resource Assessment Commission 1991; Merlan 1991; Jacobs 1994; Gelder and Jacobs 1998).

The emergence of widespread governmental and popular concern with the issue of sustainability in the 1980s provided a suitable focus for our examination of this broadly defined process of policy making and responses to it. Leading up to the 1992 UN Conference on Environment and Development in Rio, sustainability emerged as a central issue at all scales. National governments, the Business Council on Sustainable Development (Schmidheiny 1992), UN agencies, popular environmentalist movements, indigenous peoples' organisations, the non-government development agencies and many others, all sought to contribute to social reform which enhanced sustainability. In some places this has led to critical review of existing decision making systems. Questions about regimes controlling resources have led to major efforts in policy reform which provide a forum for new professional practice in resource management. Some interventions used sustainability to enhance more open access to resources, or more security over terms of resource exploitation (Gardner 1993), but more wide ranging exercises in reform and review are worth attention.

The New Zealand experience

In many ways, New Zealand's comprehensive late 1980s review of failing resource management and planning systems, and introduction of a comprehensive Resource Management Act in 1991 deals with a wide range of 'issues, objectives, policies, methods and results' (NZ Ministry for the Environment 1994), and specifically targets sustainability. The explicit purpose of the act is to promote sustainable management of natural and physical resources, which is taken to mean managing natural and physical resources in a way that enables communities to provide for their wellbeing; avoiding, remedying or mitigating any adverse effects of activities in communities; and avoiding, remedying or mitigating adverse effects of communities' activities.

The introduction of the Resource Management Act represented an enormous reform effort. One hundred and sixty-seven separate pieces of legislation relating to natural resources were revoked, and there was simultaneous reform of the mining legislation with passing of a new Crown Minerals Act. The new Act specified powers and responsibilities for all levels of government, greatly simplified processes for allocation and use of natural resources, clarified accountability and decision making structures and removed inconsistencies and contradictions which had evolved in the fragmented and piecemeal system that had developed in New Zealand since 1840, when the Treaty of Waitangi was signed between Queen Victoria and a number of Maori Chiefs. Parallel to this reform was a reorganisation of local and regional government in New Zealand, which made Regional Councils 'responsible for maintaining the ecological integrity of their catchments' (Alexandra *et al.* 1994: ix). As part of this reform, over 700 statutory authorities were abolished, with their roles being transferred to the appropriate regional and district councils or the private sector (ibid). Following New Zealand's acceptance of its importance as a result of the Treaty of Waitangi Act 1985,[1] the new legislation on resource management also placed much more importance on Maori values, Maori rights and the Treaty principles, including:

- Requirements to recognise and provide for the relationship of Maori and their culture and traditions with their ancestral lands, water, sites, *waahi tapu* (sacred places) and other *taonga* (treasures);
- Requirements to have particular regard to *kaitiakitanga* (guardianship or stewardship responsibilities);
- Requirements to take into account the principles of the Treaty of Waitangi;

- Requirements to consult with *iwi* (tribes) in the preparation of plans and policies.

(Crengle 1993: 8)

In this way, contemporary Maori interests and values were entrenched in the national, regional and local systems of planning, managing and monitoring resource use and development in New Zealand. The radical legislative reform in New Zealand has been widely seen as a model for national scale reform in other jurisdictions. Clearly, the size and complexity of New Zealand in some ways lends itself to such wide-ranging reform. Without an entrenched sub-national level of sovereign governments (like Australia's and the USA's states and Canada's provinces), territorial restructuring towards a bioregional logic, with consequential facilitation of ecosystem protection and environmental management, is clearly less problematic than has been the case elsewhere. Similarly, the signing of the Treaty of Waitangi provided a reference point for orientating contemporary political decisions to the ethics and values of the indigenous populations – despite the considerable controversy over just what the terms (and the principles) of the Treaty meant.

In Australia, the absence of any instrument of dispossession or ceding of sovereignty to the crown has produced considerable difficulty in accommodating Aboriginal interests into contemporary political and resource decision making. Similarly, the unwillingness of the US system to accept the terms, principles or even legal obligations (as defined by the US Supreme Court) of numerous treaties with Native Americans, and the competing interests of Indian nations as sovereign entities, states and the national government, has mitigated against wholesale reform in the USA (see *Natural Resources and Environment* 1993; Jaimes 1992). In Canada, while there has been no parallel wholesale reform of resource management systems, the pursuit of comprehensive regional settlements of indigenous peoples' land and resource claims has led to considerable practical reform over the last two decades.

Although the impossibility of transferring a reform model directly from a unitary system to a federal system, Alexandra *et al.* argued that the Australian system was in urgent need of reform, and the New Zealand experience provided some important precedents and guidance (Alexandra *et al.* 1994). The ACF report provides an example of an deliberate effort by resource managers (even if they are not employed as professional resource managers) to influence policy development. Similar work by academics, community groups and others targeting legislative reform, development of new legislation, or particular decisions about specific circumstances, can all be seen as fitting into a 'way of doing' that targets outcomes which are more responsive to, and more accountable to social values of sustainability, justice, equity and diversity.

The USA's Surface Mining Control and Reclamation Act

In 1977, after more than a decade of pressure, the US Congress passed Public Law 95–87, the Surface Mining Control and Reclamation Act (SMCRA). This legislation provided for wide-ranging reform of the control of coal mining in the United States. Two previous Presidents had thwarted its enactment, which was also resisted by many coal companies. The new act was predicated on recognition of the extent to which:

> Many surface mining operations result in disturbances of surface areas that burden and adversely affect commerce and public welfare by destroying or diminishing the utility of land ... by causing erosion and landslides, by contributing to floods, by polluting the water, by destroying the property of citizens, by creating hazards dangerous to life and property, by degrading the quality of life in local communities, and by counteracting governmental programmes and efforts to conserve soil, water, and other natural resources.
>
> (SMCRA S.101[c])

The new legislation established a new office within the Department of the Interior, with wide-ranging responsibilities to 'protect society and the environment from the adverse effects of surface coal mining operations' (SMCRA S.102 [a]).

It also provided for a considerable amount of citizen-initiated enforcement of the provisions of the Act: S.520[a] '... any person having an interest which is or may be adversely affected may commence a civil action on his own behalf to compel compliance with this Act.'

This is unusual. In most jurisdictions, including the USA, the citizens' right to initiate legal action to enforce compliance with environmental legislation is generally severely restricted. It is more common to rely on action from chronically under-funded, under-staffed statutory authorities. Indeed, such reliance has long been a downfall of otherwise impressive legislation, as it provides an easy loophole for governments concerned with not alienating substantial economic players amongst resource companies. In this case, statutory authorities such as the Office of Surface Mining are directly accountable for their actions through the courts to US citizens whose lives and property they affect.

One result of this particular reform has been the extent to which organisations such as the Citizens Coal Council, a Washington-based lobby group with decentralised chapters around the country, has been able to play an active role in monitoring and enforcing compliance with the legislation it helped to frame.

Land management reform in Arizona: the Navajo–Hopi land dispute

A less optimistic American example can be found in the dispute over land management in the area of Arizona which is claimed by both Navajo and Hopi nations as part of their reservations. In this case, the efforts of resource companies, state and federal politicians, and a number of lawyers, to clarify ownership of coal resources in an area over which there were competing Navajo and Hopi claims has produced the largest peacetime relocation of US citizens, and a tragic and dramatic legacy of disruption, poverty and conflict. This case reflects the ways in which political, economic, cultural and environmental interests overlap and interact – often with unintended results which have lasting consequences.

In this case, complex settlement and land-use arrangements on the large Indian Reservation in Arizona occupied by both Hopi and Navajo people produced a dispute over ownership of about 1.8 million acres of land. Following the establishment of tribal governments in the two nations under the Indian Reorganization Act of 1934, under extremely dubious circumstances, legal action established a zone of joint ownership. Self-interested lawyers 'representing' both Hopi and Navajo, encouraged their 'clients' to pursue clarification of precise ownership of the areas – an action which made each lawyer a millionaire. Over a period of more than thirty years, this legal action eventually led to a partition of the lands, and wholesale relocation of the people who ended up with homes and lives on the wrong side of the lines drawn on the map. Most of those involved were Navajos.

Coal was first discovered in this area in 1909:

> The energy companies' interest in the area set the machinery of partition in gear. Without their pressure, the land dispute might have remained simply a simmering local problem between tribes, as it had for the previous seventy-five years. Some observers of the land dispute maintain that the energy companies, working in concert with lawmakers, BIA [Bureau of Indian Affairs] officials, and tribal attorneys, *created* the concept of the land dispute. They argue that the energy companies would have found it very difficult to relocate Indians from their homelands for the purpose of clearing it for strip mining. But if the Navajos were removed for another reason – to return the land to Hopi control – then the land would be cleared without any political fuss for the energy companies.
>
> (Benedek 1993: 138, emphasis added)

As it turns out, the development of Peabody Mining's Black Mesa and Kayenta mines has proceeded with considerable opposition from many Hopi traditionals – Hopis opposed to the tribal government established by the US government in the 1930s, and with considerable uncertainty about the environmental implications. Many Navajos and some Hopis work in the operations, and see the revenue they generate as crucial to self-determination. Yet in this case, many questions about the interaction of the coal issues, and the enormous human cost of partition remain unanswered.

Canada: the Great Whale Assessment Guidelines as policy innovation

We have already considered several aspects of the James Bay hydro-electric project in northern Québec. It is also worth considering the development and implementation of the environmental assessment guidelines for the Great Whale Project as an example of the policy process in resource management.

The first phase of the James Bay project, the La Grande Complex, was constructed between 1972 and 1985. It produced considerable political reform in Québec – a new treaty with the Cree and Inuit owners of the region, development of a legislative framework for impact assessment and environmental protection, and implementation of environmental and social monitoring programmes. There was also considerable institutional development of Cree and Inuit government structures, of Hydro-Québec, and of other provincial and federal authorities involved. Failure of impact assessment to predict the mercury contamination of reservoirs meant that the impact mitigation strategies adopted for the La Grande Complex were subject to critical scrutiny in the Great Whale review process. In reviewing those mitigation measures McCutcheon concluded:

> There ... seems to be consensus among both Natives and developers that most remedial mitigation measures undertaken ... were too ineffective and expensive to be worthwhile ... . By any general measure of ecological health, there is little if any difference between the mitigated and unmitigated project ... the residual (post-mitigation) impacts are for the most part, those that were predicted.
>
> (McCutcheon 1994: 65–6)

McCutcheon concluded that although consensus existed between the developers and local communities about the mitigation effort, they disagreed about the relative distribution of costs and benefits from the project:

> The significant unresolved issues ... are not ecological but ethical and practical ... . [T]he lesson for future projects is clear: remedial mitigation measures cannot be counted on to undo the ecological impacts of large projects ... . If these impacts are unacceptable prior to remedial measures, they will, in all likelihood, remain so afterward, as well.
>
> (ibid.: 70–1)

Such conclusions present governments with a significant dilemma. The project proponent asserted that the terms of the James Bay and Northern Québec Agreement of 1975 anticipated development of the entire project, and precluded rejection of the project on grounds of social or cultural impacts. On the other hand, basic human rights principles precluded the deliberate imposition of genocidal policies on indigenous peoples. The authorities responsible for evaluating the proposal recognised that it would be bad policy for governments to approve a project which demonstrably disabled a regional economy that was already operating consistently with the principles of sustainability. The dilemma was how to deal with the wider policy commitment to sustainability without precluding a project that many saw as crucial to Québec's economic future.

The policy context was further clouded by the cultural politics of identity and nationalism in contemporary Canada (Cohen 1994). McCutcheon's early analysis (1991) documents the James Bay project's importance in shaping Québecois nationalist politics, the Cree sovereignty movement and acceptance of federal environmental controls. Hydro-Québec development as a powerful Francophone institution throughout the 1960s and its close links with Québec identity and politics is parallelled in Cree politics by the importance of the post-JBNQA institutions in representing a Cree national identity, and in giving voice to claims of sovereignty, and Cree opposition to Québec separatism.

Many of these changes have been fundamental in shaping federal, provincial and tribal policy treatment of James Bay phase II, the Great Whale Project. Unprecedented co-operation between federal, provincial and tribal authorities produced impact assessment guidelines (Evaluating Committee *et al.* 1992) that are exceptional in their scope, thoroughness and perspective. These guidelines, published in August 1992, provide a glimpse of what the sorts of structures established under the New Zealand *Resource Management Act*, might look like when implemented on a wider spatial scale, and over global scale resources. They offer an approach which is holistic, inclusive and integrative, and accountable simultaneously to local cultural values and aspirations and global goals of sustainability and equity. In this approach, there is a balancing of scientific and indigenous knowledge, of development and tradition, of human and environmental values.

The Great Whale Project pushed environmental assessment procedures in Québec into the state of the art as governments, development proponents, private sector interests, citizens groups, environmentalists and native peoples sought to exercise and facilitate responsible decision making on matters of enormous scale and scope. Hydro-Québec's environmental impact study was subject to co-ordinated review by four environmental assessment review bodies. Each had specific responsibilities under the James Bay and Northern Québec Agreement, but agreed to co-ordinate and harmonise their efforts. It was anticipated that the result would be a document of about 500 pages plus supporting material that allowed the committees to review the proposal against the principal assessment criteria spelt out in the guidelines.[2] Among these criteria were requirements that the project must 'be developed in accordance with the carrying capacities of the ecosystem and human societies involved'; it must 'respect the rights of local communities to determine their future and their own societal objectives'; and 'must not endanger the durability or quality of resources that form the basis of an existing regional economy'. The guidelines also spelt out that evaluation had to be based on 'respect for the right of future generations to sustainable use of the ecosystems within the proposed project area, both for the local population and for society as a whole', and that it is essential for the EIS to 'address the combined and cumulative effects of ... impacts on entire sectors of the ecosystem, including the human societies in the proposed project area' (Evaluating Committee *et al.* 1992: 3–5). Hydro-Québec's response, published in August 1993 as a 'feasibility study', is comprised of around 5000 pages in 30 volumes in French and English, with a 300-page summary, plus maps and other material in English, French, Cree and Inuktiktot and a two-volume methodological report. In addition, the various commissions and participants have themselves issued substantial documentation, including a series of ten background papers, compilations of press materials, and a 300-page bibliography from the Great Whale Public Review Support Office which lists 2849 relevant items.

From an international perspective, the guidelines are interesting because of the extent to which they differ from similar documents for large-scale resource projects in other jurisdictions. Rather than a standardised format of chapter headings and sub-headings, the Great Whale guidelines provide both an organisational framework, and an integrative conceptual framework. The guidelines spelt out the policy rationale for discharging the governments' responsibilities. In pushing the proponent towards this integrative approach, the Commissions were adamant about Hydro-Québec's obligations:

> [These are] guidelines that the Proponent must follow in presenting the EIS on the proposed Great Whale River hydro-electric project. It is incumbent upon the Proponent to prepare a complete EIS that includes sufficient data and analyses for a complete assessment of the anticipated impacts and their repercussions.
>
> (Evaluating Committee *et al.* 1992: 2)

Also unusual from an international perspective was the extent to which local ecological knowledge and social values were seen as directly relevant to the review process:

> Local residents' knowledge of their biophysical and social milieu is essential to an adequate assessment of the impacts of a development project. Furthermore, each cultural group has its own conceptual and symbolic system that reflects the group's image of itself and of its communities, its environment and its past and future. Since this conceptual and symbolic system partly determines the group's reaction to change, it is an intrinsic element of the environment itself and must be thoroughly understood before the impacts of a development project can be assessed.
>
> (Evaluating Committee *et al.* 1992: 6–7)

The guidelines also required Hydro-Québec to provide detailed economic justification of the proposal as a responsible energy policy. This was important because many critics felt that market projections and assessments of alternatives to massive dam developments had been far from rigorous. Even the guidelines' approach to the descriptive task is innovative compared with many practices:

> The Proponent shall provide a definition of the environment in keeping with the multicultural character of the territory in which the proposed project would be built, shall identify and target analysis of valued ecosystem components, and shall indicate and justify the spatial and temporal boundaries assigned to each component. Once this has been done, the Proponent shall have outlined a portrait of the environment in which the proposed project would take place in a way that each affected cultural group could recognize, and in all cases particular emphasis shall be accorded to the interactions between ecosystems and human communities.
>
> (Evaluating Committee *et al.* 1992: 27)

Despite Hydro-Québec's earlier efforts to separate various components of the project for isolated assessment, the committees insisted on integrated evaluation of the hydro-electric development works, the supporting infrastructural works and power collector system.

Finally, and perhaps most significantly, the guidelines were absolutely clear about the approach they expected Hydro-Québec to adopt in evaluating impacts:

> In order to avoid the reductionist and compartmentalizing tendencies of an encyclopaedic approach, the impacts of the proposed project on the various components of the environment shall be evaluated in terms of five fundamental issues
>
> - health;
> - access to the territory;
> - availability of resources;
> - social cohesion;
> - respect for values.
>
> Furthermore, the combined and cumulative effects of very different types of impacts on certain elements of the ecosystems shall be evaluated. These cumulative impacts, though more complex, do not differ from other environmental impacts. As regards the source of the impacts, the Proponent shall take into consideration not only the various elements of the proposed complex but those of other projects as well, specifically including the La Grande, Churchill-Nelson and Conawapa complexes.
>
> (Evaluating Committee *et al.* 1992: 63–4)

Australian examples

Finally, an Australian example will serve to illustrate the practical dilemmas involved in targeting policy arenas as a way of improving resource management. In 1989, the West Australian government established a social impact unit (SIU), with a mandate to 'ensure that the social impacts of development proposals were addressed as part of the environmental impact assessment procedures administered by the EPA' (Beckwith 1994: 200). A new conservative government disbanded the unit in 1993, downgrading its functions into a section of the Department of Resources Development. The implication of this administrative change is a shift in the balance between regulation and encouragement of development.

When the West Australian unit was established, community expectations and industry fears both ran high. Industry groups saw it as an additional and unnecessary bureaucratic hurdle and source of delay for project approvals – even a threat to the economic viability of some projects. Community groups in areas affected by proposals saw it as a guarantee that their concerns would be addressed. Like WA's earlier flirtation with social impact assessment in the Argyle Social Impact Group (see Chapter 7), this exercise was subject to arbitrary political intervention (and eventual closure). In both cases, there was considerable mismatch between political, community and industry expectations of institutional innovations. And in both cases short-term political imperatives over-rode longer term policy objectives.

The incorporation of rigorous assessment of social impacts as part of the project evaluation and approval process is an important policy goal for achieving more equitable and just resource management outcomes. In Western Australia, dealing with social impacts has relied on the discretionary powers of the Environmental Protection Authority (EPA). Beckwith recognises that the four years in which the SIU operated saw 'steady improvement in the standard of social impact assessment in Western Australia' (1994: 205), but that 'SIA has yet to be fully integrated into the EPA's organisational culture' (1994: 210).[3]

Alternative models for managing resources

Resource management systems are not static. Policy making processes are one source of change affecting them: policies of local, regional, provincial–state–territorial and national levels of government, by intergovernmental agencies, international organisations, non-governmental organisations, and affected communities and other stakeholders have a wide range of effects on practical resource management. Public and expert participation in policy making, therefore, is a crucial arena of practical resource management.

The policies, institutions, programmes, and processes that shape resource management, however, are often fragmented, mutually inconsistent, unfair, unsustainable, inequitable, and often irrational. In many jurisdictions, the need for policy reform is obvious and urgent, yet contributions to such reform do not proceed in a vacuum. The policy context is beset by hidden and competing agendas, incompatible spatial and temporal frames of accountability and lack of mutual respect, understanding and tolerance. One can learn much by looking at reform processes in other places. If nothing else, such exercises allow us to see familiar issues in a new context, removed from the ideological baggage we bring to issues in our home territory (Jull 1992a). In such reviews, one can also identify and critically analyse alternative ways of doing things:

- New sorts of report like the Ngai Tahu regional resource management plan for the Canterbury Region in NZ (Tau *et al.* 1990) or the Great Whale River Project impact guidelines (Evaluating Committee *et al.* 1992);
- New legislative approaches (NZ Resource Management Act, US SMCRA, Canadian regional settlements);
- New decision making structures (Nunavut,[4] increased recognition of traditional ecological knowledge, the Columbia River co-management model[5]).

Having seen these issues in the context of government policies, one could undertake similar analysis of corporate policies and strategies and the policies and activities of community organisations. We have already recognised in our simplified model of dynamic resource management systems, that it is not only governments whose decisions shape resource management outcomes. Clearly, we could pursue programmes targeting corporate and community reform in much the same way as we target reform of public policy.

14 Co-management of local resources

The changing context of resource management

Resource managers are increasingly required to respond to changing societal values and government regulations. In recent years alternative approaches have emerged to facilitate higher levels of local control, influence and participation from local interests. In particular, co-management models provide a way of integrating many of the social, cultural, political and environmental issues, but it often falls short of its potential. In many situations co-management provides the dominant paradigm with an avenue for extending control into indigenous domains in what Rose (1999) refers to as 'deep colonising'.

The extent to which societal expectations have changed is clearest in relation to environmental and human rights performance. In previous eras, governments granted virtually unfettered interests in resources to corporations, and companies were able to exercise a high level of sovereign control over matters within the territorial boundaries of resource concessions. In Australia, mining leases granted in Queensland, the Northern Territory and Western Australia in the 1950s and 1960s, excised new towns' existing local government areas. Local governance was integrated into companies' industrial relations practices. In many industries, corporate control of key information on pricing, costs, technology and marketing made governments dependent on companies for data used to set policies. In-house trading, transfer pricing and technology transfers within transnational companies diminished national resource sovereignty, and its replacement with corporate sovereignty. In many regions, companies were literally given 'freedom to narrate the world' (Morrison 1993: 64) according to their own needs and priorities (Howitt 1995).

In the post-war minerals boom, national resources policies were largely limited to facilitating profitable exploitation of resources. The legacies of this period can be found in often tragic environmental conditions at unrehabilitated resource sites and reduced biodiversity in many places. Local communities' futures were woven into the unsustainable social and economic fabric produced by such systems of resource exploitation, leaving many unresolved grievances. Indigenous peoples, whose traditional property rights were ignored and transgressed by developments in this era, were often structurally excluded from access to any of the benefits flowing from exploitation of their resources. While some, such as Hugh Morgan (1987, 1991, 1993), Heilbuth and Raffaele (1993) and Howard and Widderson (1996), feel challenges to corporate sovereignty as a threat to national integrity, their position is politically and strategically indefensible.

Co-management models offer an alternative to corporate-centred approaches. Co-management can be understood as a system which combines elements of several management systems – local-level, state-level, traditional, industrial, global and so on (Berkes *et al.* 1991: 12). Co-management represents one way of doing resource management that may produce outcomes more consistent with the principles of sustainability, equity, justice and diversity.

Joint management of national parks and conservation areas

Areas of high natural heritage value are often areas in which indigenous peoples retain strong interests. In both Canada (Berg *et al.* 1993; Dearden and Berg 1993; Fenge *et al.* 1993) and Australia (Birckhead *et al.* 1992; Woenne-Green *et al.* 1994), indigenous claims over national park and wilderness areas have led to development of joint management arrangements. In Australia, since the recognition of Aboriginal ownership of important conservation and heritage areas at Uluru-Katatjuta and Kakadu in the Northern Territory (Yapp 1989; Hill and Press 1993, 1994; Toyne 1994), joint arrangements for the administration of large national parks on Aboriginal land have been refined to involve levels of local indigenous involvement in managing conservation areas. This has become widely accepted as standard, but has also been problematic in developing Aboriginal control over tourism, wildlife management and other matters (Creagh 1992; Brown 1992; Altman and Allen 1992; Moreton-Robinson and Runciman 1990).

The Uluru–Kakadu Model involves recognition of Aboriginal ownership of certain areas of high conservation value in return for agreement to lease the area back to National Parks authorities with majority indigenous membership on the park's Board of Management. The legislative foundation of this model is the Commonwealth's Aboriginal Land Rights (NT) Act 1976. In other areas of Australia, where there is no similar legislation, arrangements for local indigenous participation in the management of conservation areas remains problematic and *ad hoc*. Conventional definitions of wilderness values have reinforced popular visions of Australia as *terra nullius* (Robertson *et al.* 1991), negating Aboriginal visions of caring for country (Young *et al.* 1991; Taylor 1995), and ways of seeing environmental conditions (Rose 1988, 1996a, 1999).

Efforts to entrench the rights of traditional Aboriginal landowners (native title holders *and* legally dispossessed landowners with continuing interests in particular places) in the management of other areas has proved slow and difficult. A 1990 submission to the Western Australian government from the Gulingi Nangga Aboriginal Corporation in Derby, for example, proposed several ways of applying the principles of the Uluru Model to the Buccaneer Archipelago in the coastal zone of the west Kimberley region (Nesbitt 1992; see also Jackson 1995 on indigenous sea rights). This position was reinforced at a 1991 meeting in the East Kimberleys, which passed a resolution arguing that:

> National Parks should be under Aboriginal control, Aboriginal People should make the rules. Aboriginal People should prepare the management plans, and to have access to all areas within National Parks ... . All National Parks should be made A Class Aboriginal Reserves [under the WA Land Act 1933 and Aboriginal Affairs Planning Authority Act 1972] and Aboriginal People can look at sub-leasing them to National Parks.
> (Kimberley Land Council 1991: 7, Resolution no. 61.
> See also Resolutions 62–7)

Despite its weaknesses, the Uluru–Kakadu Model offers a standard to which indigenous people without the level of legislative support provided by the Land Rights Act aspire to. Where there is acceptance of the principles involved, co-management models for conservation areas can provide a mechanism for addressing legislative and attitudinal hurdles. To date, however, particularly in the context of the native title debates in Australia, most states have been completely unwilling to amend legislation in the necessary ways.

One of the key problems for many non-indigenous people's acceptance of a decisive indigenous influence in resource management arises from a failure to accept the right of indigenous landowners to undertake a range of traditional activities, particularly those involving harvesting resources in conservation areas (IWGIA 1991). In the case of Inuit harvesting of animals for fur, European and North American animal rights campaigns have criticised utilisation of wildlife by indigenous cultures and undertaken emotive campaigns to boycott fur products (IWGIA 1991: 25). These campaigns have undermined the maintenance of local Inuit culture, self-esteem, independence, viability and sustainability.

A second problem arises from the failure of Western corporate interests to accept the importance and legal integrity of indigenous knowledge when it has commercial value. This failure entrenches indigenous marginalization, and reinforces corporate power and corporate sovereignty. For example, three-quarters of the current prescribed medicinal drugs derived from plants have been discovered through indigenous peoples' knowledge (Gray 1991: iii).

Increased judicial, political and social acceptance of indigenous rights (native title, citizenship, human rights protection and so on), has provided a vehicle for more routine involvement of indigenous peoples in management decisions; and recognition of the utilitarian and moral value of indigenous knowledge and values has increased acceptance of indigenous presences in 'wilderness' areas. While this is a long way from co-management being accepted practice, it is a considerable advance on the situation when joint management at Uluru–Katatjuta and Kakadu was first proposed.

Alternatives in other resource sectors

Berkes *et al.* (1991) examine Canadian approaches to co-management of wildlife resources in some detail. As they define it, co-management in this setting involves some combination of 'local-level' and 'state-level' systems. State-level management is undertaken by a centralised authority, based on scientific data, and enforced by judicial and legal sanctions. In contrast, local-level wildlife management systems in the Canadian North rely on decentralised, local authority and consensus, are based on customary practice, cultural traditions and local knowledge and enforced through social sanctions (Berkes *et al.* 1991: 12).

Wildlife resources are generally defined in Western legal traditions as 'common property'. In many indigenous traditions, harvesting sites (traplines, fishing sites, hunting ranges etc) are managed socially, although they may be exploited exclusively by identified persons. Cree institutions around Hudson Bay (Ontario and Québec) have been increasingly involved in co-management arrangements:

> The benefits of greater Cree participation in and responsibility for control of local resources are likely to be both economic and non-economic. Co-management will help to reverse the erosion of traditional leadership among the Cree, and to restore these leaders to positions of greater influence. In the more strictly economic sphere, co-management is a building block for an expended, firmer foundation for the local economies of the remote Cree communities … and to the realization of Cree goals of increased self-determination and cultural economy.
>
> (Berkes *et al.* 1991: 16)

In his compelling analysis of environmentalist–Indian alliances in northern Wisconsin, Gedicks also asserts the emergence of regional resource co-management models from struggles over fishing rights and mineral development:

> Instead of the resource colony that the multinational mining and oil corporations, with active encouragement from the state of Wisconsin, is trying to impose on northern Wisconsin, *Anishinaabe Niijii* ('Friends of the Chippewa', established in 1989) wants to declare an environmental zone, to be jointly managed by the state and the Chippewa.
>
> (Gedicks 1993: 193–4)

Similarly, in the Columbia River basin in the northwest USA and British Columbia, following a long and bitter dispute about Indian rights to take fish in 'all the usual and accustomed places', as promised under many treaties, a co-management structure has emerged that brings together national governments, state and provincial governments and key tribal governments (Institute of Natural Progress 1993; see also McGinnis 1995). In this case, industrial resource management of water resources for power generation was been severely challenged as Endangered Species legislation in the USA was employed to challenge the principles used to justify large-scale hydroelecticity dams in the region, and force their costly removal in order to re-establish healthy rivers, capable of sustaining salmon and other species.

In North American mining industries, it is possible to find some systems in which co-management is emerging. In the BHP coal mines in New Mexico, the company deals directly with the Navajo Nation government, which levies royalties and business taxes, and is currently preparing for lease re-negotiations in which it is widely anticipated that the Navajo Nation will take on direct equity in the new lease operating system, and assume a more directly managerial role in operations. In this case, BHP employs a large number of Navajo both as operators and management (BHP 1993, 1994), and much of the negotiations will involve respected Navajo people against other Navajo. Already, preferential employment agreements with the Navajo have seen up to 85 per cent of BHP's New Mexico employees being Navajo. The next decade, it seems to me, is likely to see a genuine co-management arrangement for many of the local scale functions currently undertaken at these mines.

Elsewhere in the mining industry, suggestions for co-management options remain a long way off. One review, for example, suggested that its mining personnel and native community representatives:

> Generally felt that it was a little too sceptical to expect that there would be much change in the relationship between Aboriginal communities and mining companies in the short term.
>
> (Canadian Aboriginal Minerals Association 1994: i)

According to this review, education and training remain a high priority for increased participation in mining employment, but for both groups – Indians and mining personnel.

Despite this pessimism, examples can be found where co-operative approaches to resource management are emerging. For example, at the Golden Patricia mine (Ontario), development for a gold mine involved negotiation of a main agreement between the company and five local Windigo communities, covering issues such as employment, recruitment, environmental protection, provision of scholarships and apprenticeships (Shaw and Lalonde 1994). Five sub-agreements covered human resource development needs, traditional economic activities, economic and business development, social, cultural and community support, and administration, management and implementation of the agreements. The 1993 renegotiation of the agreements confirmed and expanded most of the original terms and sought to expand some aspects of the operation.

The Inuit Circumpolar Conference's Arctic Environmental Protection Strategy also points towards some form of resource co-management on the ground (Reimer 1993/94). Reimer argues that although this shift towards joint management is positive and overdue, it 'does not measure up to the Inuit vision of the North', which is reflected in a regional conservation strategy and steps towards demilitarisation, international co-operation and recognition of indigenous rights.

In New Zealand, refinement of the 1991 Resource Management Act's approach to Maori (and local-scale) resource management (see Chapter 13) has seen many opportunities for decisions to be made, and operations to be monitored in ways consistent with and acceptable to Maori values and customs (Gibbs 1994). High levels of Maori involvement in fisheries management, for example, has produced a group of Maori people who exercise considerable influence and even direct power over some sorts of decisions to do with fisheries management.

Beyond co-management

Following Notzke (1995) it is appropriate to move discussion about co-management from dealing with indigenous participation in co-management regimes as a concession by government, towards dealing with it as a constitutional (or indigenous sovereign) right. Notzke suggests a number of different approaches to co-management:

- Co-management as a result of comprehensive claims settlements (regional agreements)
- Co-management as a means of crisis resolution
- Co-management as a result of indigenous peoples' common law rights (Sparrow *v.* Regina 1986)
- Co-management of national parks
- Strategic co-management
- Co-management as a constitutional right.

Most co-management seeks to incorporate indigenous and other local groups into a system of resource management in which resources are defined and managed consistently with the dominant paradigm. Notzke's final two categories move beyond this constraint and offer a way of seeing things that may overcome the impasse that emerged from the confrontation between indigenous rights and development prerogatives. In the Canadian context, Notzke suggests that:

> The last two decades have witnessed a gradual transformation in the ideas of social justice and environmental consciousness on the part of the

mainstream society, and concurrently an increased degree of politicization of aboriginal people. Together these trends not only gave rise to the formulation and settlement of comprehensive claims, but also to a changing approach to environmental and socio-economic impact assessment. Aboriginal people are discovering that this Euro-Canadian device can indeed serve as a very useful tool for empowerment and result in effective impact management and in adaptive management procedures for resource development on aborignal land.

(Notzke 1995: 205–06)

This has led towards the development of regional environmental councils, some with statutory and some with negotiated powers, with representation from a wide range of stakeholders and responsibility for moving towards co-operative management of the environmental consequences of industrial resource systems. Such circumstances have led to the development of some interesting models involving 'strategic co-management'. In Australia, the Cape York Land Use Agreement (O'Faircheallaigh 1996c; Farley *et al.* 1997; Teehan 1997), for example, represents a strategic agreement between a number of non-statutory parties (in that case it was Aborigines, environmentalists and pastoralists) who have decided it is in *all* their strategic interests to reach some form of co-management arrangement over resources in which each has some sort of interest. Other agreements such as the Zapopan Agreement at Mt Todd and the McArthur River Agreement at Borroloola (O'Faircheallaigh 1996a; Strapp 1994; Craig *et al.* 1996; Howitt 1997b; Jawoyn Association 1995, 1997a, b) reflect strategic approaches to co-management arrangements. Similarly, we could develop models for regional agreements between Aboriginal groups and mining, forestry, tourism or pastoral industry groups to provide clear mechanisms for *how* to address concerns in resource-rich areas where indigeonous people are currently left out of decision making – or reduced to archaeological relics!

One of the lessons to be learnt from experience in strategic co-management, however, is that the first regional agreement that is needed when moving towards strategic co-management is a regional agreement between Aboriginal groups. In seeking to change power relations through co-management processes, it is important to realise that the beneficiaries of existing arrangements will seek to exploit any weakness within the indigenous position.

Divide and conquer

For the resource companies, the first rule of negotiating (and they almost can't help themselves on this) is often 'divide and conquer'. I was once sitting down with a mining company man talking about the language we could use in trying to get the company to recognise that they had really been 'part of the damage' done to the Aboriginal community on the doorstep of their mine. I wanted to talk about 'negotiating' a change in the relationship and he shook his head. 'You have to remember', he said to me. 'On my side, the corporate warriors think that a "negotiation" is something you *win* – not something that helps resolve a problem. And my boss's preferred starting position in *any* negotiation – even if it's with someone in his own company – is with his foot firmly on his opponent's throat'. The easiest way to get the other side's boots back on your throat is to offer yourselves up one at a time to be bought of cheaply – or to hold out for too much, when everybody else is ready to be bought off.

Strategic co-management agreements obviously take time to develop – but the windows of opportunity that arise in dealing with resource companies are often short-lived. Resource companies will make indigenous stakeholders think time is always shorter than it is, but the marketplace is demanding, and opportunities do pass. Waiting for constitutional recognition, or ideal market circumstances to achieve an ideal outcome may well see real opportunities for improvements bypassed or delayed. This is an area where values and judgement come into play and differences of opinion will abound.

In Canada, constitutional processes have led to new perspectives on issues of the constitutional status of indigenous peoples, and the sorts of rights that might accrue to such peoples (Crawford 1988). The constitutional recognition of Nunavut, the extension of common law rights to water, wildlife and other resources, and the recognition that indigenous peoples and nations have a constitutional right to some level of self-government and self-determination are all suggestive of a move towards constitutional recognition of a right to co-management of resources and environmental systems. Constitutional debate in Canada has, for example, continued in the context of a statement which is unlikely to be changed in future discussion:

> The exercise of the right of self-government includes the authority of the duly constituted legislative bodies of Aboriginal peoples, each with its own jurisdiction:
>
>   (b) to develop, maintain and strengthen their relationship with their lands, waters and environment so as to determine and control their development as peoples according to their own values and priorities and ensure the integrity of their societies.
>
> (quoted in Notzke 1995: 207)

This leads towards recognition of exclusive rights in resource management systems, as well as shared rights and responsibilities, and towards specific constitutional recognition (see also McHugh 1996). It also leads towards consideration of the nature and limitations of negotiation as a strategy for achieving better resource management (Chapter 4). Co-management models, such as those emerging in North America, particularly Canada, and Australia, particularly in the area of conservation reserves and wildlife management, are ways of doing resource management that offer means of generating local-scale win–win scenarios for the futures in resource localities, but for many indigenous groups, indigenous management of such areas would be a better result than co-management.

Part VI

From theory to praxis

15 Sustainability, equity and optimism

The professional practice of resource management occurs in contested terrain where issues of sustainability, human rights and social justice are in constant tension with economic imperatives and technical sophistication. To conclude this book, I want to talk about optimism. I'd also like to draw inspiration from the words of a song that says more powerfully than I can, something about the issues tackled in this book. We commenced our exercise in rethinking resource management with Leon Rosselson's song about the English Civil War. This time, the words come from North America's recent colonial history. In her song, 'My Country 'Tis of thy People You're Dying', Buffy Sainte Marie reflects on the contemporary relevance of dispossession and marginalisation of American Indians in American society. Her words are as relevant and powerful today, at the beginning of the twenty-first century and more than 500 years after Columbus' misreading of the landscape commenced the Eurocentric history of the world, and a few years after the quincentenary of Columbus' voyage, as it was when she wrote them over twenty years ago:

My Country 'Tis of thy People You're Dying

Now that your big eyes are finally opened.
Now that you're wondering, 'How must they feel?'
Meaning them that you've chased cross America's movie screens;
Now that you're wondering, 'How can it be real?'
That the ones you've called colorful, noble and proud
In your school propaganda,
They starve in their splendour.
You ask for my comment, I simply will render:
My country 'tis of thy people you're dying.

Now that the longhouses breed superstition.
You force us to send our children away
To your schools where they're taught to despise their traditions
Forbid them their languages;
Then further say that American history really began

When Columbus set sail out of Europe,
And stress that the nations of leeches who conquered this land
Were the biggest, and bravest, and boldest, and best.
And yet where in your history books is the tale
Of the genocide basic to this country's birth?
Of the preachers who lied?
How the Bill of Rights failed?
How a nation of patriots returned to their earth?
And where will it tell of the Liberty Bell
As it rang with a thud over Kinzua mud?
Or of brave Uncle Sam in Alaska this year?
My country 'tis of thy people you're dying.

Hear how the bargain was made for West,
With her shivering children in zero degrees.
'Blankets for your land' – so the treaties attest.
Oh well, blankets for land, that's a bargain indeed.
And the blankets were those Uncle Sam had collected
From smallpox diseased dying soldiers that day.
And the tribes were wiped out
And the history books censored
A hundred years of your statesmen
Have thought, 'It's better this way'.
But a few of the conquered have somehow survived
And their blood runs the redder
Though genes have been paled.
From the Grand Canyon's caverns
To Craven's sad hills
The wounded, the losers, the robbed sing their tale.
From Los Angeles County to upstate New York,
The white nations fatten while other grow lean.
Oh the tricked and evicted they know what I mean:
My country 'tis of thy people you're dying.

The past it just crumbled; the future just threatens
Our lifeblood is shut up in your chemical tanks,
And now here you come, bill of sale in your hand
And surprise in your eyes, that we're lacking in thanks
For the blessings of civilisation you've brought us
For the lessons you've taught us;
The ruin you've wrought us;
Oh see what our trust in America got us.
My country 'tis of thy people you're dying.

Now that the pride of the sires receives charity.
Now that we're harmless and safe behind laws.
Now that my life's to be known as your heritage.
Now that even the graves have been robbed.
Now that our own chosen way is your novelty.
Hands on our hearts
We salute you your victory:
Choke on your blue white and scarlet hypocrisy.
Pitying the blindness that you've never seen -
That the eagles of war whose wings lent you glory,
They were never no more than buzzards and crows:
Pushed the wrens from their nest;
Stole their eggs; changed their story.
The mockingbird sings it;
It's all that she knows.
'Oh but what can I do?', say a powerless few.
With a lump in your throat and a tear in your eye:
Can't you see that their poverty's profiting you?
My country 'tis of thy people you're dying.

Reaching a conclusion

This book is partly aimed at supporting the professional education of resource managers. As with any educational enterprise, learning to be a resource manager does not finish with graduation. University learning does not make a person educated in any final sense; one is never completely educated, and (as employers will remind readers who apply for positions in real-world resource management systems), a conceptual toolkit needs the addition of real-world experience before it begins to be really adequate. It is easy to forget this as one emerges from the academy brimming with learning and enthusiasm. I am reminded of a salutary lesson taught to me by an unschooled bushman in the forests of northern New South Wales some years ago. Mr Lloyd was telling me of 'a smart university feller who came up to study emus':

> Gee he was smart. What he knew about emus was enough to fill a whole book. But what he didn't know was enough to fill a bloody library – but he couldn't see that.

Over the years, as I've taught the course on which this book is based, I've thought long and hard about what to say in conclusion. Initially, I thought the last words needed to provide a profound summation; a final convincing explanation; a revelatory insight. But after dealing with the concerns, aspirations and insights of several hundred students who have completed the course, it now seems clear to me that the most important challenge isn't to provide profundity, but to justify optimism.

At the turn of the century, we face the prospects of a new millenium with a widely pervasive mood of social pessimism. In the words of another song-writer, 'everything put together sooner or later falls apart' (Paul Simon 1971). This recognition of fragility extends across the personal and the political – relationships, communities, political programmes, ecological balances, faith and for many even personal sanity. Buffy Sainte Marie's powerful song, too, contains an element of pessimism; yet she has continued her creative endeavours, creating new visions of possibilities, contributing to cultural renewal and change. Like many artists, she sees a basis for pessimism, but never fully accepts the argument in favour of desperation. So in this final chapter, I want to explore some of the foundations for optimism in resource management and the wider struggle for social justice.

Since this song was written, the justification for pessimism may seem to have grown. Certainly, many of the issues and circumstances we've examined in this course are a source of desperation for many people. Certainly, for many Aboriginal people, caught in a web of alienation, poverty, doubt, racism and powerlessness, pessimism has long been close to the surface. The stories of Aboriginal people who have died in custody in Australian police cells and prisons in recent years starkly illustrate the depth of desperation for many (E. Johnston 1991; Dodson 1991; Langton *et al.*, 1990). If we take the news from many parts of the world at face value, the news from Siberia, from Amazonia, from Azania, from Burma, from Somalia, from Kurdestan, from East Timor, from Haiti, from Kosovo … we see desperation multiplied many times. Yet we also hear from indigenous people and the marginalised and dispossessed, an extraordinary determination. In the bicentenary of invasion, Aboriginal groups around Australia shouted 'We have survived', and continued the process of survival and renewal.

In my own case, I have been privileged to come to political and professional maturity working closely with such people. Around them, involved in their struggles, pessimism in a privileged young whitefeller has seemed an indulgence – an unnecessary and unjustifiable luxury – a travesty of everything they struggle for and to which I wanted to contribute.

In the field of resource management, we constantly straddle an interface between human misery and ecological fragility. It is easy to take a vantage point in which pessimism overwhelms the basis for action; or in which pessimism justifies abdication of personal and collective responsibility – if nothing can change the system, why shouldn't I get in for my cut too? In the face of genuinely global crises, in the face of extraordinarily complex local and regional crises, what can one person really do?

In this final chapter, then, I want to engage in a little self-reflection and auto-critique in order to explore the basis for my own optimism, and to lay foundations for others to identify a basis for optimism in their own professional practices. I believe that optimism is perhaps the most important conceptual tool a resource manager can develop. I've had the privilege of having as my teachers many Aboriginal people whose optimism was never naïve, never superficial, never simple and rosy, and often extraordinarily humble. Their optimism comes from clarity about preferred futures, and trust in humane values and the ability of other people, including their children and future generations, to maintain integrity; to continue their struggle. It is something of their teaching of me that I hope to pass on to you.

A personal journey

When I was working in the Kimberleys in 1980, I found myself enmeshed in the events at Noonkanbah. Because I was still working on interviews with some of the companies involved, the Aboriginal people I was working with decided they would prefer it if I didn't stand on the picket line with them, and continued to provide an analysis of the bigger picture. Left behind, I spent the day writing two songs,[1] which I performed a couple of days later for a group of Aboriginal women, just after the convoy had pushed its way on to Aboriginal land at Noonkanbah. After I finished singing, there was no clapping, no discussion, just silence. I was worried that maybe I'd done the wrong thing, but one of the women came up to me and told me how much they appreciated the songs. 'Just when we were beginning to feel down', she said,' that song has made us feel strong again. Put it on a tape for me'. A few days later, I found myself in the dirt at Noonkanbah, sitting in a circle of old men with my mandolin. I'd been told that these men really only liked traditional music and not to expect too much. I sang the song, and again there was a long silence. One of the men spoke to a young boy in Walmajarri, and he disappeared, returning a few minutes later with a tape recorder, a Slim Dusty tape, and some sticky tape. Ceremoniously the older man covered the tabs on the cassette and put the tape into a cassette recorder, pressed the record button, and said, 'Sing it again'. Again, pessimism seemed like an irrelevance amongst these people Hawke describes as real heroes of the Australian nation (Hawke and Gallagher 1989).

As an educator, I guess I find another basis for optimism. Effective education is always an 'aha-experience' – you are never really the same after you have learned something. My work as an educator draws on a whole range of influences, but particularly on the work of Freire (1972a,b, 1976). In most of my teaching, I have been engaged not so much in teaching a subject or content as in a broader process of personal and community development. Whether it was in my two teacher school in the heart of New South Wales' conservative New England, or my one-teacher demountable on a commune on the New South Wales north coast, or a year-four classroooms with 13 language backgrounds in suburban Sydney, there was always more to do than just the narrow curriculum tasks.

Those classrooms were always linked to complex and challenging communities. I find it hard to claim that I caused any changes – although things certainly changed. All I can say is that I contributed to those changes. For me, that is enough. I don't see much value in seeking monuments to our individual work as a mark of our professional or human contributions. All any of us can do is what we are able to do and what we have the opportunities to do. Everything I have lived through, however, compels me to say that this is the driving imperative for action. If we are capable of contributing to humane change, to contributing to the creation of more preferable futures – and when all's said and done, that is one of the characteristics of humans – then we are obliged to do so.

This book has aimed to confront readers with issues that would challenge you in some way. For some readers, the material and events discussed, the issues addressed might have actually changed how you 'see' things. Much of my own quite unshakeable optimism comes from my experience as an educator, working with students and others who respond to this material in thoughtful and challenging ways. If students can change in the course of a few months of exposure to ideas about 'new ways of seeing', 'new ways of thinking' and 'new ways of doing', then there continues to be a basis for optimism in the ability of wider human communities to pursue alternatives to the probable futures which can be projected from current trends. It is not an optimism based on faith in an inevitable triumph of 'good' over 'evil', but an optimism rooted in experience. Inevitability, therefore, has nothing to do with it. We are constantly confronted with the material foundations for both realities – both the pessimistic reading and the optimistic reading. We are all able to interpret the trend projections that emphasise the vulnerability of the future, and there is always a basis and opportunity for active engagement in the change processes needed to produce alternative outcomes. Core concepts needed to shape these preferred alternatives abound in the literature and realities reviewed in this book. The notions of sustainability, equity, empowerment and participation are all building blocks in a professional practice for resource managers which address rather than reinforce the sorts of issues we've discussed. Glimpses of optimistic readings in many places have been woven into the account reported here. In trying to pull together the disparate strands of discourse and experience, my own conclusion is that it is only through situated engagement – the hard work of dealing with justice *in situ* – that we can really achieve this. We need to rethink resource management, but ultimately the challenges to students, teachers and professionals alike are not philosophical or theoretical. They are complex and ongoing issues of practice, ethics and experience. They demand critical thought, critical vision and coherent personal and collective praxis.

Notes

1 Worlds turned upside down

1 Song lyric by Leon Rosselson, © Leon Rosselson. Recorded on 'Rosselsongs', Fuse Records CFCD007. Reproduced with kind permission of Leon Rosselson.

2 Song lyric by Leon Rosselson, © Leon Rosselson. Recorded on 'Rosselsongs', Fuse Records CFCD007. Reproduced with kind permission of Leon Rosselson.

3 Marx (1954 [1887]: 703–4) quotes my own ancestor William Howitt as saying 'The barbarities and desperate outrages of the so-called Christian race, throughout every region of the world, and upon every people they have been able to subdue, are not paralleled by those of any other race, however fierce, untaught, and however reckless of mercy and of shame, in age of the earth'.

4 While the systemic excesses of European imperialism and its appeals to European superiority may be relegated to a past history (see Blaut 1993), the overt racism of many fundamentalist political movements at the turn of the twenty-first century, and the continued human rights abuses that characterise political processes in many parts of the contemporary world, suggests that violent repression of difference, dissent and diversity remains acceptable as a basis for social policy in some quarters – and is tolerated by the community of nation states in most circumstances.

5 In his Preface to *A Contribution to the Critique of Political Economy*, Marx wrote: At a certain stage of their development, the material productive forces of society come into conflict with the existing social relations of production, or ... the property relations within which they have been at work hitherto. From forms of development of the productive forces these relations turn into their fetters ... . No social order ever perishes before all the productive forces for which there is room in it have developed; and new, higher relations of production never appear before the material conditions of their existence have matured in the womb of the old society (Marx 1975 [1859]).

6 It is important to recognise that the Aboriginal concept of 'country' discussed in the preceding extract by Deborah Bird Rose not only incorporates much more than the standard English term 'country', but also incorporates more than 'land'. Not only does it extend to include 'sea country' within the traditional estates of coastal peoples, but it also encompasses the whole of the relevant estate, its history and geography, the animate and inanimate materials upon and within it, and the complex unities between living, past and future generations linked to it. In other words, these relationships to country must be understood as including peoples' collective 'ownership' of their traditional estates in the fullest possible sense. For further discussion of this point see Jackson (1995), Rose (1996a) and Sharp (1996).

2 The problem of 'seeing'

1   This quotation is a line from the American Indian anthem 'My country, 'tis of thy people you're dying' by Buffy Sainte-Marie.
2   Song lyric by Leon Rosselson, © Leon Rosselson. Recorded on 'Rosselsongs', Fuse Records CFCD007. Reproduced with kind permission of Leon Rosselson.

3 Complexity in resource management systems

1   I am grateful to Dinny Smith, a colleague in the mining industry, for the idea of these issues as potential 'showstoppers'. His work in cross-cultural development in a major Australian mining company has emphasised the importance of getting resource management professionals to 'see' the importance of avoiding social and cultural misunderstandings. For a more conventional academic assessment, see Colin Filer's work on the way in which failure to address these issues lays 'time bombs' in the development path of resource projects in Papua New Guinea (1990).
2   Harvey (1996: 48–57) claims that the principles of dialectics can be summarised in eleven principles. Ollman's first feature is encompassed in Harvey's third principle; the second in Harvey's fourth principle; the third in Harvey's sixth principle, and the final one in Harvey's eighth and ninth principles. Harvey's discussion adds to Ollman's four features a degree of space–time awareness, a more detailed exploration of the implications of flow ontologies versus thing ontologies, and an engagement to some extent with educational and environmental issues. Ollman's approach is preferred here because it highlights issues of importance to the task at hand.
3   Swyngedouw refers to at least the following: scalar narratives; scalar levels and perspectives; spatial scale as something that is produced; scalar spatial configurations; scale as the arena and the moment where socio-spatial power relations are contested and compromises negotiated and regulated; mechanisms of scale transformation and transgressions; jumping of scales; scales as 'nested'; scale mediating between co-operation and competition, between homogenization and differentiation, and between empowerment and disempowerment; scale reconfiguration; scale-defined institutions or levels of governance; scaling; significant new institutional or regulatory scales; scale-produced tensions; nested scales; rescalings; scale transgressions; scale politics; a nested set of related and interpenetrating spatial scales; a profound rearticulation of scales; upscaling; and emancipatory; decidedly scaled politics; and empowering politics of scale.
4   For a summary of the situation at Gove see Howitt (1992) and Williams (1987).

4 Beyond negotiation

1   Perhaps the Yolngu-speaking areas of northeast Arnhem Land are most obvious here because of their profile through advocates such as Manduwuy Yunupingu (1994). Other areas, however, such as Cape York, the Kimberleys, Jawoyn Country and the Ngaanyatjarra and Pitjantjatjara lands all retain coherent practices that can be understood as constituting sovereignty. In the Torres Strait, although acceptance of a managerial agenda has been one of the conditions imposed by the Commonwealth, there has been a lot of effort put in to re-membering indigenous knowledge–power systems since the 1992 High Court decision in the Murray Island case.

5 Reading landscapes

1   Relph's paper (1989) refers to nineteenth-century French geographer and anarchist Elisee Reclus' description of humanity as 'the conscience of the earth' to suggest that this sort of

responsibility extends to each one of us as people, not just to certain professionals: 'It implies that each one of us has a reponsibility for the earth and for environmental implications of human actions, both our own and those of others' (Relph 1989: 158; see also Fleming 1992).

2   A lot of relevant work on sense of place and the meaning of place can be found in the work of cultural geographers. For comprehensive reviews of themes and issues see Cosgrove (1978, 1983, 1987, 1992) and Anderson and Gayle (1992). See also recent work in anthropology (for example Hirsch and O'Hanlon 1995).

3   See also the work of Hugh Brody (1981) and Deborah Bird Rose (1988, 1996, 1999).

6 Ethics for resource managers

1   I should, of course, point out here that values are not arbitrarily chosen on an individual basis, but are socially and culturally constructed. Personal values are overdetermined by an extraordinary range of structural and circumstantial determinants. In many ways, we are all part of broad-scale value shifts over long time scales. Even in the short-term and particularly in the West, we rarely completely reflect our parents' values; we rarely accept unchallenged the institutional values (for example, church, state, political party and other affiliations) of previous generations, and have generally sought ways of challenging them. In other societies where tradition is valued more highly than innovation, values shifts are probably slower, but nevertheless, even in the most conservative societies (as all the great literatures tell us), crucial change in the moral order is often a product of the actions of those who challenge the commonly-held values of the dominant society.

2   I use the term alienation here in the Marxist sense addressed for example by Ollman (1976) and referring to the complex process by which workers become disconnected from their products and the means of producing value from their own labour power.

3   The people involved in these decisions were the Aboriginal people who make up the Executive Committee of the Napranum Aboriginal Corporation. Until it was wound up in mid-1994, these people also comprised the Aboriginal executive members of Weipa Aborigines Society, to whom the original review process was directly accountable.

4   The 60 000 word manuscript was also submitted to University of Queensland Press for consideration, and to Pat Dodson, Chairperson of the Council for Aboriginal Reconciliation, with a request that he consider writing a preface for a published version of the document.

5   On the Comalco side, there was concern about the effect of publication on the Wik Claim, in which Comalco was a respondent. Within Comalco, there has been much debate about the best response to this legal action, which includes areas of rich bauxite in the southern section of the Comalco lease. On the one hand, those who recognise that the company and the Aboriginal communities will need to maintain long-term good neighbour relations well beyond the conclusion of the case have argued for the maintenance and continued development of 'good neighbour' programmes such as continued involvement in NAC. On the other hand, the very substantial financial and strategic demands of defending the Wik Claim, the constraints put on interaction with Wik people and other potential claimants by legal strategies, and the anger generated by claims which, if successful might constrain Comalco's future mine development, have led some to suggest that defence of the claim should be given priority over all other work in Aboriginal public relations.

8 Recognition, respect and reconciliation

1   At the 1996 federal election, Ms Hanson stood as the endorsed Liberal Party candidate for the seat of Oxley, which was held for the Australian Labor Party by the Attorney-General Michael Lavarche. The seat had previously been held by the former leader of the ALP, then

Governor-General Bill Hayden. Ms Hanson was expelled from the Liberal Party prior to the election because of anti-Aboriginal racist commentaries during the election campaign, and was elected as an independent. In the same period, the endorsed ALP candidate for the seat of Kalgoorlie in Western Australia was also expelled from his party for continued racist comments, but was also elected to the Parliament. Hanson formed a new political party ('Pauline Hanson's One Nation Party'), and became a focus of enormous media attention with anti-Aboriginal and anti-multiculturalism statements, criticism of the 'Asianization' of Australia and strong statements against national migration policies. Langton (1997) provides an Aboriginal perspective on the new 'wedge' politics that Hanson represents.

2   Mabo and Others *v.* The State of Queensland [no. 2] (1992) 175 CLR 1 (see Bartlett 1993); The Wik Peoples *v.* The State of Queensland and others and The Thayorre Peoples *v.* The State of Queensland and others (see Hiley 1997; Bachelard 1997).

3   Reynolds (1996: xi) convincingly draws 'a direct line' from the Blackburn judgement in the Gove Land Rights case, through a Privy Council decision of 1889 (Cooper *v.* Stuart 14 AC 1889 291) to the observations made by Sir Joseph Banks 'from the quarter-deck of the Endeavour in 1770', and points also to an 1836 NSW Supreme Court decision (R *v.* Murrell) which was cited in a 1976 (R *v.* Wedge 1976 NSWLR 581) as having settled the legal status of Aboriginal interests with the statement 'although it might be granted that on first taking possession of the Colony, the Aborigines were entitled to be recognized as free and independent, yet they were not in such a position with regard to strength to be considered free and independent tribes. They had no sovereignty' (cited in Reynolds 1996: 7).

4   For the text of the decision and detailed commentary, see Bartlett 1993a. For further discussion see (*inter alia*) Council for Aboriginal Reconciliation (1994), Stephenson and Ratnapala (1993), Rowse (1993) and Sharp (1992, 1996).

5   In their initial response to the Mabo decision, the state government in Western Australia attempted to convert unextinguished native title rights to specified statutory use rights under the Land (Titles and Traditional Usage) Act 1993, which was rejected by the High Court in a legal challenge (see Bartlett 1995).

6   This was in fact the position advocated by the new Minister for Aboriginal Affairs in the Howard Coalition Government during vigorous debate over reconciliation, native title and self-determination in early 1996.

7   In recent years I have been privileged to work with many Aboriginal communities on the impacts of mining projects on their lives and futures. The work reported in the following 'snapshots' draws heavily on work circulated in a number of discussion papers and community-based reports (Howitt 1991c, 1992a, 1992b, 1994, 1995, 1997d, 1998b; Howitt and Jackson 2000). In framing these snapshots, I have also drawn on the work of research students who have worked with me over recent years (for example du Cros 1996; Norris 1996; Kealy 1996; Suchet 1996; Jackson 1995, 1996, 1997).

8   *Balanda* is the Yolngu-matha word referring to outsiders. It is widely translated as 'whitefellers', but reflects the long interaction between Yolngu and outside influences as it is derived from the Macassan rendition of 'Hollander'.

9   This review was undertaken by the author and circulated as an unpublished report under the title 'Part of the Damage? A review of the relationship between Comalco and the Weipa Aborigines Society'. Despite the intention of the WAS Executive Committee to publish it, the report remained unpublished because its was seen as damaging to Comalco. The company shredded copies of the report, only a small number of which remain in circulation in the community. The WAS Executive, and later the NAC Executive, agreed not to pursue publication in return for undertakings from senior management of Comalco regarding the transition process envisaged for the development of NAC. Many of these undertakings were not honoured by the company, but some positive changes did arise from these discussions (see below). The ethical dimensions of this situation were discussed in detail in Chapter 6 (pp. 181–84, above).

10   NAC is incorporated under the Commonwealth Aboriginal Councils and Associations 1983 and must have constitutional changes approved by the Canberra-based Registrar of

Aboriginal Corporations. It took the registrar over sixteen months and many follow-up actions from NAC to finally approve this change in February 1996.

11 The author was part of the Economic and Social Impact Assessment consultancy team, along with Professor Ciaran O'Faircheallaigh (Griffith University) and Dr Annie Holden (ImpaxSIA Consultants). The report produced for this study remains confidential as the negotiations were concluded only in April 2001.

12 For accounts of the historical background of the Noonkanbah communities see for example Hawke and Gallagher (1989: 62–99) and Kolig (1987, 1990).

13 See Jackson (1996).

14 It is worth noting that when Western Australia was granted independence in 1889, its constitution included a provision that removed administration of Aboriginal affairs from the government and placed it under direct control of an Aborigines Protection Board accountable directly to the Governor. It also required devotion of a fixed percentage of state revenues to Aboriginal affairs and the work of the Aborigines Protection Board – a provision 'that had not applied to any of the other colonies in Australia when they were granted self-government' (Hawke and Gallagher 1989: 42).

15 The story of the Bunuba resistance is one of the heroic episodes of Australian history. Jandamarra, a complex figure in Bunuba myth and white history, mounted a coherent and strategically brilliant guerrilla campaign against police and pastoral intrusions into the Bunuba heartlands north of Noonkanbah commencing with an ambush of a police patrol in 1894 and ending with Jandamarra's death in 1897. Pederson and Woorunmurra's compelling account of this period (1995) is an accessible and rewarding resource for those interested in better understanding the roots of contemporary indigenous grievances in Australia.

16 It is worth noting that this 1968 decision did not require immediate institution of adequate cash wages for Aboriginal labour, but proposed a slow phasing in of award wages, with maintenance of a 'slow worker' clause in the Pastoral Industry Award providing a continuing loophole for unequal wages even after the decision. For discussion of Aboriginal employment conditions in the pastoral industry and the effects of the Pastoral Industry Award decision see Stevens (1974, 1981), Hardy (1968), Rowse (1987) and Reynolds (1992).

17 The 1967 Referendum was an enormously important symbolic gesture of reconciliation from white Australians towards inclusion of indigenous Australians in Australian society. Although technically it neither gave indigenous Australians voting rights, nor included them in census counts, both these matters are widely attributed to the referendum, which was overwhelmingly endorsed in most parts of Australia (see Pearson 1994; Bandler 1989).

18 In the 1950s, the core of the Strelley Mob organised the first major strike among Aboriginal pastoral workers (see Stuart 1959).

19 For background information on the Argyle Diamonds project and its impacts on Aboriginal interests, see (among others) Dixon and Dillon (1990), Langton (1983), Coombs *et al.* (1989), Howitt (1989) and Dillon (1991).

20 As discussed below, the report of the ASIG review was never made public. This commentary on aspects of the review is drawn from discussions with participants in the review process and materials provided by WA government staff.

21 The AIAS (now AIATSIS) was commissioned by the Minister for Aboriginal Affairs in March 1977 to monitor the social impacts of uranium mining in the region. Work commenced in 1978. Regular reports were published by the Commonwealth government, with a consolidated report published in 1984 (AIAS 1984). AIAS also published a set of guidelines for research on social impacts of uranium mining on Aboriginal people (AIAS 1980a) and a 'knowledge directory' (AIAS 1980b). There was also a very substantial literature produced by this project and the people involved in the research (see e.g. Appendix VIII of AIAS 1984) and significant academic discussion of the project (see Kesteven 1986; Williams 1986, among others). The SIUM project was also influential on the East Kimberley Impact Assessment Project discussed above.

22 The following section is a revision of Howitt (1998b).

23   This point is echoed by Canada's Royal Commission on Aboriginal Peoples, which identified four compelling reasons to 'work out fair and lasting terms of co-existence with Aboriginal people':
   • Canada's claim to be a fair and enlightened society depends on it.
   • The life chances of Aboriginal people, which are still shamefully low, must be improved.
   • Negotiation, as conducted under the current rules, has proved unequal to the task of settling grievances.
   • Continued failure may well lead to violence.
                            (Royal Commission on Aboriginal People 1996: Ch1, p1)
24   While the Jawoyn Association's innovative negotiations with the operators of the Mt Todd project were widely praised at the time, the downturn in gold prices in 1998 and turmoil in Asian economies led to the closure of the mine, bankruptcy for the operating company, dismissal of the workforce and potentially substantial shortfalls in rehabilitation funds set aside to repair environmental damage at the site.
25   I gratefully acknowledge the role of my colleagues Ciaran O'Faircheallaigh (Griffith University) and Annie Holden (ImpaxSIA Consultants, Brisbane) in developing this perspective.

9   Dependent nations or sovereign governments?

1   Figures are from the 1990 US Census, quoted from Navajo Nation (1993: 7).
2   Parts of BHP's operations are also on the Mountain Ute reservation.
3   Within the Navajo Nation, the local Chapter House is the equivalent of a local government area.
4   Will Collette, Citizens Coal Council, interview in Washington DC, 30 June 1994.

10   Indigenous rights or states' rights

1   The term Sami is variously spelled as Saami, Sami, Sámi and Same. I have adopted the former as my standard spelling, but maintained others in quotations.
2   Björklund's paper investigates the inapplicability of Hardin's 'tragedy of the commons' (1968) to Sami management of coastal fisheries in northern Norway. For further discussion of this case and its wider relevance, see e.g. Sharp (1996).
3   The following discussion draws principally on Paine's report, commissioned as part of the Sami legal case against the Alta project (Paine 1982). For further discussion of Sami reindeer pastoralism as a resource management system, see also Björklund (1988), Kvist (1991) and Paine (1992).
4   Sami in Swedish territory did not face the same sorts of pressures to abandon Sami language as those living within Norway, although this was a result of assumptions about the inappropriateness of 'normal' education for Sami.
5   Paine (1991) provides an overview of issues of stigmatisation of Sami identity in Norway and the development of Sami 'nationalism'. See also Aarseth (1993), Aikio (1993), Niia (1991) and Nysto (1991), among others.

11   Diversity and world order

1   This proposition parallels McDowell's feminist response to the challenges facing institutional radical geography in the 1990s (McDowell 1992, see also Howitt 2000).
2   This phrase comes from Wagner (1991).
3   This contrast between 'autonomy' and 'power' relies on the work of Galtung referred to in Chapter 3, above.

4 A good example of this can be seen in the case of the Gulingi Nangga submission for application of a modification of the Uluru Model to a proposed marine park in the Buccaneer Archipelago of the West Kimberley coast (Nesbitt 1992).

12 Social impact assessment

1 Similar points could be made in relation to WA. See for example Coombs *et al.*, and recent media coverage of disputes over resource development at Marandoo and Yakabindie. Earlier disputes at Noonkanbah and Argyle raised similar issues, but these were not addressed in any substantial way in the early 1980s. In the case of Argyle, the ongoing monitoring group ASIG (Argyle Social Impact Group) was disbanded by the WA Government without adequate consultation in 1990. The report of an inquiry into the operations of ASIG has never been published, although indications are that it would have supported continuation of ASIG with improved performance on impact monitoring and mitigation (see Chapter 8).
2 For example, if one considers the interventions in the longstanding and complex conflict between Navajo and Hopi Indians (and within both the Navajo and Hopi nations) over a rich area of coal-bearing land in the southwest USA (see Chapter 9).
3 The Comalco ESIA report was completed in early 1997.
4 The dispute over uranium mining at Jabiluka gained worldwide attention when senior traditional owner Yvonne Margarula and her colleague Jaqui Katona were awarded the Goldman Environmental Prize in April 1999. For information on this dispute see the following URLs: http://aucwa.iinet.net.au/internet/net-jabiluka.html; http://www.mirrar.net/; http://www.sea-us.org.au/jabiluka/jabiluka.html.

13 Policy arenas

1 The NZ Parliament passed an Act in 1975 which allowed Maori to take action to implement the terms and principles of the Treaty of Waitangi in relation to actions after 1975. In 1985, the Lange Government allowed the Treaty of Waitangi Tribunal, established under the 1975 legislation, to consider grievances back to the 1840 signing of the Treaty. For some background see for example Kaiwharu (1989), Crengle (1993) and Tau *et al.* (1990).
2 Scientific Co-ordinator of the Great Whale Public Review Support Office interview in Montréal, 28 June 1994.
3 For discussion of a similar unit in the Queensland bureaucracy, see Dale and Lane (1995) and Dale *et al.* (1997).
4 On Nunavut, see Weller (1988), Canadian Arctic Resource Committee (1993), Ivanitz (1997) and also http://www.nunavut.com/home.html.
5 On the Columbia River, see for example Wood (1996), Gooding (1997), Pyle (1995) and McGinnis (1995).

15 Sustainability, equity and optimism

1 The songs were 'The Noonkanbah Scabs' and 'The Road to Noonkanbah', published in Stringybark and Greenhide (1981; 2.6: 15–23); they were also released by Larrikin Records ('Noonkanbah!', Riss.002).

Bibliography

Aarseth, Bjørn (1993) *The Sámi: Past and Present.* (Translated from Norwegian by Jean Aase) Norskfolkemuseum, Oslo.

Adams, Philip (1980) CRA and the Aborigines. *The Age*, 11 October, 24.

Agnew, John (1993) Representing space: space, scale and culture in social science, in Duncan, James and Ley, David (eds) *Place/Culture/Representation.* Routledge, London and New York: 251–71.

Agrawal, Arun (1995a) Dismantling the divide between indigenous and scientific knowledge. *Development and Change* 26, 413–39.

—— (1995b) Indigenous and scientific knowledge: some critical comments. *Indigenous Knowledge and Development Monitor* 3(3); URL http://www.nufficcs.nl/ciran/ikdm/3–3/articles/agrawal.html

AIAS (Australian Institute of Aboriginal Studies) (1981) *Social Impact of Uranium Mining on Aborigines of the Northern Territory.* AGPS, Canberra.

Aikio, Marjut (1989) Forced shift of environment; the language ecological orientation. A case study of a Norwegian-Finnish reindeer Sámi group, in Broadbent, Noel D. (ed.) *Nordic Perspectives on Arctic Cultural and Political Ecology Including a Statement of Principles and Priorities in Arctic and Northern Research.* (No. 9.) Miscellaneous Publications, Center for Arctic Cultural Research, Umea University: 105–19.

Aikio, Pekka (1987) Experiences drawn from the Finnish Sámi Parliament, in IWGIA (ed.) *Self Determination and Indigenous Peoples: Sámi Rights in Northern Perspectives.* (IWGIA Document, 58). International Work Group for Indigenous Affairs, Copenhagen: 91–102.

Alexander, Anne and van Dijk, Johan (1996) Scientific knowledge and indigenous perceptions of area, weight and space. *Indigenous Knowledge and Development Monitor* 4(3); http://www.nufficcs.nl/ciran/ikdm/4–3/articles/dijk.html

Alexandra, Jason; Wishart, Felicity; Fisher, Tim; Rosenbaum, Helen and Japp, Ian (1994) *New Zealand Legislates for Sustainable Development – Lesson for Australia: A Brief Review of New Zealand's Resource Management Act.* Australian Conservation Foundation, Fitzroy, Victoria.

Allen, Mark (1989) Native American control of tribal natural resource development in the context of the Federal Trust and tribal self-determination. *Boston College Environmental Affairs Law Review* 16, 875–95.

Althusser, Louis (1969) *For Marx.* Penguin, Harmondsworth. (Originally published in French in 1965 by Maspero).

Altman, Jon (1983) *Aborigines and Mining Royalties in the Northern Territory*. Australian Institute of Aboriginal Studies, Canberra.

Altman, Jon and Allen, Linda (1992) Living off the land in national parks: issues for Aboriginal Australians, in Birckhead, Jim; de Lacy, Terry and Smith, Laurajane (eds) *Aboriginal Involvement in Parks and Protected Areas* Aboriginal Studies Press, Canberra: 117–36.

Altman, Jon and Smith, Diane E. (1994) *The Economic Impact of Mining Moneys: the Nabarlek case, Western Arnhem Land*. CAEPR Discussion Paper, 63/1994. Centre for Aboriginal Economic Policy Research, ANU, Canberra.

Anderson, Benedict (1983) *Imagined Communities: Reflections on the Origin and Spread of Nationalism*. Verso, London.

—— (1992) The New World Disorder, *24 Hours Supplement*, February: 40–46.

Anderson, Kay and Gayle, Fay (eds) (1992) *Inventing Places: Studies in Cultural Geography*. Longman, Melbourne.

Aplin, Graeme (1998) *Australians and their Environment: an Introduction to Environmental Studies*. Oxford University Press, Melbourne.

Armour, Audrey (1992) The challenge of assessing social impacts. *Social Impact* 1(4) 6–8.

Atwood, Bain (1996) Mabo, Australia and the end of history, in Atwood, Bain (ed) *In the Age of Mabo: History, Aborigines and Australia*. Allen & Unwin, Sydney: 100–16.

Australia: House of Representatives Select Committee on Grievances of Yirrkala Aborigines Arnhem Land Reserve (1963) *Report*. Australian Government Publishing Service, Canberra.

Australia: House of Representatives Standing Committee on Aboriginal Affairs (1974) *Report on the Present Conditions of Yirrkala People*. Parliamentary Papers, 227/74. Parliament of the Commonwealth of Australia, Canberra.

Australia: Human Rights and Equal Opportunity Commission (1997) *Bringing Them Home: National Inquiry into the Separation of Aboriginal and Torres Strait Islander Children from their Families*. HREOC, Sydney.

Australian Environment Council (1984) *Guide to Environmental Legislation and Administrative Arrangements in Australia*. AGPS, Canberra.

Australian Institute of Aboriginal Studies (1984) *Aborigines and Uranium: Consolidated Report on the Social Impact of Uranium Mining on the Aborigines of the Northern Territory*. Australian Institute of Aboriginal Studies, Canberra.

Australian National Parks and Wildlife Service and Kakadu National Park Board of Management (1991) *Kakadu National Park Plan of Management*. Commonwealth of Australia, Canberra.

Australian Petroleum Exploration Association (APEA) (1981) *Aboriginal Communities and Petroleum Exploration and Development*, August.

Aviation Planning Consultants (1989) *Air Commute Feasibility for Comalco Weipa Operations*. Comalco, Brisbane.

Awashish, Philip (1988) The stakes for the Cree of Québec, in Vincent, Sylvie and Bowers, Garry (eds) *Baie James et Nord Québécois: Dix Ans Après/James Bay and Northern Québec: Ten Years After*. Recherches amérindiennes au québec, Montréal: 42–5.

Bachelard, Michael (1997) *The Great Land Grab: What every Australian should know about Wik, Mabo and the Ten-Point Plan*. Hyland House, South Melbourne.

Back, William Douglas and Taylor, Jeffrey S. (1980) Navajo water rights: pulling the plug on the Colorado River? *Natural Resources Journal* 20(1) 71–90.

Ball, Desmond (1980) *A Suitable Piece of Real Estate: American Installations in Australia*. Hale and Iremonger, Sydney.

Bandler, Faith (1989) *Turning the Tide: A Personal History of the Federal Council for the Advancement of Aborigines and Torres Strait Islanders*. Aboriginal Studies Press, Canberra.

Banks, Glenn (1996) Compensation for mining: benefit or time-bomb? The Porgera gold mine, in Howitt, Richard; Connell, John and Hirsch, Philip (eds) *Resources, Nations and Indigenous Peoples: Case Studies from Australasia, Melanesia and Southeast Asia*. Oxford University Press, Sydney: 223–35.

Banks, Glenn and Ballard, Chris (eds) (1997) *The Ok Tedi Settlement: Issues, Outcomes and Implications*. Pacific Policy Papers 27. National Centre for Development Studies and Resource Management in Asia Pacific, Australian National University, Canberra.

Barnet, Richard J. and Müller, Ronald E. (1974) *Global Reach: The Power of the Multinational Corporations*. Simon & Schuster, New York.

Bartlett, Richard (ed.) (1993a) *The Mabo Decision: Commentary by R.H. Bartlett and the Full Text of the Decision*. Butterworth, Sydney.

Bartlett, Richard (1993b) How native title at common law upsets the dominance of resource titles: resource security and Mabo, in *The Challenge of Resource Security: Law and Policy* (ed: Gardner, A.). Federation Press, Leichhardt NSW: 118–32.

—— (1995) Racism and the W.A. Government: The High Court decision. *Aboriginal Law Bulletin* 3(73), 8–9.

—— (1996) Dispossession by the National Native Title Tribunal. *Western Australian Law Review* 26, 108–37.

BBC Consultant Planners and Environmental Affairs Pty Ltd (1994) *Social Impact Assessment: A Report*. (Report prepared for the Review of Commonwealth Environmental Impact Assessment). Commonwealth Environment Protection Agency, Canberra.

Beattie, Scott (1997) Is mediation a real alternative to law? Pitfalls for Aboriginal participants. *Australian Dispute Resolution Journal* 8(1) 57–69.

Beauregard, Robert (1989) Between modernity and post-modernity: the ambiguous position of US planning. *Environment and Planning D: Society and Space* 7: 381–95.

Beckwith, Jo Ann (1994) Social impact assessment in Western Australia at a crossroads. *Impact Assessment* 12(2) 199–213.

Bell, Colin (1978) Studying the locally powerful, in Bell, Colin and Encel, Sol (eds) *Inside the Whale: Ten Personal Accounts of Social Research*. Pergamom, Sydney: 14–40.

Benedek, Emily (1993) *The Wind Won't Know Me: a History of the Navajo–Hopi Land Dispute*. Vintage, New York.

Benterrak, K., Muecke, S. and Roe, (1984) *Reading the Country: Introduction to Nomadology*. Fremantle Arts Centre Press, Fremantle

Berdoulay, V. (1978) The Vidal–Durkheim debate, in Ley, D. and Samuels, M. (eds) *Humanistic Geography: Prospects and Problems*. Croom Helm, London: 77–90.

—— (1989) Place, meaning, and discourse in French language geography, in Agnew, J. and Duncan, J. (eds) *The Power of Place: Bringing Together the Geographical and Sociological Imaginations*. Unwin Hyman, Boston: 124–39.

Berg, Lawrence; Fenge, Terry and Dearden, Philip (1993) The role of aboriginal peoples in national park designation, planning and management in Canada, in Dearden, Philip and Rollins, Rick (eds) *Parks and Protected Areas in Canada: Planning and Management.* Oxford University Press, Toronto: 225–55.

Berger, John (1973) *Ways of Seeing.* BBC and Penguin, London.

Berger, Thomas R. (1981) Public inquiries and environmental assessment, in Clark (ed.) *Environmental Assessment in Australia and Canada.* Westwater, Vancouver: 377–400

—— (1977) *Northern Frontier Northern Homeland: the Report of the Mackenzie Valley Pipeline Inquiry* (2 vols) James Lorimer, Toronto.

—— (1988) *Northern Frontier Northern Homeland: the Report of the Mackenzie Valley Pipeline Inquiry.* Revised edn. Douglas & McIntyre, Vancouver.

—— (1991) *A Long and Terrible Shadow: White Values, Native Rights in the Americas.* Douglas & McIntyre and the University of Washington Press, Vancouver and Seattle.

—— (1994) The independent review of the Sardar Sarovar projects 1991–1992. *Impact Assessment* 12(1) 3–18.

Bergman, Ingela (1991) Spatial structures in Saami cultural landscapes, in Kvist, Roger (ed.) *Readings in Saami History, Culture and Language II.* (Umeå University Center for Arctic Cultural Research, Miscellaneous Publications, 12.) University of Umeå Center for Arctic Cultural Research, Umeå, Sweden: 59–68.

Berkes, Fikret; George, Peter and Preston, Richard J. (1991) Co-management: the evolution in theory and practice of the joint administration of living resources. *Alternatives* 18(2) 12–18.

Bernard, H. Russell (1988) *Research Methods in Cultural Anthropology.* Sage, Newbury Park, CA.

Bhaba, Homi K. (1994) *The Location of Culture.* Routledge, London and New York.

BHP (1993) Navajo returns badlands to good use. *BHP Review,* 71(3) 19–21.

—— (1994) Building a future, keeping tradition. *BHP Review,* 72(1) 21–4.

—— (1995) BHP and Ok Tedi: are you getting the full picture? *Good Weekend* (*Sydney Morning Herald* and *Melbourne Age* supplement) 1999 (4 November).

Birch, Charles (1990) *On Purpose.* University of NSW Press, Sydney.

Birckhead, J; De Lacy, T. and Smith, L. (eds) (1992) *Aboriginal Involvement in Parks and Protected Areas.* Aboriginal Studies Press, Canberra

Bjorklund, Ivar (1991) Property in common, common property or private property? Norwegian fishery management in a Sami coastal area. *North Atlantic Studies* 3(1), 41–5.

Bjorklund, Ivar and Brantenberg, Terje (1981) Alta-Kautokeinoutbyggingen, reindrift og samisk kultur. *Ottar* 129, 16–31.

Blackburn, J. (1970) Milirrpum and others *v.* Nabalco Pty Ltd and another. *Federal Law Reports* 17(10) 141–294.

Blainey, G. (1963) *The Rush that Never Ended.* Melbourne University Press, Melbourne.

Blaut, James M. (1993) *The Colonizer's Model of the World: Geographical Diffusionism and Eurocentric History.* Guildford Press, New York and London.

Blay, J. (1984) *Part of the Scenery.* McPhee Gribble/Penguin, Fitzroy, Victoria.

Blishen, B.R.; Lockhart, A.; Craib, and Lockhart, E. (1979) *Socio-economic Impact Model for Northern Development.* (2 vols) Department of Indian and Northern Affairs, Ottawa.

Bodley, John H. (1990) *Victims of Progress*. 3rd edn. Mayfield, Palo Alto.

Boothroyd, Peter (1995) The need for retrospective impact assessment: the megaprojects example, in Gagnon, Christiane (ed.) *Evaluation des impacts sociaux: vers un développement viable?* Groupe de recherche et d'intervention régional, Université du Québec à Chicoutimi, Chicoutimi: 43–64.

Boothroyd, Peter; Knight, Nancy; Eberle, Margaret; Kawaguchi, June and Gagnon, Christiane (1995) The need for retrospective impact assessment: the megaprojects example. *Impact Assessment* 13(3) 253–71.

Borring, Jan; Hveem, Britt; Lindal, Asmund and Sunde, Helge (1981) *ALTA-bilde/ ALTA pictures: 12 Years' Struggle for the Alta-Kautokeino Watercourse*. PAX forlag a.s., Oslo.

Bouma, Gary D. (1996) *The Research Process*. 3rd edn. Oxford University Press, Melbourne.

Bradley, John (1988) *Yanyuwa Country: The Yanyuwa people of Borroloola Tell the History of Their Land*. Greenhouse Publications, Richmond, Vic.

Branch, K.; Hooper, D.A; Thompson, J. and Creighton, J. (1984) *Guide to Social Assessment: a Framework for Assessing Social Change*. Westview Press, Boulder, CO.

Brantenberg, Odd Terje (1985) The Alta-Kautokeino Conflict, Saami reindeer herding and ethnopolitics, in Brosted, J.; Dahl, J.; Gray, A.; Gullov, H. C.; Henriksen, G.; Jorgensen, J. B. and Kleivan, I. (eds) *Native Power*. Universitetsforlaget, Bergen: 23–48.

—— (1991) Norway: constructing indigenous self-government in a nation state, in Jull, Peter and Roberts, Sally (eds) *The Challenge of Northern Regions*. North Australian Research Unit, Darwin: 66–128.

Brascoupé, Simon (1997) Strengthening traditional economies and perspectives, in *Sharing the Harvest: The Road to Self-Reliance*. Report of the National Round Table on Aboriginal Economic Development and Resources for the Royal Commission on Aboriginal Peoples (For Seven Generations/Pour sept générations). CD-ROM. Libraxus, Ontario.

Bright, April (1994) Burn grass, in Rose, Deborah Bird (ed.) *Country in Flames: Proceedings of the 1994 Symposium on Biodiversity and Fire in North Australia*. Department of the Environment, Sport and Territories and North Australia Research Unit, Canberra and Darwin: 59–62.

Broadbent, Noel D. (1989) The Chernobyl accident and reindeer herding in Sweden, in Broadbent, Noel D. (ed.) *Readings in Saami History, Culture and Language*. (Umea University Center for Arctic Cultural Research, Miscellaneous Publications, 7.) Umea University Center for Arctic Research, Umea, Sweden: 127–42.

Brody, Hugh (1981) *Maps and Dreams: Indians and the British Columbia Frontier*. Douglas & McIntyre, Vancouver.

Brösted, Jens (1987) Sámi rights and self-determination, in IWGIA (ed.) *Self Determination and Indigenous Peoples: Sámi Rights in Northern Perspectives*. (IWGIA Document 58). International Work Group for Indigenous Affairs, Copenhagen, 155–72.

Brower, David (1978) Wilderness and the wilderness within us. *Habitat International* 5(6)

Brown, A.J. (ed.) (1992) *Keeping the Land Alive: Aboriginal Peoples and Wilderness Protection in Australia*. Environmental Defender's Office, Wilderness Society, Sydney.

Bryan, R. (1987) The state and the internationalisation of capital. *Journal of Contemporary Asia* 17: 253–75.

Bryson, I. and Howitt, R. (1994) *A Preliminary Bibliography on Mabo and Native Title*. Unpublished paper, Macquarie University.

Buckley, K. and Wheelwright, T. (1988) *No Paradise For Workers*. Oxford University Press, Melbourne.

Builders' Labourers' Federation (1975) *Builders' Labourers' Song Book*. Australian Building Construction and Employees' and Builders' Labourers' Federation, Melbourne.

Burdge, Rabel J (1994a) Community needs assessment and techniques, in Burdge, Rabel J. (ed.) *A Conceptual Approach to Social Impact Assessment: Collection of Writings by Rabel J Burdge and Colleagues*. Social Ecology Press, Middleton, Wisconsin: 197–212.

—— (1994b) Needs assessment surveys for decision makers, in Burdge, Rabel J. (ed.) *A Conceptual Approach to Social Impact Assessment: Collection of Writings by Rabel J Burdge and Colleagues*. Social Ecology Press, Middleton, Wisconsin: 187–95.

Burdge, Rabel J and Vanclay, Frank (1996) Social impact assessment: a contribution to the state-of-the-art series. *Impact Assessment* 14(1) 59–86.

Burger, Julian (1990) *Gaia Atlas of First Peoples: a Future for the Indigenous World*. Penguin, Harmondsworth.

Butler, Judith (1998) Merely cultural. *New Left Review* 227: 33–44.

Campbell, Petra (1990) *Death of the Soviet Amazon*. One TV Productions. 55 minutes [videorecording].

—— (1991) Doomed deerherders of the Soviet Amazon. *Sydney Morning Herald* (12 January) 32.

Canadian Aboriginal Minerals Association (CAMA) (1994) *Aboriginal Community and Mineral Industry Perspectives: Mineral Business Opportunities –Issues and Recommendations*. CAMA, Ottawa.

Canadian Arctic Resources Committee (1993) Creating Nunavut and breaking the mold of the past. *Northern Perspectives* (Special issue on Nunavut Agreement) 21(3) 1–19.

Canby, William C., Jr (1988) *American Indian Law in a Nutshell*. West Publishing, St Paul, MN.

Cant, Garth; Overton, John and Pawson, Eric (eds) (1993) *Indigenous Land Rights in Commonwealth Countries: Dispossession, Negotiation and Community Action*. Department of Geography, University of Canterbury and the Ngai Tahu Maori Trust Board for the Commonwealth Geographical Bureau, Christchurch.

Carter, Novia (1981) SIA: new wine in old bottles, in Tester and Mykes (eds) *SIA: Theory, Method and Practice*. Detselig, Calgary: 5–12.

Castillon, David A. (1992) *Conservation of Natural Resources: a Resource Management Approach*. William C. Brown, Dubuque, IA.

Caulfield, R.A. (1992) Alaska's subsistence management regimes, *Polar Record*, 28 (164) 23–32.

Chaliand, G. (ed.) (1989) *Minority Peoples in the Age of Nation States*. Pluto Press, London.

Chandrakanth, M.G. and Romm, Jeff (1991) Sacred forests, secular forest policies and people's actions. *Natural Resources Journal* 31(4) 741–56.

Charest, Paul (1982) Hydroelectric dam construction and the foraging activities of eastern Quebec Montagnais, in Leacock, Eleanor and Lee, Richard (ed.) *Politics and History in Band Societies.* Cambridge University Press and Editions de la Maison des Sciences de l'Homme, Cambridge and Paris: 413–26.

Charta 79 (ed.) (1982) *The Sami People and Human Rights.* Charta 79, Oslo.

Chase, Athol (1990) Anthropology and impact assessment: development pressures and indigenous interests in Australia. *Environmental Impact Assessment Review* 10: 11–23.

Chomsky, Noam (1993) *Year 501: the Conquest Continues.* Verso, London.

—— (1994) *World Orders Old and New.* Columbia University Press, New York.

Christensen, Will (1981) *The Wangaya Way: Tradition and Change in a Reserve Setting.* PhD dissertation, Department of Anthropology, University of Western Australia, Nedlands.

—— (1990) Aborigines and the Argyle Diamond Project, in Dixon, Rod A. and Dillon, Michael C. (eds) *Aborigines and Diamond Mining: The Politics of Resource Development in the East Kimberley Western Australia.* University of Western Australia Press, Nedlands: 29–39.

Christie, Michael (1990) Aboriginal science for the ecologically sustainable future. *Batchelor Journal of Aboriginal Education* November: 56–68.

—— (1992) Grounded and ex-centric knowledges: exploring Aboriginal alternatives to Western thinking. Paper presented to the Conference on Thinking, Townsville, July 1992.

Christie, Michael and Perrett, Bill (1996) Negotiating resources: language, knowledge and the search for 'secret English' in northeast Arnhem Land, in Howitt, Richard; Connell, John and Hirsch, Philip (eds) *Resources, Nations and Indigenous Peoples: Case Studies from Australasia, Melanesia and Southeast Asia.* Oxford University Press, Melbourne: 57–65.

Christophersen, Christine and Langton, Marcia (1995a) Allarda! *Arena Magazine* June–July: 28–32.

—— (1995b) Ranger's contaminated water: downstream Aborigines say 'NO'. Edited version of a paper presented to the Australian Conservation Foundation Women and the Environment Conference, Melbourne, March 1995.

Christophersen, S. (1989) On being outside 'The Project'. *Antipode* 21 (2) 83–9.

Churchill, Ward (1992) The earth is our mother: struggles for American Indian land and liberation in the contemporary Univted States, in Jaimes, M.A. (ed.) *The State of Native America: Genocide, Colonization and Resistance.* South End Press, Boston: 139–88.

—— (1995) *Since Predator Came: Notes from the Struggle for American Indian Liberation.* Aigis Publications, Littleton, CO.

Churchill, Ward (ed.) (1988) *Critical Issues in Native North America.* (IWGIA Document 62). International Work Group on Indigenous Affairs, Copenhagen.

Ciacca, John (1988) The practical and philosophical stakes, in Vincent, Sylvie and Bowers, Garry (eds) *Baie James et Nord Québécois: Dix Ans Après/James Bay and Northern Québec: Ten Years After.* Recherches amérindiennes au québec, Montréal: 39–41.

Cobo, Jose R. Martinez (1986) *Study of the Problem of Discrimination Against Indigenous Populations.* United Nations Document E/CN.4/Sub.2/1986/7.

Coggins, George Cameron and Donley, James (1996) Natural resources development on Native American Indian reservations in the United States, in Meyers, Gary D. (ed.) *The Way Forward: Collaboration and Co-operation 'in Country'*. 2nd edn. National Native Title Tribunal, Perth: 90–101.

Cohen, Barri (1994) Technological colonialism and the politics of water. *Cultural Studies* 8(1) 32–55.

Committee on Guidelines and Principles for Social Impact Assessment (1994) *Guidelines and Principles for Social Impact Assessment*. (NOAA Technical Memorandum NMFS-F/SPO–16 May 1994). US Department of Commerce, National Oceanic and Atmospheric Administration and National Marine Fisheries Service, Washington DC.

Conacher, Arthur (1988) Resource development and environmental stress: EIA and beyond in Australia and Canada. *Geoforum* 19(3) 339–52.

Connell, John (1991) Compensation and conflict: the Bougainville Copper Mine, Papua New Guinea, in Connell, J. and Howitt, R. (eds) *Mining and Indigenous Peoples in Australasia*. Sydney University Press, Sydney: 54–75.

Connell, John and Howitt, Richard (1991a) Mining, dispossession and development, in Connell, J. and Howitt, R. (eds) *Mining and Indigenous Peoples in Australasia*, Sydney University Press, Sydney: 1–17.

Connell, John and Howitt, Richard (eds) (1991b) *Mining and Indigenous Peoples in Australasia*. Sydney University Press, Sydney.

Conrad, Joseph (1955 [1926]) Geography and some explorers, in *Tales of Hearsay and Last Essays: Collected works of Joseph Conrad* vol. 18. Dent & Sons, London, 1–21.

—— (1995 [1917]) *Heart of Darkness*. Penguin, Harmondsworth.

Cook, E. (1982) The consumer as creator: a criticism of faith in limitless ingenuity. *Energy Exploration and Exploitation* 1: 189–201

Coombs, H.C.; McCann, H.; Ross, H. and Williams, N.M. (eds) (1989) *Land of Promises: Aborigines and Development in the East Kimberley*. Centre for Environmental Studies, ANU and Aboriginal Studies Press, Canberra.

Cosgrove, D. (1978) Place, landscape and the dialectics of cultural geography. *Canadian Geographer* 22 (1) 66–72

—— (1983) Towards a radical cultural geography: problems of theory. *Antipode* 15(1) 1–11.

—— (1987) New directions in cultural geography. *Area* 19(2) 95–101.

—— (1992) Orders and a new world: cultural geography 1990–91. *Progress in Human Geography* 16(2) 272–80.

Council for Aboriginal Reconciliation (1993) *Exploring for Common Ground: Aboriginal Reconciliation and the Australian Mining Industry*. Australian Government Publishing Service, Canberra.

Coveney, Peter and Highfield, Roger (1995) *Frontiers of Complexity: the Search for Order in a Chaotic World*. Faber & Faber, London.

Cox, Gary (1994) *Better Communities Through Social Impact Assessment*. Best Practice Paper No. 4. NSW Office on Social Policy, Sydney.

CRA Ltd (1997) *Aboriginal People and Mining*. CRA Ltd Discussion Paper, CRA, Melbourne, February 1997.

Crabb, Peter (1984) Hydroelectric power in Newfoundland, Tasmania and the South Island of New Zealand, in Goldsmith, Edward and Hilyard, Nicholas (eds) *The Social and Environmental Effects of Large Dams, Volume 2: Case Studies.* Wadebridge Ecological Centre, Camelford, UK: 55–68.

Craig, Donna (1988) SIA and policy making: the relationship between SIA, EIA and the decision process, in Hindmarsh, R.A.; Hundloe, T.J.; McDonald, G.T. and Rickson, R.E. (eds) *Papers on Assessing the Social Impacts of Development.* Institute of Applied Environmental Research, Griffith University, Brisbane: 60–77.

—— (1990) Social impact assessment: politically oriented approaches and applications. *Environmental Impact Assessment Review* 10: 37–54.

Craig, Ehrlich; Ross, Helen and Lane, Marcus (1996) *Northern Land Council: Indigenous Participation in Commonwealth Environmental Impact Assessment.* Environmental Protection Agency, Canberra.

Cramér, Tomas (1994) Saami rights cleansing in Scandanavia. *Indigenous Affairs* 1994/4: 52–5.

Crang, (1992) The politics of polyphony: reconfigurations in geographical authority. *Environment and Planning D: Society and Space* 10: 527–49.

Crawford, James (ed.) (1988) *The Rights of Peoples.* Clarendon Press, Oxford.

Creagh, Carson (1992) Looking after the land at Uluru. *Ecos* 71(Autumn) 6–13.

Crengle, Diane (1993) *Taking into Account the Principles of the Treaty of Waitangi: Ideas for Implementation of Section 8 Resource Management Act 1991.* New Zealand Ministry for the Environment, Wellington.

Crough, Greg and Christophersen, Christine (1993) *Aboriginal People in the Economy of the Kimberley Region.* North Australia Research Unit, Australian National University, Darwin.

Cummings, Barbara J. (1990) *Dam the Rivers, Damn the People.* Earthscan Publications, London.

d'Abbs, Peter and Jones, Trish (1996) Gunbang ... or ceremonies? Combating alcohol misuse in the Kakadu/West Arnhem region. A report prepared for the Gunbang Action Committee by the Menzies School of Health Research, July.

Dahl, Jens (1992) Development of indigenous and circumpolar peoples' rights, in Lyck, Lise (ed.) *Nordic Arctic Research on Contemporary Problems.* Nordic Arctic Research Forum, Copenhagen: 183–9.

Dale, Allan (1991) *Aboriginal Access to Land Management Funding and Services – Case Studies: Kowanyama, Aurukun, Woorabinda and Trelawney.* Australian National Parks and Wildlife Service, Canberra.

Dale, Allan and Lane, Marcus (1995) Queensland's social impact assessment unit: its origins and prospects. *Queensland Planner* 35(3) 5–10.

Dale, Allan; Chapman, Peter and McDonald, Morag (1997) Social impact assessment in Queensland: why practice lags behind legislative opportunity. *Impact Assessment* 15(2) 159–79.

DASET (Department of the Arts, Sport, the Environment and Territories) (1991) *Australian National Report to the United Nations Conference on Environment and Development.* AGPS, Canberra.

Davidson, R. (1987) *Travelling Light.* Methuen, Sydney.

Davies, Jocelyn (1991) *Aboriginal Access to Funds and Services for Land Management – Case Studies: West Coast and Eyre Peninsula, South Australia.* Australian National Parks and Wildlife Service, Canberra.

Davis, Leon (1996) *Redefining Australia*. Luncheon address to Australian Business in Europe, August 1996.

Davis, Shelton and Mathews, R (eds.) (1976) *The Geological Imperative: anthropology and development in the Amazon Basin of South America*. Anthropology Resource Center, Boston.

Davis, Steven and Prescott, Victor (1992) *Aboriginal Frontiers and Boundaries*. Melbourne University Press, Melbourne.

De Walt, B. and Pelto, A. (eds) (1985) *Micro and Macro Levels of Analysis in Anthropology: Issues in Theory and Research*. Westview, Boulder, CO.

Dearden, Philip and Berg, Lawrence (1993) Canada's national parks: a model of administrative penetration. *Canadian Geographer* 37(3) 194–211.

Delaney, David and Leitner, Helga (1997) The political construction of scale. *Political Geography* 16(2) 93–7.

Deloria, Vine Jr (1988) *Custer Died for Your Sins: an Indian Manifesto*. Revised edn. University of Oklahoma Press, Norman and London.

Denniston, Derek (1994) Defending the land with maps. *World Watch* January/ February: 27–31.

Denzin, Norman K. and Lincoln, Yvonna S. (eds) (1998) *The Landscape of Qualitative Research: Theories and Issues*. Sage, Thousand Oaks, CA.

Deurden, Frank and Kuhn, Richard G. (1996) The application of geographic information systems by First Nations and Government in Northern Canada. *Cartographica* 33(2) 49–62.

Deutsche, R. (1991) Boys town. *Environment and Planning D: Society and Space* 9: 5–30.

Diamond, Albert (1988) The costs of implementing the Agreements, in Vincent, Sylvie and Bowers, Garry (eds) *Baie James et Nord Québécois: Dix Ans Après/James Bay and Northern Québec: Ten Years After*. Recherches amérindiennes au québec, Montréal: 115–18.

Dicken, Peter (1998) *Global Shift: Transforming the World Economy*. 3rd edn. Paul Chapman, London.

Dillon, Michael (1990) Social impact at Argyle: genesis of a public policy, in Dixon, Rod and Dillon, Michael (eds) *Aborigines and Diamond Mining: the Politics of Resource Development in the East Kimberley Western Australia*. University of Western Australia Press, Nedlands: 130–54.

—— (1991) Interpreting Argyle: Aborigines and diamond mining in northwest Australia, in Connell, John and Howitt, Richard (eds) *Mining and Indigenous Peoples in Australasia*. University of Sydney Press, Sydney: 138–52.

Dixon, Rod and Dillon, Michael (eds) (1990) *Aborigines and Diamond Mining: the Politics of Resource Development in the East Kimberley, Western Australia*. University of Western Australia Press, Nedlands.

Dodson, Michael (1994a) The end of the beginning: re(de)finding Aboriginality. The Wentworth Lecture. *Australian Aboriginal Studies* 1994(1) 2–13.

—— (1994b) Towards the exercise of indigenous rights: policy, power and self-determination. *Race and Class* 35(4) 65–76.

—— (1995) *Aboriginal and Torres Strait Islander Social Justice Commissioner: Native Title Report July 1994–June 1995*. Human Rights and Equal Opportunity Commission, Sydney.

—— (1996) *Aboriginal and Torres Strait Islander Social Justice Commissioner: Native Title Report July 1995–June 1996*. Human Rights and Equal Opportunity Commission, Sydney.

Dodson, Patrick (1991) *Royal Commission into Aboriginal Deaths in Custody: Background Issues in Western Australia Report* (2 vols) AGPS, Canberra.

Dodson, Patrick; Altman, Jon; Yunupingu, Galarrwuy; Cooper, Victor; Hicks, John; Jones, Neville; Johnston, Arthur; Jones, Mandy; Jackson, Andrew; Roeger, Steve; Scott, Doug and Carbon, Barry (1997) *Kakadu Region Social Impact Study: Report of the Study Advisory Group*. Supervising Scientist, Canberra.

Donovan, F. (1986) *An Assessment of the Social Impact of Argyle Diamond Mines on the East Kimberley Region*, East Kimberley Working Paper 11, Centre for Resources and Environmental Studies, ANU, Canberra.

Dove, J.; Miriung, T. and Togolo, M. (1974) Mining bitterness, in Sack, Peter G. (ed.) *Problem of Choice: Land in Papua New Guinea's Future*. Australian National University Press, Canberra: 181–9.

Dowling, Robyn (2000) Power, subjectivity and ethics in qualitative research, in Hay, Iain (ed.) *Qualitative Research Methods in Geography*. Oxford University Press, Melbourne: 23–49.

Downie, Paul (1990) Far East a potential bonanza. *Australian* (12 February) 17.

Drache, Daniel and Perin, Roberto (eds) (1992) *Negotiating with a Sovereign Québec*. James Lorimer, Toronto.

Driver, Felix (1992) Geography's empire: histories of geographical knowledge. *Environment and Planning D: Society and Space* 10: 23–40.

du Cros, Dominique (1996) Regaining ground: Aboriginal land management in the Upper Hunter Valley. BSc (Hons) Thesis, School of Earth Sciences, Macquarie University, Sydney.

Duff, Alan (1990) *Once Were Warriors*. Tandem Press, Auckland.

Duncan, J. and Duncan, N. (1988) (Re)reading the landscape, *Environment and Planning D: Society and Space*, 6: 117–26.

Duncan, James and Ley, David (eds) (1993) *Place/Culture/Representation*. Routledge, London and New York.

Dwyer, Peter (1994) Modern conservation and indigenous peoples: in search of wisdom. *Pacific Conservation Biology* 1: 91–7.

Eidheim, Harald (1985) Indigenous peoples and the State: the Saami case in Norway, in Brosted, Jens; Dahl, Jens; Gray, Andrew; Gullov, Hans Christian; Henriksen, Georg; Jorgensen, Jorgen B. and Kleivan, Inge (eds) *Native Power*. Universitetsforlaget, Oslo: 155–71.

Ekins, (1992) *A New World Order: Grassroots Movements for Global Change*. Routledge, London.

Ekins, Paul and Max-Neef, Manfred (eds) (1992) *Real-life Economics: Understanding Wealth Creation*. Routledge, London and New York.

Ellis, Elisabeth (1998) Reaching Out/Reaching In/Reaching Across: Identity and the spatial politics of disablement and enablement. BA(Hons) thesis, School of Earth Sciences, Macquarie University, Sydney.

Escobar, Arturo (1992a) Imagining a post-development era? Critical thought, development and social movements. *Social Text* 31/32: 20–56.

—— (1992b) Planning, in Sachs, Wolfgang (ed.) *The Development Dictionary*. Zed Press, London: 132–45.

ESD Working Groups. (1991) *Final Report – Executive Summaries*. AGPS, Canberra.

Evaluating Committee, Kativik Environmental Quality Commission, Federal Review Committee North of the 55th Parallel and Federal Environmental Assessment Review Panel. (1992) *Guidelines: Environmental Impact Statement for the Proposed Great Whale River Hydro-electric Project.* Great Whale Public Review Support Office, Montréal.

Fagan, Robert H. and Bryan Richard (1991) Australia and the changing global economy: background to social inequality in the 1980s, in O'Leary, J. and Sharp, R. (eds) *Inequality in Australia: Slicing the Cake.* Heinemann, Melbourne: 7–31.

Fagan, Robert H. and Webber, Michael (1994) *Global Restructuring: the Australian Experience.* Oxford University Press, Melbourne.

Farley, Rick; McRae, Tony and Lane, Patricia (1997) Outlook for regional development: opportunities for agreements. Paper presented at the North Australia Regional Outlook Conference, Darwin, September 1997. URL: http://www.nntt.gov.au/4825642b003b24e5/

Fearnside, Philip M. (1994) The Canadian feasibility study of the Three Gorges Dam proposed for China's Yangzi River: a grave embarrassment to the impact assessment profession. *Impact Assessment* 12(1) 21–57.

Feit, Harvey (1985) Legitimation and autonomy in James Bay Cree responses to hydro-electric development, in Dyck, Noel (ed.) *Indigenous Peoples and the Nation-State: 'Fourth World' Politics in Canada, Australia and Norway.* Social and Economic Papers ed. Vol. 14. Institute of Social and Economic Research, Memorial University of Newfoundland, St Johns: 27–66.

Feit, Harvey A (1986) Hunting and the quest for power: the James Bay Cree and Whitemen in the Twentieth Century, in Morrison, R. Bruce and Wilson, C. Roderick (eds) *Native Peoples: the Canadian experience.* McClelland and Stewart, Toronto: 171–207.

Feitelson, E. (1991) Sharing the globe: the role of attachment to place. *Global Environmental Change* 1: 396–406.

Fenge, Terry (1993) National parks in the Canadian Arctic: the case of the Nunavut Land Claim Agreement, in Cant, Garth; Overton, John and Pawson, Eric (eds) *Indigenous Land Rights in Commonwealth Countries.* Proceedings of a Commonwealth Geographical Bureau Workshop held in Christchurch, February 1992: Department of Geography and Ngai Tahu Maori Trust Board, Christchurch: 195–207.

Filer, Colin (1990) The Bougainville Rebellion, the mining industry and the process of social disintegration in Papua New Guinea. *Canberra Anthropology* 13(1) 1–39.

—— (1993) The policy and methodology of social impact mitigation in the mining industry. Paper presented to the 20th Waigani Seminar, University of PNG, Port Moresby, September 1993.

Finlayson, Julie and Dale, Alan (1996) Negotiating indigenous self-determination at the regional level: experiences with regional planning, in Sullivan, Patrick (ed.) *Shooting the Banker: Essays on ATSIC and Self-Determination.* North Australia Research Unit, Darwin: 70–88.

Finsterbusch, Kurt (1995) In praise of SIA – a personal review of the field of social impact assessment: feasibility, justification, history, methods, issues. *Impact Assessment* 13(3) 229–52.

Firey, W. (1960) *Man, Mind and Land: a Theory of Resource Use.* Free Press, Glencoe, IL.

Fisher, R. and Ury, W. (1991) *Getting to Yes: Negotiating an Agreement Without Giving In*. 2nd edn. Business Books, London.

Fisk, Milton (1993) Poststructuralism, difference, and Marxism. *Praxis International* 12(4) 323–40.

Fitzgerald, T. (1974) *The Contribution of the Mineral Industry to Australian Welfare*. AGPS, Canberra.

Fitzsimmons, Margaret (1989) The matter of nature. *Antipode* 21(2) 106–20.

Fjellstrøm, Phebe (1987) Cultural- and traditional-ecological perspectives in Saami religion, in Ahlback, Tore (ed.) *Saami Religion*. Donner Insititute in Religious and Cultural History, Åbo, Finland: 34–45.

Flannery, Tim (1994) *The Future Eaters: an Ecological History of the Australasian Lands and People*. Reed, Melbourne.

Fleming, Marie (ed.) (1988) *The Geography of Freedom: the Odyssey of Elisee Reclus*. Black Rose, Montréal and New York.

Fleras, Augie and Elliot, Jean Leonard (1992) *The 'Nations Within': Aboriginal-State Relations in Canada, the United States, and New Zealand*. Oxford University Press, Toronto.

Folch-Serra, Mireya. (1993) David Harvey and his critics: the clash with disenchanted women and post-modern discontents. *Canadian Geographer* 37(2) 176–84.

Formby, John (1977) *Environmental Impact Assessment Procedures in Australia*, CRES Working Paper, CRES, ANU, Canberra.

—— (1987) *The Australian Government's Experience with Environmental Impact Assessment*, CRES, ANU, Canberra.

Fothergill, Anthony (1992) Of Conrad, cannibals, and kin, in Gidley, Mick (ed.) *Representing Others: White Views of Indigenous Peoples*. University of Exeter Press, Exeter, 37–59.

Foucault, Michel (1980a) Questions on geography, in Gordon, Colin (ed.) *Michel Foucault: Power/Knowledge*. Harvester Press, Brighton, Victoria: 63–77.

—— (1980b) Truth/Power, in Gordon, Colin (ed.) *Michel Foucault: Power/Knowledge*. Harvester Press, Brighton, Victoria.

Fox, J. (1992) The problem of scale in community resource management. *Environmental Management* 16 (3) 289–97

Fox, R. W.; Kelleher, G.G. and Kerr, C.B. (1977a) *Ranger Uranium Environmental Inquiry First Report*. AGPS, Canberra.

—— (1977b) *Ranger Uranium Environmental Inquiry Second Report*. AGPS, Canberra.

Fraser, Nancy (1989) Foucault on power, in *Unruly Practices: Power, Discourse and Gender in Contemporary Social Theory*. University of Minnesota Press, Minneapolis: 17–34.

—— (1995) From redistribution to recognition? Dilemmas of justice in a 'post-socialist' age. *New Left Review* 212: 68–93.

—— (1997a) *Justice Interruptus: Critical Reflections on the 'Post-Socialist' Condition*. Routledge, New York and London.

—— (1997b) A rejoinder to Iris Young. *New Left Review* 223: 126–9.

—— (1998) Heterosexism, misrecognition and capitalism: a response to Judith Butler. *New Left Review* 228: 140–49.

Freire, Paulo (1972a) *Pedagogy of the Oppressed*. Penguin, Harmondsworth.

—— (1972b) *Cultural Action for Freedom*. Penguin, Harmondsworth.

—— (1976) *Education: The Practice of Freedom*. Writers & Readers Cooperative, London.

Fundación Sabiduría Indígena (FSI) and Kothari, Brij (1997) Rights to the benefits of research: compensating indigenous peoples for their intellectual contribution. *Human Organization* 56(2) 127–37.

Funk, Ray (1985) The Mackenzie Valley Pipeline Inquiry in retrospect, in Derman, W. and Whiteford, S. (eds) *Social Impact Analysis and Development Planning in the Third World*. Westview Press, Boulder, CO: 119–32.

Gagnon, Christine; Hirsch, Phillip and Howitt, Richard (1993) Can SIA empower communities? *Environmental Impact Assessment Review* 13: 229–53.

Gale, Fay (1983) Kakadu National Park: tourist paradise or environmental conflict? *Proceedings of the Royal Geographical Society of Australasia, South Australian Branch* 83: 1–11.

Gale, Richard and Miller, Marc L. (1985) Professional and public resource management arenas: forests and marine fisheries. *Environment and Behaviour* 17(6) 651–78.

Galligan, Brian and Lynch, Georgina (1992) Integrating conservation and development: Australia's Resource Assessment Commission and the testing case of Coronation Hill, in Moffatt, Ian and Webb, Ann (eds) *Conservation and Development Issues in North Australia*. North Australia Research Unit, Darwin: 239–49.

Galtung, Johan (1973) *The European Community: an Emerging Superpower*. Allen & Unwin, London.

—— (1980) *The True Worlds: a Transnational Perspective*. Free Press, New York.

—— (1989) The fourth political wave. *Social Alternatives* 8(1) 31–3.

Gamble, Donald J. (1978) The Berger Inquiry: an impact assessment process. *Science* 199(3 March), 946–952.

—— (1986) Crushing of cultures: western applied science in northern societies. *Arctic* 39(1) 20–23.

Gardner, Alex (ed.) (1993) *The Challenge of Resource Security: Law and Policy*. Federation Press, Leichhardt, NSW.

Gardner, J. and Roseland, M. (1989a) Thinking globally: the role of sustainability in sustainable development. *Alternatives,* 16.3: 26–35.

Gardner, J. and Roseland, M. (1989b) Acting locally: community strategies for equitable sustainable development, *Alternatives* 16(3) 36–48.

Gedicks, Al (1982) The new resource wars. *Raw Materials Report* 1(2) 8–13.

—— (1993) *The New Resources Wars: Native and Environmental Struggles against Multinational Corporations*. Black Rose, Montréal.

Geisler, C.G.; Green, R.; Usner, D. and West, R. (eds) (1982) *Indian SIA: the Social Impact Assessment of Rapid Resource Development on Native Peoples*. Natural Resources Research Laboratory, University of Michigan.

Gelder, Ken and Jacobs, Jane (1998) *Uncanny Australia: Sacredness and Identity in a Post-Colonial Nation*. Melbourne University Press, Melbourne.

Gerritsen, Rolf and MacIntyre, Martha (1991) Dilemmas of distribution: the Misima gold mine, Papua New Guinea, in Connell, John and Howitt, Richard (eds) *Mining and Indigenous Peoples in Australasia*. Sydney University Press, Sydney: 34–53.

Gibbs, Leah (1999) Gold Country: colonialism and mining in the Eastern Goldfields, Western Australia. BSc (Hons) thesis, Department of Human Geography, Macquarie University, Sydney.

Gibbs, Nici (1994) *Enabling Sustainable Communities: A Strategic Policy Paper*. New Zealand Ministry for the Environment, Wellington.

Gibson, Katherine (1984) Industrial reorgansiation and coal production in Australia 1860–1982). *Australian Geographical Studies* 22: 221–42.

—— (1990a) *'Hewers of Cake and Drawers of Tea': Women on the Central Queensland Coalfields.* ERRRU Working Paper No. 2. Economic and Regional Restructuring Research Unit, University of Sydney

—— (1990b) Australian coal in the global context: a paradox of efficiency and crisis. *Environment and Planning A* 22: 2629–46.

Gibson, Katherine D and Horvath, Ronald J (1983) Aspects of a theory of transition within the capitalist mode of production. *Environment and Planning D: Society and Space* 1, 121–138.

Gibson, Katherine and Horvath, Ronald (1984) Abstraction in Marx's method. *Antipode* 16.1: 12–25.

Gibson-Graham, J. K. (1996) *The End of Capitalism (As We Knew It); A Feminist Critique of Political Economy.* Blackwell, Cambridge.

Gibson-Graham, J. K.; Resnick, Stephen A. and Wolff, Richard D. (eds.) (2000) *Class and Its Others.* University of Minnesota Press, Minneapolis/London.

Godoy, R. (1984) Mining: anthropological perspectives. *Annual Review of Anthropology* 14: 199–217.

Gómez-Peña, G. (1992) The New World (B)order: a work in progress. *Third Text* 21: 71–9.

Gondolf, E.W. and S.R. Wells (1986) Empowered native community: modified SIA. *Environmental Impact Assessment Review* 6: 373–83.

Gooding, Susan S. (1997) Imagined spaces – storied places: a case study of the Colville Tribes and the evolution of treaty fishing rights. *Droits et Cultures* 33(1) 53–95.

Gorrie, Peter (1990) The James Bay power project. *Canadian Geographic* 110(1), 20–31.

Graham, Julie (1988) Post-modernism and Marxism. *Antipode* 20(1) 60–66.

—— (1990) Theory and essentialism in Marxist geography. *Antipode* 22(1) 53–66.

—— (1991) Fordism/post-Fordism: the second cultural divide? *Rethinking Marxism* 4(1) 39–58.

—— (1992) Anti-essentialism and overdetermination – a response to Dick Peet. *Antipode* 24(2) 141–56.

Graham, Julie and St Martin, Kevin (1990) Knowledge and the 'localities' debate: meditations on a theme by Cox and Mair. *Antipode* 22(2) 168–74.

Grand Council of the Crees (1995) *Sovereign Injustice: Forcible Inclusion of the James Bay Crees and Cree Territory into a Sovereign Québec.* Grand Council of the Crees, Nemaska, Québec.

Gray, Andrew (1991) *Between the Spice of Life and the Melting Pot: Biodiversity Conservation and its Impact on Indigenous Peoples.* (IWGIA Document 70). International Working Group on Indigenous Affairs, Copenhagen.

Gregory, R. (1976) Some implications of rapid growth of the mineral sector. *Australian Journal of Agricultural Economics* 20(2) 71–91.

Griffin, David R. and Falk, Richard (eds) (1993) *Post-modern Politics for a Planet in Crisis: Policy, Process and Presidential Vision.* State University of New York Press, New York.

Hadjimichalis, Costis (1994) Global-local social conflicts: examples from Southern Europe, in Amin, Ash and Thrift, Nigel (eds) *Globalization, Institutions, and Regional Development in Europe.* Oxford University Press, Oxford: 239–56.

Haggett, Peter (1965) *Locational Analysis in Human Geography*. Edward Arnold, London.

Hardy, Frank (1968) *The Unlucky Australians*. Thomas Nelson Limited, Melbourne.

Harman, Elizabeth J. (1981) Ideology and mineral development in Western Australia 1960–1980 (Chapter 9), in Harman, E.J. and Head, B. (eds) *State Capital and Resources in the North and West of Australia*. University of Western Australia Press, Perth: 167–96.

Harman, W. (1989) Reclaiming traditional wisdom for the needs of modern society. Paper presented to the Indigenous Science Conference, University of Calgary, July 1989.

Harvey, David (1968) *Explanation in Geography*.

—— (1973) *Social Justice and the City* (recently reissued).

—— (1974) Population, resources and the ideology of science. *Economic Geography* 50(3) 256–77.

—— (1982) *The Limits to Capital*. Blackwells, Oxford.

—— (1984) On the history and present condition of geography: an historical materialist manifesto. *Professional Geographer* 36(1) 1–11.

—— (1985) *The Urbanization of Capital: Studies in the History and Theory of Capitalist Urbanization*. The Johns Hopkins University Press, Baltimore.

—— (1989) *The Condition of Post-modernity*, Blackwells, Oxford

—— (1992) Social justice, post-modernism and the city. *International Journal of Urban and Regional Research* 16(4) 588–601.

—— (1993a) From space to place and back again: reflections on the condition of postmodernity, in Bird, Jon; Curtis, Barry; Putnam, Tim; Robertson, George and Tickner, Lisa (eds) *Mapping the Futures: Local Cultures, Global Change*. Routledge, London: 3–29.

—— (1993b) The nature of environment: the dialectics of social and environmental change, in Miliband, Ralph and Panitch, Leo (eds) *Socialist Register 1993: Real Problems, False Solutions*. Merlin Press, London: 1–51.

—— (1996) *Justice, Nature and the Geography of Difference*. Blackwells, Oxford.

—— (1997) Considerations on the environment of justice. Paper presented to the Conference on Environmental Justice: Global Ethics for the 21st Century, University of Melbourne, 1–3 October 1997.

Harvey, Nick (1998) *Environmental Impact Assessment: Procedures, Practice, and Prospects in Australia*. Oxford University Press, Melbourne.

Harwood, Alison (1997) Putting protest in its place: power, protest and participation – Gulf Aboriginal people and the Century Mine proposal. BA(Hons) thesis, School of Earth Sciences, Macquarie University, Sydney.

Hawke, Steven and Gallagher, Michael (1989) *Noonkanbah, Whose Land, Whose Law*. Fremantle Arts Centre Press, Fremantle.

Hay, Iain (ed.) (2000) *Qualitative Research Methods in Geography*. Oxford University Press, Melbourne.

Healey, Patsy (1997) *Collaborative Planning: Shaping Places in Fragmented Societies*. Macmillan, London.

Hebbert, Michael (1990) Regionalist aspirations in contemporary Europe. *Plural Societies* 21: 5–16.

Heilbuth, B. and Raffaele, B. (1993) Time to stop the war against mining. *Readers Digest* 143 (857) 19–26.

Herbert, Xavier (1975) *Poor Fellow My Country*. Harper Collins, Melbourne.

Hiley, Graham (ed.) (1997) *The Wik Case: Issues and Implications*. Butterworths, Sydney.

Hill, Christopher (1972) *The World Turned Upside Down: Radical Ideas during the English Revolution*. Temple Smith, London.

Hill, M.A. and Press, A.J. (1993) Kakadu National Park: an experiment in partnership. *The Australian Quarterly* 65 (4: The politics of Mabo) 23–33.

Hill, M.A. and Press, A.J. (1994) Kakadu National Park: an experiment in partnership, in Goot, Murray and Rowse, Tim (eds) *Make a Better Offer: the Politics of Mabo*. Pluto Press, Leichhardt: 23–33.

Hirsch, Eric and O'Hanlon, Michael (eds) (1995) *The Anthropology of Landscape: Perspectives on Place and Space*. Clarendon Press, Oxford.

Hirsch, Philip (1993) *Political Economy of Environment in Thailand*. Journal of Contemporary Asia Publishers, Manila.

Hobson, George (1992) Traditional knowledge is science. *Northern Perspectives* 20(1) 2.

Högman, Johan Fr (1989) The collision of cultures in northern Europe, in Müller-Wille, Ludger (ed.) *Social Change and Space: Indigenous Nations and Ethnic Communities in Canada and Finland*. Department of Geography, McGill University, Montreal: 35–8.

Holden, Annie and O'Faircheallaigh, Ciaran (1995a) *Economic and Social Impacts of Silica Mining at Cape Flattery*. Griffith University, Brisbane.

—— (1995b) *The Mapoon People and the Skardon Kaolin Project: An Economic and Social Impact Assessment*. Griffith University, Brisbane.

—— (1996) *The Economic and Social Impact on Aboriginal People of the ALSPAC Ely Bauxite Project*. Griffith University, Brisbane.

Hollingsworth Dames and Moore (1991) *The Century Project: Initial Advice Statement*. HDM, Brisbane.

Hooson, David (ed.) (1994) *Geography and National Identity*. Blackwells, Oxford.

Horvath, Ron. (1991) Combining the sociological, historical and geographical imaginations: an historical factorial ecology of global development. Paper presented to Economic and Regional Restructuring Research Unit Seminar, University of Sydney, November 1991.

Howard, Albert and Widdowson, Frances (1996) Traditional knowledge threatens environmental assessment. *Options Politiques* November: 34–6.

Howitt, Richard (1978) The management strategies of the aluminium companies with interests on Cape York Peninsula 1955–1978. BA(Hons) thesis, Department of Geography, University of Newcastle, NSW.

—— (1979) *Beyond the Geological Imperative: A Study of the Social Impact of the Management Strategies of the Aluminium Companies with Interests on Cape York Peninsula 1955–1978*. Research Papers in Geography, 23, Department of Geography, University of Newcastle, Newcastle, NSW.

—— (1989a) Social impact assessment and resource development: issues from the Australian experience. *Australian Geographer* 20(2) 153–66.

—— (1989b) Whatever happened to Amax: restructuring of a global resources corporation. *JAPE*.

—— (1989c) A different Kimberley: Aboriginal marginalisation and the Argyle Diamond Mine. *Geography* 74(3), 232–8.

—— (1990a) *'All They Get is the Dust': Aborigines, Mining and Regional Restructuring in WA's Eastern Goldfields.* ERRRU Working Paper No. 1. Economic and Regional Restructuring Research Unit, University of Sydney.

—— (1990b) Learning from cross-cultural perspectives on old and new environmental wisdom. Unpublished paper, Department of Geography, University of Sydney.

—— (1991a) Aborigines and restructuring in the mining sector: vested and representative interests. *Australian Geographer* 22(2) 117–19.

—— (1991b) 'A world in a grain of sand': towards a reconceptualision of geographical scale. *Australian Geographer* 24(1) 33–44.

—— (1991c) Aborigines and gold mining in Central Australia, in Connell, J. and Howitt, R. (eds) *Mining and Indigenous Peoples in Australiasia.* Sydney University Press, Sydney.

—— (1992a) *Aborigines, Mining and Regional Restructuring in Northeast Arnhem Land.* ERRRU Working Paper No. 10. Economic and Regional Restructuring Research Unit, University of Sydney.

—— (1992b) Weipa: industrialisation and indigenous rights in a remote Australian mining area. *Geography* 77(3) 223–35.

—— (1992c) The political relevance of locality studies: a remote Antipodean viewpoint. *Area* 24(1) 73–81.

—— (1993a) Social impact assessment as 'applied peoples' geography'. *Australian Geographical Studies* 31(2) 127–40.

—— (1993b) Marginalisation in theory and practice: a brief conceptual introduction, in Howitt, R. (ed.) *Marginalisation in Theory and Practice.* ERRRU Working Paper No. 12. Economic and Regional Restructuring Research Unit, University of Sydney.

—— (1994a) *Participation, Empowerment and Intervention in Locality Studies: Methodological and Ethical Issues in Locality Research.* Paper presented at the Institute of Australian Geographers Annual Conference, Magnetic Island, Queensland, September 1994.

—— (1994b) Resource conflicts at Coronation Hill, in Hirsch, Philip and Thoms, Martin (eds) *Australasian Geography for the 1990s.* Department of Geography, University of Sydney: 6–11.

—— (1995) *Developmentalism, Impact Assessment and Aborigines: Rethinking Regional Narratives at Weipa.* NARU Discussion Paper 24. North Australia Research Unit, Darwin.

—— (1996) Resources, nations and indigenous peoples, in Howitt, R., Connell, J. and Hirsch, P. (eds) *Resources, Nations and Indigenous Peoples: Case Studies from Australasia, Melanesia and Southeast Asia.* Oxford University Press, Melbourne: 1–30.

—— (1997a) Terra Nullius no more? Changing Australian geographies through negotiation. Paper presented at the Inaugural International Conference in Critical Geography, University of British Columbia and Simon Fraser University, Vancouver, August 1997.

—— (1997b) *The Other Side of the Table: Regional Agreements and the Culture of Mining Companies.* AIATSIS Regional Agreements Discussion Papers 3. Native Title Research Unit, Australian Institute of Aboriginal and Torres Strait Islander Studies, Canberra.

—— (1997c) Getting the scale right: the geopolitics of regional agreements. *Northern Analyst* 2: 15–17.

—— (1997d) *Aboriginal Social Impact Issues in the Kakadu Region*. School of Earth Sciences, Macquarie University, Sydney.

—— (1998a) Scale as relation: musical metaphors of geographical scale. *Area* 30(1), 49–58.

—— (1998b) Recognition, reconciliation and respect: steps towards decolonisation? *Australian Aboriginal Studies* 1998/1, 28–34.

—— (1999) Indigenous rights and regional economies: rethinking the building blocks. Paper presented to Rethinking Economy: Alternative Accounts, Australian National University, Canberra (9 August 1999).

—— (2000) For whom do we teach? the paradox of excellence. *Journal of Geography in Higher Education* 24(3), 317–23.

Howitt, Richard and Crough, Greg (1996) Australia, in Campbell, Bonnie and Ericsson, Magnus (eds) *Restructuring in Global Aluminium*. Mining Journal Books, London: 175–207.

Howitt, Richard and Douglas, John (1983) *Aborigines and Mining Companies in Northern Australia*. Alternative Publishing Co-op, Chippendale.

Howitt, Richard and Hutchinson, Frank (1987) *Education for Peace K–6*. NSW Department of Education, Sydney.

Howitt, Richard and Jackson, Sue (1998) Some things *do* change: indigenous rights, geographers and geography in Australia. *Australian Geographer* 29(2) 155–73.

—— (2000) Social impact assessment and linear projects, in Goldman, Laurence R. (ed.) *Social Impact Analysis: An Applied Anthropology Manual*. Berg, Oxford: 257–94.

Hulme, Keri (1994) *The Bone People*. 10th anniversary ed. Picador, London.

Hunter, Ernest M. (1993) *Aboriginal Health and History: Power and Prejudice in Remote Australia*. Cambridge University Press, Cambridge.

Hydro-Québec (1993) *Grande-Baleine Complex. Highlights and Overview Documents* compiled as an information kit by Hydro-Québec Communications et Relations publiques, Québec.

Hyndman, David (1991) Zipping down the fly on the Ok Tedi project, in Connell, John and Howitt, Richard (eds) *Mining and Indigenous Peoples in Australasia*. Sydney University Press, Sydney: 76–90.

Indigenous Knowledge and Development Monitor (1996a) Reactions to Agrawal's 'Indigenous and scientific knowledge: some critical comments'. *Indigenous Knowledge and Development Monitor* 4 (1). http://www.nufficcs.nl/ciran/ikdm/4–1/articles/agrawal.html

—— (1996b) A sequel to the debate. *Indigenous Knowledge and Development Monitor* 4(2). URL http://www.nufficcs.nl/ciran/ikdm/4–2/articles/

Institute for Natural Progress (1992) In usual and accustomed places: contemporary American Indian fishing rights struggles, in Jaimes, M. A. (ed.) *The State of Native America: Genocide, Colonization and Resistance*. South End Press, Boston: 217–39.

Interorganizational Committee on Guidelines and Principles for Social Impact Assessment (1994) *Guidelines and Principles for Social Impact Assessment*. (NOAA Technical Memorandum NMFS–F/SPO–16 May 1994) US Department of Commerce, National Oceanic and Atmospheric Administration and National Marine Fisheries Service, Washington DC.

Ivanitz, Michele (1997) *The Emperor Has No Clothes: Canadian Comprehensive Claims and their Relevance to Australia.* Land, Rights, Laws: Issues of Native Title, Regional Agreements Papers 4. Native Titles Research Unit, Australian Institute of Aboriginal and Torres Strait Islander Studies, Canberra.

IWGIA (1990) *Indigenous Peoples of the Soviet North.* International Working Group on Indigenous Affairs, Copenhagen.

—— (1991) *Arctic Environment: Indigenous Perspectives.* (IWGIA Document 69). International Working Group on Indigenous Affairs, Copenhagen.

Jackson, Richard T. (1991) Not without influence: villages, mining companies and government in Papua New Guinea, in Connell, J. and Howitt, R. (eds) *Mining and Indigenous Peoples in Australasia.* Sydney University Press, Sydney: 18–33.

Jackson, Sue (1995) The water is not empty: cross-cultural issues in conceptualising sea space. *Australian Geographer* 26(1) 87–96.

—— (1996) *When History meets the New Native Title Era at the Negotiating Table: a Case Study in Reconciling Land Use in Broome, Western Australia.* North Australia Research Unit, Darwin.

—— (1997) A disturbing story: the fiction of rationality in land use planning in Aboriginal Australia. *Australian Planner* 34(4) 221–6.

Jacobs, Jane (1994) 'Shake 'im this country': the mapping of the Aboriginal sacred in Australia – the case of Coronation Hill, in Jackson, Peter and Penrose, Jan (eds) *Constructions of Race, Place and Nation.* University of Minnesota Press, Minneapolis: 100–18.

—— (1996) *Edge of Empire: Post-colonialism and the City.* Routledge, London and New York.

—— (1997) Resisting reconciliation: the social geographies of (post)colonial Australia, in Pile, Steve and Keith, Michael (eds) *Geographies of Resistance.* Routledge, London and New York: 203–18.

Jacobs, P. and Chatagnier, H. (1985) *Environement Kativik Environment.* Proceedings of Kativik Environment Conference, Environment & Inuit Life in N Quebec, 1984, Kativik Regional Government.

Jacobs, Peter (1988) Towards a network of knowing and of planning in northern Canada, in Wonders, William C. (ed.) *Knowing the North: Reflections on Tradition, Technology and Science.* Boreal Institute of Northern Studies, University of Alberta, Edmonton: 51–60.

Jacobs, Peter and Mulvihill, Peter (1995) Ancient lands: new perspectives. Towards multi-cultural literacy in landscape management. *Landscape and Urban Planning* 32: 7–17.

Jaimes, M. Annette (ed.) (1992) *The State of Native America: Genocide, Colonization and Resistance.* South End Press, Boston.

James Bay Mercury Committee (1992) *Report of Activities.* Fifth annual report of the Committee for 1991–1992.

Jawoyn Association (1989) *'Just Sweet Talk': The Jawoyn Response to the Coronation Hill Joint Venture Draft Environmental Impact Statement. A Submission to the Minister for Arts, Sport, the Environment, Tourism and Territories.* Jawoyn Association, Katherine, NT.

—— (1994) *Rebuilding the Jawoyn Nation: Approaching Economic Independence.* Green Ant, Darwin.

—— (1997a) *Jawoyn Mining Policy: A Path to Independence.* Jawoyn Association, Katherine, NT.

—— (1997b) *Nyarrang nyan-burrk bunbun yunggaihmih: 'We're moving ahead'.* *Jawoyn Association Five Year Plan.* Jawoyn Association, Katherine.

Jhappan, C. Radha (1992) Global community? Supranational strategies of Canada's aboriginal peoples. *Journal of Indigenous Studies* 3(1) 59–97.

—— (1990) Indian symbolic politics: the double-edged sword of publicity. *Canadian Ethnic Studies* 22(3) 19–39.

Johansen, Bruce and Maestas, Roberto (1979) *Wasi'chu: The Continuing Indian Wars.* Monthly Review Press, New York and London.

Johnson, Colin (1983) *Doctor Wooreddy's Prescription for Enduring the Ending of the World.* Hyland House, Melbourne.

Johnson, T. (1991) Caring for the Earth: new activism among Hopi traditionals. *Amicus Journal* Winter 91: 22–7.

Johnston, E. (1991) *Royal Commission into Aboriginal Deaths in Custody, National Report* (5 vols) AGPS, Canberra.

Johnston, R. (1989) *Environmental Problems: Nature, Economy and State.* Belhaven Press, London.

Jonas, Andrew (1994) The scale politics of spatiality. *Environment and Planning D: Society and Space* 12(3) 257–64.

Jull, Peter (1991a) *Australian Nationhood and Outback Indigenous Peoples.* NARU Discussion Paper 1. North Australia Research Unit, Darwin.

—— (1991b) Indigenous survivors in Australia and Canada. *Alternatives* 18(2) 28–33.

—— (1992a) *The Constitutional Culture of Nationhood, Northern Territories and Indigenous Peoples.* NARU Discussion Paper 6. North Australia Research Unit, Darwin.

—— (1992b) *A Guide for Australian Research into Northern Regions and Indigenous Policy in North America and Europe.* NARU Discussion Paper 3. North Australian Research Unit, Darwin.

Kaiwharu, Hugh (ed.). (1989) *Waitangi: Maori and Pakeha Perspectives on the Treaty of Waitangi.* Oxford University Press, Auckland.

Kakadu Board of Management and Australian Nature Conservation Agency (1996) *Kakadu National Park Draft Plan of Management.* Kakadu Board of Management and Australian Nature Conservation Agency, Jabiru.

Kaliss, Tony (1997) What was the 'other' that came on Columbus's ships? An interpretation of the writing about interaction between Northern Native peoples in Canada and the United States and the 'other'. *Journal of Indigenous Studies* 3(2) 27–42.

Kanaaneh, Moslih (1997) The 'anthropologicality' of indigenous anthropology. *Dialectical Anthropology* 22: 1–21.

Katz, Cindi (1994) Playing the field: questions of fieldwork in geography. *Professional Geographer* 46(1) 67–72.

Kealy, Venessa (1996) Imagined spaces: interpreting perceptions of place and regulation of spaces through the processes of normalisation and reconciliation at Weipa. B.A. (Hons) Thesis, School of Earth Sciences, Macquarie University.

Kean, John; Richardson, Garry and Trueman, Norma (eds) (1988) *Aboriginal Role in Nature Conservation: Emu Conference June 7–9, 1988.* South Australian National Parks and Wildlife Service, Adelaide.

Kelly, Philip F. (1997) Globalization, power and the politics of scale in the Philippines. *Geoforum* 28(2) 151–71.

Kesteven, Sue (1986) The project to monitor the social impact of uranium mining on Aboriginal communities of the NT. *Australian Aboriginal Studies* 1/1986: 43–5.

Kevans, Denis (1985) *Ah, White Man, Have You Any Sacred Sites?* Denis Kevans, Sydney.

Kimberley Land Council and Warinarri Resource Centre (1991) *The Crocodile Hole Report*. Kimberley Land Council, Derby.

Kleivan, Helge (1978) Incompatible patterns of land use: the controversy over hydro-electric schemes in the heartland of the Sami people. *IWGIA Newsletter* 20–21, 56–80.

Knudtson, Peter and Suzuki, David (1992) *Wisdom of the Elders*. Allen & Unwin, Sydney.

Kobayashi, Audrey (1989) A critique of dialectical landscape, in Kobayashi A. and Mackenzie S. (eds) *Remaking Human Geography*. Unwin Hyman, Boston: 164–83.

Kobayashi, Audrey (1994) Coloring the field: gender, 'race', and the politics of field-work. *Professional Geographer* 46(1) 73–80.

Kohen, James (1995) *Aboriginal Environmental Impacts*. UNSW Press, Sydney.

Kolig, Erich (1987) *The Noonkanbah Story*. University of Otago Press, Dunedin.

—— (1990) Government policies and religious strategies: fighting with myth at Noonkanbah, in Tonkinson, Robert and Howard, Michael (eds) *Going it Alone: Prospects for Aboriginal Autonmy – Essays in Honour of Ronald and Catherine Berndt*. Aboriginal Studies Press, Canberra: 234–52.

Kuhn, Richard G. and Duerden, Frank (1996) A review of traditional environmental knowledge: an interdisciplinary Canadian perspective. *Culture* 16(1) 71–84.

Kvist, Roger (1991) Saami reindeer pastoralism as an indigenous resource manage-ment system: the case of Tuorpon and Sirkas, 1760–1860. *Arctic Anthropology* 28(2), 121–34.

Lafitte, Gabriel (1995) Big ugly Australian: Ok Tedi, BHP and the PNG élite. *Arena Magazine* 19 (October–November) 18–19.

Lane, Marcus B.; Dale, Allan; Ross, Helen; Hill, Alan and Rickson, Roy (1990) *Social Impact of Development: An Analysis of the Social Impact of Development on Aborig-inal Communities in the Region*. A report for the Resource Assessment Commission for the Kakadu Conservation Zone Inquiry. AGPS, Canberra.

Lane, Marcus B.; Ross, Helen and Dale, Allan (1997) Social impact research: inte-grating the social, political and planning paradigms. *Human Organization* 56(3) 302–10.

Lane, Marcus and Chase, Athol (1996) Resource development on Cape York Penin-sula: marginalisation and denial of indigenous perspectives, in Howitt, Richard; Connell, John and Hirsch, Philip (eds) *Resources, Nations and Indigenous Peoples: Case Studies from Australasia, Melanesia and Southeast Asia*. Oxford University Press, Sydney: 172–83.

Lane, Marcus and Yarrow, David (1998) Development on native title land – inte-grating the right to negotiate with the impact assessment process. *Environmental and Planning Law Journal* 15(2), 147–55.

Langton, Marcia (1983) A national strategy to deal with exploration and mining on or near Aboriginal land, in Peterson, N. and Langton, M. (eds) *Aborigines, Land and Land Rights*. Australian Institute of Aboriginal Studies, Canberra: 385–402.

—— (1995) What do we mean by wilderness? Wilderness and *terra nullius* in Austra-lian art. Paper presented to the Sydney Institute, 12 October 1995.

—— (1996) No future in a return to racial paternalism. *Australian* (18 April) 13.

—— (1997) Pauline as the thin edge of the wedge, in Adams, Philip (ed.) *The Retreat from Tolerance: a Snapshot of Australian Society.* ABC Books, Sydney: 86–107.

—— (1998) *Burning Questions: Emerging Environmental Issues for Indigenous Peoples in Northern Australia.* Centre for Indigenous Natural and Cultural Resource Management, Northern Territory University, Darwin.

Langton, Marcia; Ah Mat, Leslie; Schaber, Evelyn; Thomas, Merle; Moss, Bonita; Mackinolty, ; Tilton, Edward and Spencer, Lloyd (1990) *Too Much Sorry Business. The Submission of the Northern Territory Aboriginal Issues Unit of the Royal Commission into Aboriginal Deaths in Custody to Commissioner Elliot Johnston, QC.* July, Royal Commission into Aboriginal Deaths in Custody, Alice Springs.

Lanning, G. and Mueller, M. (1979) *Africa Undermined: Mining Companies and the Underdevelopment of Africa.* Penguin, Harmondsworth.

Lasko, Nils (1987) The importance of indigenous influence on the system of decision-making in the nation-state, in IWGIA (ed.) *Self Determination and Indigenous Peoples: Sami Rights and Northern Perspectives.* (IWGIA Document, 58.) IWGIA, Copenhagen: 73–84.

Laurence, Robert (1993) American Indians and the environment: a legal primer for newcomers. *Natural Resources and Environment* 7(4), 3–6, 48–50.

Lawrence, David (1996) *Kakadu: the Making of a National Park.* ANCA, Canberra.

Le Guin, Ursula (1988) *Always Coming Home.* Grafton, London.

—— (1989) *Dancing at the Edge of the World: Thoughts on Words, Women, Places.* Victor Gollancz, London.

Lea, John (1984) Housing needs and social demands in the special development projects of the Northern territory, in Drakakis-Smith, David (ed.) *Housing in the North: Policies and Markets.* ANU, North Australia Research Unit, Darwin: 125–57.

—— (1987) Resilience and socially sustainable growth: insights from the Northern Territory, in Parker, Paul (ed.) *Resource Development and Local Government: Policies for Growth, Decline and Diversity.* (Developments in Local Government 5.) Office of Local Government, Canberra: 51–8.

Lea, John and Zehner, Robert A.B. (1985) Democracy and planning in a small mining town: the governance transition in Jabiru, NT, in Loveday, Peter and Wade-Marshall, Deborah (eds) *Economy and People in the North.* NARU, Darwin: 225–47.

—— (1986) *Yellowcake and Crocodiles.* Allen & Unwin, Sydney.

Lefebvre, Henri (1991) *The Production of Space.* Blackwells, Oxford (trans. D. Nicholson-Smith).

Leftwich, Adrian. (1983) *Redefining Politics: People, Resources and Power.* Methuen, London.

Levitus, Robert (1991) The boundaries of Gagadju Association membership: anthropology, law and public policy, in Connell, John and Howitt, Richard (eds) *Mining and Indigenous Peoples in Australasia.* University of Sydney Press, Sydney: 153–68.

Levitus, Robert and Aboriginal Project Committee (1997) *Kakadu Region Social Impact Study: Report of the Aboriginal Project Committee.* Supervising Scientist, Canberra.

Levy, M.A., Keohane, R.O. and Haas, P.M. (1992) Institutions for the earth: promoting international environmental protection. *Environment* 34(4) 12–17, 29–36.

Libby, Ronald T. (1989) *Hawke's Law: the Politics of Mining and Aboriginal Land Rights in Australia.* University of Western Australia, Nedlands.

Lipietz, Alain (1996) Geography, ecology, democracy. *Antipode* 28(3) 219–28.

Little Bear, Leroy; Boldt, M. Enno and Long, J. Anthony (eds) (1984) *Pathways to Self-determination: Canadian Indians and the Canadian State.* University of Toronto Press, Toronto.

Luxembourg, Rosa (1963) *The Accumulation of Capital.* Routledge Kegan Paul, London.

Mailhot, José (1994) *Traditional Ecological Knowledge: the Diversity of Knowledge Systems and their Study* (trans. Axel Harvey). 2nd edn. (Great Whale Environmental Assessment 4.) Great Whale Public Review Support Office, Montréal.

Marcus, G. (1986) Contemporary problems of ethnography in the modern world system, in Clifford and Marcus (eds) *Writing Culture: The Poetics and Politics of Ethnography.* University of California Press, Berkeley: 165–93.

Marcus, G. and Fischer, M.M.J. (1986) A crisis of representation in human sciences, in Marcus, G.E. and Fischer, M.M.J.(eds) *Anthropology as Cultural Critique: An Experimental Moment in the Human Sciences.* University of Chicago Press, Chicago, 7–16.

Mares, (1992) Why a series on the new world order? *24 Hours Supplement,* February: 4–9.

Marika, R. (1989) A celebration of the knowledge of native people from around the world. *Yutana Dhawu* August : 8–13.

Marx, Karl ([1851] 1975) Preface to a contribution to the critique of political economy, in Livingston, Rodney and Benton, Gregory (eds) *Karl Marx: Early Writings.* Penguin in association with New Left Review, Harmondsworth: 424–8.

—— (1954) [1887]: *Capital,* vol. 1. Progress Publishers, Moscow.

—— (1975 [1859]) Preface to a contribution to the critique of political economy, in *Karl Marx: Early Writings.* Penguin, Harmondsworth: 424–8.

Massey, Doreen (1973) Towards a critique of industrial location theory. *Antipode* 5(3) 33–9.

—— (1984a) Introduction, in Massey, D. and Allen, J. (eds) *Geography Matters! A Reader.* Cambridge University Press, Cambridge: 1–11.

—— (1984b) *Spatial Structures of Production.* Methuen, London.

—— (1993a) Questions of locality. *Geography* 78(2), 142–9.

—— (1993b) Power-geometry and a progressive sense of place, in Bird, J. *et al.* (eds) *Mapping the Futures: Local Cultures, Global Change.* Routledge, London: 59–69.

—— (1994) Double articulation: a 'place in the world', in Bammer, A. (ed.) *Displacements – Cultural Identities in Question.* Bloomington and Indianapolis, Indiana: 110–19.

Max-Neef, Manfred (1992) Development and human needs, in Ekins, Paul and Max-Neef, Manfred (eds) *Real-Life Economics: Understanding Wealth Creation.* Routledge, London and New York: 197–214.

Maxwell, James; Lee, Jennifer; Briscoe, Forrest; Stewart, Ann and Suzuki, Tatsujiro (1997) Locked on course: Hydro-Québec's commitment to mega-projects. *Environmental Impact Assessment Review* 17: 19–38.

Maybury-Lewis, David (1992) *Millenium: Tribal Wisdom and the Modern World.* Viking, New York.

McClellan, J.R.; Fisk, J. and Jonas, W. (1985) *The Report of the Royal Commission into British Nuclear Tests in Australia* (2 vols). Australian Government Publishing Service, Canberra.

McCutcheon, Sean (1991) *Electric Rivers: The Story of the James Bay Project.* Black Rose, Montréal.

McCutcheon, Sean (1994) *Mitigation at the La Grande Complex: a Review*. Background Paper No. 8. Great Whale Public Review Support Office, Montréal.

McDowell, Linda (1994) Multiple voices: speaking from inside and outside 'The Project'. *Antipode* 24(1) 56–72.

McGinnis, Michael V. (1995) On the verge of collapse: The Columbia river system, wild salmon and the Northwest Power Planning Council. *Natural Resources Journal* 35(1) 63–92.

McHugh, Paul G. (1991) Constitutional myths and the Treaty of Waitangi. *The New Zealand Law Journal* (September) 316–19.

McHugh, Paul G. (1996) The legal and constitutional position of the crown in resource management, in Howitt, Richard; Connell, John and Hirsch, Philip (eds) *Resources, Nations and Indigenous Peoples: Case Studies from Australasia, Melanesia and Southeast Asia*. Oxford University Press, Melbourne: 300–16.

McLaughlin, Dene and Niemann, Grant (1984) The exploration and mining process in relation to Aboriginal lands in the Northern Territory. Paper presented to the AIMM Conference, Darwin.

McNabb, Steven (1993) Contract and consulting anthropology in Alaska. *Human Organization* 52(2) 216–24.

McTaggart, Robin and Kemmis, Stephen (eds) (1988) *The Action Research Reader*. 3rd edn. Deakin University Press, Victoria, Australia.

Meadows, D.H, Meadows, D.L, Randers J. and Behrens, W. (1972) *The Limits to Growth*, Universe Books, New York

Meares, Peter. (ed.) (1992) Whatever happened to the new world order? Special supplement to *24 Hours*, February 1992.

Mercer, D. (1991) *'A Question of Balance': Natural Resource Conflict Issues in Australia*. Federation Press, Leichhardt.

Merlan, Francesca (1991) The limits of cultural constructionism: the case of Coronation Hill. *Oceania* 61: 341–52.

Michael, John (1996) Making a stand: standpoint epistemologies, political positions: proposition 187. *Telos* 108: 93–103.

Michalenko, G. (1981) The social assessment of corporation, in Tester and Mykes (eds) *SIA: Theory, Method and Practice*. Detselig, Calgary: 168–79.

Milliken, Robert (1986) *No Conceivable Injury: The Story of Britain and Australia's Atomic Cover-up*. Penguin, Ringwood, Victoria.

Minichiello, Victor; Aroni, Rosalie; Timewell, Eric and Alexander, Loris (1995) *In-depth Interviewing: Principles, Technique, Analysis*. 2nd edn. Addison Wesley Longman, Melbourne.

Mitchell, Bruce (1989) *Geography and Resource Analysis*. 2nd edn. Longman, New York.

Moody, R. (ed.) (1988) *Indigenous Voice* (2 vols) IWGIA and Zed Books, London.

Moody, Roger (1991) *Plunder!* PARTIZANS/CAFCA, London and Christchurch.

Moreton-Robinson, A. and Runciman, C. (1990) Land rights in Kakadu: self management or domination. *Journal for Social Justice Studies*, special edition series 3 (Contemporary Race Relations) 77–90.

Morgan, Hugh (1987) Mining – still the backbone of Australia. *Mining Review* (May) 16–20.

—— (1991) Reflections on Coronation Hill. *Australian and World Affairs* 10 (Spring) 19–29.

—— (1993) The heritage threat to land use: trouble for pastoral and mining industries from heritage powers. *Mining Review* 17(5) 26–8.

Morris, Meaghan. (1992) The man in the mirror: David Harvey's 'condition' of postmodernity. *Theory, Culture and Society* 9: 253–79.

Morrison, Andrea (ed.) (1997) *Justice for Natives: Searching for Common Ground.* McGill University Press, Montréal.

Morrison, Toni (1993) *Playing in the Dark: Whiteness and the Literary Imagination.* Picador, London.

Muecke, Stephen (1992) Marginality, writing and education. *Cultural Studies*, 6(2) 261–70.

—— (1993) Studying the other: a dialogue with a post-grad. *Cultural Studies* 7(2) 324–9.

Müller-Wille, Ludger (ed.) (1989) *Social Change and Space: Indigenous Nations and Ethnic Communities in Canada and Finland.* Department of Geography, McGill University, Montreal.

Muzaffar, I. (1992) The new world order: gold or god? In: Whatever happened to the new world order? Special supplement to *24 Hours*, February: 28–33.

Myers, F. (1986) *Pintupi Country, Pintupi Self: Sentiment, Place and Politics among Western Desert Aborigines.* Smithsonian Institute, Washington and AIAS, Canberra.

Nader, Laura (1964) Perspectives gained from fieldwork, in *Horizons of Anthropology.* 2nd edn. Aldine, Chicago: 187–98.

—— (1974) Up the anthropologist – perspective gained from studying up, in Hymes, Dell (ed.) *Reinventing Anthropology.* Vintage Books, New York, 284–311.

Nash, J. (1979) *We Eat the Mines and the Mines Eat Us*, Columbia University Press, New York.

Nast, Heidi (1994) Opening remarks on 'Women in the Field'. *Professional Geographer* 46(1) 54–66.

Natural Resources and Environment (1993) Developing resources on American Indian lands: special issue. *Natural Resources and Environment* 7(4) 1–64.

Navajo Nation (1993) *Overall Economic Development Plan 1992–93.* Navajo Nation Division of Economic Development, Window Rock, AZ.

—— (1994) *Navajo Nation Fax 93: A Statistical Abstract of the Navajo Nation.* Division of Economic Development, Navajo Nation, Window Rock, AZ.

Nesbitt, Bradley (1992) Aboriginal 'joint' management of Northwest Kimberley Conservation Reserves: achievable under existing legislation but is there the political will? In: Birckhead, Jim; deLacy, Terry and Smith, Laurajane (eds) *Aboriginal Involvement in Parks and Protected Areas.* AIATSIS Report Series. Aboriginal Studies Press, Canberra: 251–61.

New Zealand Conservation Authority: Te Whakahaere Matua Atawhai O Aotearoa (1994) *Maori Customary Use of Native Birds, Plants and other Traditional Materials.* New Zealand Conservation Authority: Te Whakahaere Matua Atawhai O Aotearoa, Wellington, New Zealand.

New Zealand Ministry for the Environment. (1994) *Issues, Objectives, Policies, Methods and Results under the Resource Management Act.* Working Paper No 1. July 1994. NZ Ministry for the Environment, Wellington.

Niia, Lars Petter (1991) Saami culture: the will of the Saami, in Kvist, Roger (ed.) *Readings in Saami History, Culture and Language II.* Umeå University Center for Arctic Cultural Research, Miscellaneous Publications, 12. University of Umeå Center for Arctic Cultural Research, Umeå, Sweden: 149–58.

Norgaard, R. (1992) Sustainability as intergenerational equity: economic theory and environmental planning, *Environmental Impact Assessment Review*, 12 (1/2) 85–124

Norris, Beth (1996) Local culture or global coal: aboriginal struggles to protect and nurture their cultural heritage in the Upper Hunter Valley. BA (Hons) Thesis, School of Earth Sciences, Macquarie University, Sydney.

Nottingham, Isla (1990) Social impact reporting: a Maori perspective – the Taharoa Case. *Environmental Impact Assessment Review* 10: 175–84.

Notzke, Claudia (1994) *Aboriginal Peoples and Natural Resources in Canada.* Captus Press, North York, Ontario.

—— (1995) A new perspective in aboriginal natural resource management: co-management'. *Geoforum* 26(2) 187–209.

Nystö, Sven-Roald (1991) Economic development as seen from a Sami viewpoint, in Seyersted, Per (ed.) *The Arctic: Canada and the Nordic Countries.* Nordic Association for Canadian Studies, Lund: 35–40.

O'Faircheallaigh, Ciaran (1986) The economic impact on Aboriginal communities of the Ranger Project: 1979–1985. *Australian Aboriginal Studies* 2(2), 2–14.

—— (1987) *Mine Infrastructure and Economic Development in North Australia.* NARU, Darwin.

—— (1988) Land rights and mineral exploration: the Northern Territory experience. *The Australian Quarterly* 60(1, Autumn), 70–84.

—— (1991) Resource exploitation and indigenous people: towards a general analytical framework, in Jull, Peter and Roberts, Sally (eds) *The Challenge of Northern Regions.* North Australia Research Unit, Darwin: 228–71.

—— (1995) *Mineral Development Agreements Negotiated by Aboriginal Communities in the 1990s.* CAEPR Discussion Paper 85/95 ed. Centre for Aboriginal Economic Policy Research, ANU, Canberra.

—— (1996a) Negotiating with resource companies: issues and constraints for Aboriginal communities, in Howitt, Richard; Connell, John and Hirsch, Philip (eds) *Resources, Nations and Indigenous Peoples: Case Studies from Australasia, Melanesia and Southeast Asia.* Oxford University Press, Melbourne: 184–201.

—— (1996b) *Making Social Impact Assessment Count: A Negotiation-based approach for Indigenous Peoples.* Aboriginal Politics and Public Sector Management Research Papers 3. Centre for Australian Public Sector Management, Griffith University, Brisbane.

—— (1996c) Indigenous land use agreements: achieving negotiated outcomes, in Meyers, Gary D. (ed.) *The Way Forward: Collaboration and Co-operation 'in Country'.* 2nd edn. Proceedings of the Indigenous Land Use Agreements Conference, Darwin 26–29 September 1995. National Native Title Tribunal, Perth: 216–21.

O'Reilly, James (1988) The role of the courts in the evolution of the James Bay Hydroelectic Project, in Vincent, Sylvie and Bowers, Garry (eds) *Baie James et Nord Québécois: Dix Ans Après/James Bay and Northern Québec: Ten Years After.* Recherches amérindiennes au québec, Montréal: 30–8.

O'Tuathail, Gearoid (1993) The East–West conflict? Japan and the Bush adminsitration's 'New World Order'. *Area* 25 (2) 127–35.

Ollman, Bertell (1976) *Alienation: Marx's Conception of Man in Capitalist Society.* 2nd edn. Cambridge University Press, Cambridge.

—— (1990) Putting dialectics to work: the process of abstraction in Marx's method. *Rethinking Marxism* 3(1) 26–74.

—— (1993) *Dialectical Investigations.* Routledge, New York.

Omara-Ojungu, (1991) *Resource Management in Developing Countries.* Longman, London.

Osherenko, Gail (1992) Human/nature relations in the Arctic: changing perspectives. *Polar Record* 28(167) 277–84.

—— (1993) Using peripheral vision in the northern sea route: assessing impacts on indigenous peoples, in Simonsen, Henning (ed.) *Proceedings from the Northern Sea Route Expert Meeting, Tromso, October 1992.* Fridtjof Nansen Institute, Lysaker, Norway: 115–32.

Owens, Nancy J. (1978) Can tribes control energy development? In: Jorgensen, Joseph G.; Clemmer, Richard O.; Little, Ronald L.; Owens, Nancy J. and Robbins, Lynn A. (eds) *Native Americans and Energy Development.* Anthropology Resource Center, Cambridge Mass.: 49–62.

Paine, Robert (1982) *Dam a river, damn a people? Saami (Lapp) livelihood and the Alta/Kautokeino hydro-electric project and the Norwegian parliament.* (IWGIA Document 45). International Work Group for Indigenous Affairs, Copenhagen.

—— (1992) Social construction of the 'Tragedy of the Commons' and Saami reindeer pastoralism. *Acta Borealia* 9(2), 3–20.

Palinkas, Lawrence; Downs, Michael A.; Petterson, John S. and Russell, John (1993) Social, cultural and psychological impacts of the Exxon Valdez oil spill. *Human Organization* 52(1) 1–13.

Palmer, Ian (1988) *Buying Back the Land: Organisational Struggle and the Aboriginal Land Fund Commission.* Aboriginal Studies Press, Canberra.

Pardy, Rob; Parsons, Mike; Siemon, Don and Wigglesworth, Ann (1978) *Purari: Overpowering PNG?* International Development Action and Purari Action Group, Fitzroy, Victoria.

Parry, Bronwyn (1994) *International Conventions and the Empowerment of Small Nations: The Case of the United Nations Convention on the Law of the Sea III.* ERRRU Working Papers, 15. Economic and Regional Restructuring Research Unit, Departments of Economics and Geography, University of Sydney, Sydney.

—— (1996) Hunting the gene-hunters – the role of status, nepotism and chance in securing and structuring interviews with corporate élites. Paper for the IBG Conference, Glasgow, Scotland, 3 January 1996.

Payne, R.J. and Graham, R. (1984) Non-hierarchical alternatives in northern resource management. *Etudes Inuit Studies* 8(2) 117–30.

Pearson, Noel (1994) Mabo: towards respecting equality and difference, in Anon (ed.) *Voices from the Land: 1993 Boyer Lectures.* ABC Books, Sydney: 89–101.

—— (1996) The concept of native title at common law, in Northern and Central Land Councils (ed.) *Land Rights Past, Present and Future: Conference Papers.* Northern and Central Land Councils, Casuarina: 118–23.

—— (1997) The concept of native title at common law, in Yunupingu, Galawurruy (ed.) *Our Land Is Our Life: Land Rights – Past Present and Future.* University of Queensland Press, Brisbane: 150–61.

Pederson, Howard and Woorunmurra, Banjo (1995) *Jandamarra and the Bunuba Resistance.* Magabala Books, Broome.

Penn, Alan (1985) The James Bay and Northern Quebec Agreement: natural resources, public lands and the implementation of a native land claim settlement. Paper prepared for the Royal Commission on Aboriginal People. Alan Penn, Montreal.

People's Grand Jury (1977) *The Amax War Against Humanity: a Case Study of a Multinational Corporation's Robbery of Namibian Copper and Cheyenne Coal.* People's Grand Jury on Land and Human Exploitation, Washington.

Perry, John (ed.) (1989) *Doing Fieldwork: Eight Personal Accounts of Social Research.* Deakin University Press, Geelong, Victoria.

Pile, Steve and Keith, Michael (eds) (1997) *Geographies of Resistance.* Routledge, London and New York.

Pollin, Robert (1981) The multinational mineral industry in crisis. *Monthly Review* 31(11) 25–38.

Press, Tony; Lea, David; Webb, Ann and Graham, Alistair (eds) (1995) *Kakadu: Natural and Cultural Heritage Management.* Australian Nature Conservation Agency and North Australia Research Unit, Darwin.

Pretes, Michael and Robinson, Michael (1989) Beyond boom and bust: a strategy for sustainable development in the North. *Polar Record* 25(153) 115–20.

Professional Geographer (1994) Special issue on 'Women in the field'. 46(1).

Puddicombe, Stephen (1991) Realpolitik in Arctic Quebec: why Makivik Corporation won't fight this time. *Arctic Circle* Sept–Oct 1991, 14–21.

Pudup, Mary Beth (1988) Arguments within regional geography. *Progress in Human Geography* 12(3) 369–90.

Pyle, Michael T (1995) Beyond fish ladders: dam removal as a strategy for restoring America's rivers. *Stanford Environmental Law Journal* 14(1) 97–143.

Québec Provincial Government (1997) *James Bay and Northern Québec Agreement and Complementary Agreements.* Les Publications du Québec, Sainte-Foy.

Radio National (1995a) BHP and the controversial Ok Tedi compensation bill. Radio National transcripts: *The Business Report* Friday 18 August 1995.

—— (1995b) BHP accused of supporting undemocratic legislation in Papua New Guinea. Radio National transcripts: *The Business Report* Friday 11 August 1995.

—— (1995c) The bottom line at BHP. Radio National transcripts: *The Business Report* Friday 22 September 1995.

—— (1995d) BHP, Ok Tedi and Justice Cummins' contempt of court ruling. Radio National Transcripts: *The Law Report* Tuesday 26 September 1995.

Redclift, Michael (1992) Sustainable development and global environmental change. *Global Environmental Change* 2(1) 32–42.

Reimer, Chester (1993/1994) Moving toward co-operation: Inuit circumpolar policies and the Arctic Environmental Protection strategy. *Northern Perspectives* 21(4) 22–6.

Relph, E. (1989) Responsive methods, geographical imagination and the study of landscapes, in Kobayashi, A. and Mackenzie, S. (eds) *Remaking Human Geography.* Unwin Hyman, Boston: 149–63.

Renwick, W. (ed.) (1991) *Sovereignty and Indigenous Rights: The Treaty of Waitangi in International Contexts.* Victoria University Press, Wellington.

Resnick, Stephen and Wolff, Richard (1987) *Knowledge and Class: a Marxian Critique of Political Economy.* University of Chicago Press, Chicago.

—— (1992) Reply to Richard Peet. *Antipode* 24(2) 131–40.

Resource Assessment Commission (1991) *Kakadu Conservation Zone Inquiry: Final Report* (2 vols) AGPS, Canberra.

—— (1992) *Forest and Timber Inquiry: Final Report* (3 vols) Australian Government Publishing Service, Canberra.

—— (1993) *Coastal Zone Inquiry: Final Report*. Australian Government Publishing Service, Canberra.

Reynolds, Henry (1987) *The Law of the Land*. Penguin, Ringwood.

—— (1992) Mabo and pastoral leases. *Aboriginal Law Bulletin* 2(59), 8–10.

—— (1996) *Aboriginal Sovereignty*. Allen & Unwin, Sydney.

—— (1999) *Why Weren't We Told? A Personal Search for the Truth about our History*. Viking, Ringwood, Victoria.

Richardson, B. and Boer, B. (1991) *Regional Agreements in Australia and Canada: Strategies for Aboriginal Self-determination and Control*. Environmental Law Centre, School of Law, Macquarie University.

Richardson, Benjamin J.; Craig, Donna and Boer, Ben (1995) *Regional Agreements for Indigenous Lands and Cultures in Canada*. North Australia Research Unit, Darwin.

Richardson, Boyce (1976) *Strangers Devour The Land*. New York.

Rickson, R; Hundloe, T; McDonald, G and Burdge, R, (eds). (1990) Social impacts of development: putting theory and methods into practice, special issue. *Environmental Impact Assessment Review* 37(1–2).

Robbins, Lynn A. (1978) Energy Developments and the Navajo Nation, in Jorgensen, Joseph G.; Clemmer, Richard O.; Little, Ronald L.; Owens, Nancy J. and Robbins, Lynn A. (eds) *Native Americans and Energy Development*. Anthropology Resource Center, Cambridge Mass.: 35–48.

Roberts, J. (1975) *The Mapoon Story by the Mapoon People – Mapoon, Book One*. International Development Action, Fitzroy, Victoria.

Roberts, J.; Parsons, M. and Russell, B. (1975) *The Mapoon Story According to the Invaders: Church Mission, Queensland Government and Mining Company – Mapoon, Book Two*. International Development Action, Fitzroy, Victoria.

Roberts, J. P. (1981) *Massacres to Mining: the Colonisation of Aboriginal Australia*. Dove Communications, Melbourne.

Roberts, J. P. and McLean, D. (1975) *The Cape York Aluminium Companies and the Native Peoples: Comalco, RTZ, Kaiser, CRA, Alcan, Billiton, Pechiney, Tipperary – Mapoon, Book Three*. International Development Action, Fitzroy, Victoria.

Robertson, Margaret, Vang, Kevin and Brown, A. J. (1992) *Wilderness in Australia: Issues and Options – A Discussion Paper*. Australian Heritage Commission, Canberra.

Robinson, Mike; Garvin, Terry and Hodgson, Gordon (1994) *Mapping How We Use our Land using Participatory Action Research*. Arctic Institute of North America, University of Calgary, Calgary.

Rodda, A. (1991) *Women and the Environment*. Zed Books, London

Rodman, Margaret C. (1992) Empowering place: multilocality and multivocality. *American Anthropologist* 94(3) 640–56.

Rogers, Raymond A. (1995) *The Oceans Are Emptying: Fish Wars and Sustainability*. Black Rose, Montréal.

Rose, Bruce (1992) *Aboriginal Land Management Issues in Central Australian*. Central Land Council, Alice Springs.

Rose, Deborah Bird (1988) Exploring an Aboriginal land ethic. *Meanjin* 47(3) 378–87.

—— (1996a) *Nourishing Terrains: Australian Aboriginal views of Landscape and Wilderness*. Australian Heritage Commission, Canberra.

—— (1996b) *Indigenous Customary Law and the Courts: Post-modern Ethics and Legal Pluralism*. NARU Discussion Papers 2/1996. North Australia Research Unit, Darwin.

—— (1999) Indigenous ecologies and an ethic of connection, in Low, N. (ed.) *Global Ethics and Environment*. Routledge, London: 175–87.

Ross, Helen (1990a) Progress and prospects in Aboriginal SIA. *Australian Aboriginal Studies* 1: 11–17.

—— (1990b) Community social impact assessment: a framework for indigenous peoples. *EIA Review* 10: 185–93.

—— (1991) The East Kimberley impact assessment project. *Interdisciplinary Science Reviews* 16(3), 1–10.

—— (1999a) A social impact and negotiation model for the mining industry and indigenous Australians. Paper presented to International Symposium on Society and Natural Resource Management, Brisbane, Australia, July 1999.

—— (1999b) New ethos – new solutions: indigenous negotiation of co-operative environmental management agreements in Washington state. *Australian Indigenous Law Reporter* 4(2) 1–28.

Rosser, Bill (1987) *Dreamtime Nightmares*. Penguin, Ringwood, Victoria.

Rowse, Tim (1987) 'Were you ever savages?' Aboriginal insiders and pastoralists' patronage. *Oceania* 58(2, December), 81–99.

—— (1993) *After Mabo: Interpreting Indigenous Traditions*. Melbourne University Press, Melbourne.

Royal Commission on Aboriginal Peoples (1996) *People to People, Nation to Nation: Highlights from the Royal Commission on Aboriginal Peoples*. Minister of Supply and Services, Ottawa.

Ruiz, Lester Edwin J. (1988) Theology, politics, and the discourses of transformation. *Alternatives* 13: 155–76.

Rural Landholders for Coexistence (1998) *Talking Common Ground: Negotiating Agreements with Aboriginal People*. Rural Landholders for Coexistence, Sydney.

Rushdie, Salman (1992) *Imaginary Homelands: Essays and Criticism 1981–1991*. Granta, London.

Russell, Julian and Gailey, Tony (1989) *Barefoot Economist – Visionaries: Small Solutions to Enormously Large Problems*. 220 Productions. 55 minutes [videorecording].

Sachs, Wolfgang (ed.) (1992) *The Development Dictionary: a Guide to Knowledge as Power*. Zed Books, London.

Sadler, B. (1989) National parks, wilderness preservation and native peoples in Northern Canada. *Natural Resources Journal* 29: 185–204.

Saffu, Yaw (1992) The Bougainville crisis and politics in Papua New Guinea. *The Contemporary Pacific* 4(2) 325–43.

Sagers, Matthew J. and Kryukov, Valeriy (1993) The hydrocarbon processing industry in West Siberia. *Post-Soviet Geography* 34(2), 127–52.

Said, Edward (1978) *Orientalism*. Routledge & Kegan Paul, London.

Salisbury, Richard (1977) A prism of perceptions: the James Bay hydro electricity project, in Wallman, Sandra (ed.) *Perceptions of Development*. Cambridge University Press, Cambridge: 172–90.

Sallenave, John (1994) Giving traditional ecological knowledge its rightful place in environmental impact assessment. *Northern Perspectives* 22(1) 16–19.

Sandercock, Leonie (1995) Voices from the Borderlands: a meditation on a metaphor. *Journal of Planning Education and Research* 14: 77–88.

Sanders, Douglas E. (1981) International law aspects of the effects on the Sami people of the hydro-electric project on the Alta River. Opinion prepared in Vancouver, 20 September 1981.

Sardar, Z (1992–93) Lies, damn lies and Columbus: the dynamics of constructed ignorance. *Third Text* 21: 47–56.

Sayer, A. (1984) *Method in Social Science: A Realist Approach*. Hutchinson, London.

—— (1989) The 'new' regional geography and problems of narrative. *Environment and Planning D: Society and Space* 7(3) 253–76.

—— (1991) Behind the locality debate: deconstructing geography's dualisms. *Environment and Planning A*, 23(2) 283–308.

Schell, M. (1991) Interview. *Massachusetts* 2,3 (Spring) 9.

Schmidheiny, Stephan (1992) *Changing Course: A Global Business Perspective on Business Development and the Environment*. MIT Press, Cambridge, MA.

Schoenberger, Erica (1991) The corporate interview as a research method in economic geography. *Professional Geographer* 43(2) 180–89.

Schumacher, E.F. (1974) *Small is Beautiful: Economics as if People Mattered*. Sphere Books, London.

Scott, Colin (1995) Encountering the whiteman in James Bay Cree narrative history and mythology. *Aboriginal History* 19(2) 21–40.

Seabrook, Jeremy (1993) *Victims of Development: Resistance and Alternatives*. Verso, London.

Sharp, Nonie (1992) Scales from the eyes of justice. *Arena* 99(100) 55–61.

—— (1994) Native title: the reshaping of Australian identity. *Arena* 3: 115–47.

—— (1996) *No Ordinary Judgement: Mabo, The Murray Islanders' Land Case*. Aboriginal Studies Press, Canberra.

Shaw, David G. (1992) The Exxon Valdez oil spill: ecological and social consequences. *Environmental Conservation* 19(3) 253–8.

Shaw, Larry and Lalonde, Michael A. (1994) Working together: an overview of the Golden Patricia Agreement. Paper presented to the Aboriginal Mineral Development Conference, June 1994.

Shea, Courtney W. (1994) Coal mining and the environment: does SMCRA give regulators appropriate enforcement tools? *Natural Resources and Environment* Spring: 17–20, 58–60.

Shiva, V. (1988) *Staying Alive: Women, Ecology and Development*. Zed Books, London.

—— (1991) World Bank cannot protect the environment. *Third World Resurgence* 14/15: 22–3.

—— (1992) Does the new world order have trees? *24 Hours Supplement*, February: 34–9.

Silvern, Steven E. (1999) Scales of justice: law, American Indian treaty rights and political construction of scale. *Political Geography* 18: 639–68.

Sivuaq, Paulussi (1988) The extinguishment clause in the Agreement, in Vincent, Sylvie and Bowers, Garry (eds) *Baie James et Nord Québécois: Dix Ans Après/James Bay and Northern Québec: Ten Years After*. Recherches amérindiennes au québec, Montréal: 57–8.

Smith, L. Graham (1993) *Impact Assessment and Sustainable Resource Management*. Longman, Harlow.

Smith, Michael Peter (1994) Can you imagine? Transnational migration and the globalisation of grassroots politics. *Social Text* 39: 15–34.

Smith, Neil (1984a) *Uneven Development: Nature, Capital and the Production of Space.* Basil Blackwell, London.

—— (1984b) Deindustrialization and regionalization: class alliance and class struggle. *Papers of the Regional Science Association* 54: 115–28.

—— (1987) Dangers of the empirical turn: some comments on the CURS intiative. *Antipode* 19(1) 59–68.

—— (1988a) Regional adjustment or restructuring. *Urban Geography* 9(3) 318–24.

—— (1988b) The region is dead! Long live the region! *Political Geography Quarterly* 7(2) 141–52.

—— (1992) Geography, difference and the politics of scale, in Doherty, Joe; Graham, Elspeth and Malek, Mo (eds) *Postmodernism and the Social Sciences.* Macmillan, London: 57–79.

—— (1994) Homeless/global: scaling places, in Bird *et al.* (eds) *Mapping the Futures: Local Cultures, Global Change.* Routledge, London: 87–119.

Smith, Neil and Ward, Dennis (1987) The restructuring of geographical scale: coalescence and fragmentation of the northern core region. *Economic Geography* 63(2) 160–82.

Smith, Neil and Godlewska, Anne (1994) Critical histories of geography, in Smith, Neil and Godlewska, Anne (eds) *Geography and Empire.* IBG Special Publications 30. Blackwell and the Institute of British Geographers, Oxford: 1–8.

Smith, Nigel; Alvim, Paulo; Homma, A.; Falesi, I. and Serrao, A. (1991) Environmental impact on resource exploitation in Amazonia. *Global Environmental Change* 1(4) 313–20.

Soja, E. (1989) *Post-modern Geographies: The Reassertion of Space in Critical Social Theory.* Verso, London

Solzhenitsyn, Alexander (1974) *The Gulag Archipelago.* Collins/Fontana, New York.

Spoehr, A. (1956) Cultural differences in the interpretation of natural resources, in Thomas, W.L. (ed.) *Man's Role in Changing the Face of the Earth.* Wenner-Gren Foundation, National Science Foundation and Chicago University Press, Chicago: 93–102.

Stake, Robert E. (1995) *The Art of Case Study Research.* Sage, Thousand Oaks, CA.

Stanner, W.E.H. (1979) *White Man Got No Dreaming: Essays 1938–1973.* ANU Press, Canberra.

Stayner, Richard (1992) *Local and Regional Impacts of Mining and Associated Developments in Australia.* Rural Development Centre, University of New England, Armidale.

Steinlien, Oystein (1989) The Sami Law: a change of Norwegian government policy toward the Sami minority? *The Canadian Journal of Native Studies* 9(1), 1–14.

Stephenson, Margaret A. and Ratnapala, Suri (eds) (1993) *Mabo: A Judicial Revolution: The Aboriginal Land Rights Decision and its Impact on Australian Law.* University of Queensland Press, Brisbane.

Stevens, Frank (1969) Weipa: the politics of pauperization. *The Australian Quarterly* 41(3), 5–25.

—— (1974) *Aborigines in the Northern Territory Cattle Industry.* ANU Press, Canberra.

—— (1981) Equal wages for Aborigines, in Stevens, Frank (ed.) *Black Australia.* Alternative Publishing Cooperative, Sydney: 63–105.

Stevenson, M. (1992) Columbus and the war on indigenous peoples. *Race and Class* 33(3) 27–45.

Stewart, G. (1991) Australian miners pay the price for ignoring Aboriginal needs. *Australian Journal of Mining* 6(61) 24–5.

Stockbridge, M.E.; Gordon, B.; Nowicki, R. and Paterson, N. (1976) *Dominance of Giants: A Shire of Roebourne Study*. Department of Social Work, University of Western Australia, Nedlands.

Stockdale. Pro-growth, limits to growth and a sustainable development synthesis. *Society and Natural Resources* 2: 163–76.

Storey, K. and Shrimpton, M. (1992) Workshop on the use of long distance commuting by the Mining Industry (proposal and pre-conference materials). Institute of Social and Economic Research, Memorial University of Newfoundland, St Johns and Rural Development Centre, University of New England, Armidale.

Stormo, Rune and Solem, Eva (1981) Sami rights and the Alta River. *Northern Perspectives* 9(3), 2–5.

Storper, Michael (1988) Big structures, small events and large processes in economic geography. *Environment and Planning A* 20(2) 165–85.

Strapp, Terry (1994) Native Title Act case study: the Zapopan Mt Todd Project and the Jawoyn Aboriginal community. Paper presented to the Blake Dawson Waldron Conference, 'Working with the Native Title Act', Sydney, 16–17 May 1994.

Stuart, Donald (1959) *Yandy*. Australasian Book Society, Sydney.

Suchet, Sandra (1994) Rekindling culture through resources: aboriginal resource management strategies and aspirations at Weipa. BA(Hons) Thesis, School of Earth Sciences, Macquarie University, Sydney.

—— (1999) Situated engagement: a critique of wildlife management and post-colonial discourse. PhD dissertation, Department of Human Geography, Macquarie University, Sydney.

Sullivan, M. and King, D. (1992) Assessing the environmental impact of development aid programmes. *Development Bulletin* April: 12–16

Sullivan, Patrick (1995) *Beyond Native Title: Multiple Land Use Agreements and Aboriginal Governance in the Kimberley*. NARU Discussion Paper 89/1995. North Australia Research Unit, Darwin.

—— (1997) *Regional Agreements in Australia: an Overview Paper*.(Land, Rights, Laws: Issues of Native Title 17). Australian Institute of Aboriginal and Torres Strait Islander Studies, Canberra.

Sullivan, Patrick (ed.) (1996) *Shooting the Banker: Essays on ATSIC and Self-Determination*. North Australia Research Unit, Darwin.

Suzuki, D. (1992) A personal foreword: the value of native ecologies, in Knudstin, and Suzuki, D. (eds) *Wisdom of the Elders*. Allen and Unwin, Sydney: xxi–xxxv.

Svensson, Tom G. (1984) The Sami and the nation state: some comments on ethnopolitics in the northern fourth world. *Etudes Inuit Studies* 8(2), 158–66.

—— (1986) Ethnopolitics among the Sami in Scandanavia: a basic strategy toward local autonomy. *Arctic* 39(3), 208–15.

—— (1988) Patterns of transformations and local self-determination: ethnopower and the larger society in the north: the Sami case, in Dacks, Gurston and Coates, Ken (eds) *Northern Communities: the Prospects for Empowerment*. Boreal Institute for Northern Studies Occasional Papers, 25. Boreal Institute for Northern Studies, University of Alberta, Edmonton: 77–89.

Swyngedouw, Erik (1997) Neither global nor local: 'glocalization' and the politics of scale, in Cox, Kevin R. (ed.) *Spaces of Globalization: Reasserting the Power of the Local*. The Guildford Press, New York and London: 137–66.

Tann, T. (ed.) (1991) Final report of Conference on Aboriginal interests in national parks and reserves for nature conservation in Western Australia held at Millstream-Chichester National Park, August 1990: Aboriginal Affairs Planning Authority, Perth.

Tanzer, M. (1992) After Rio. *Monthly Review* 44(6) 1–11.

Tatz, Colin (1982) *Aborigines and Uranium and Other Essays*. Heinemann Educational, Richmond, Victoria.

—— (1998) The reconciliation 'bargain'. *Melbourne Journal of Politics* 25 (special issue on Reconciliation) 1–8.

—— (1999) *Genocide in Australia*. Australian Institute of Aboriginal and Torres Strait Islander Studies, Canberra.

Tau, Te Maire; Goodall, Anake; Palmer, David and Tau, Rakiihia (1990) *Te Whakatau Kaupapa: Ngai Tahu Resource Management Strategy for the Canterbury Region*. Aoraki Press, Wellington.

Tauli-Corpuz, V. (1993) An indigenous peoples' perspective on environment and development. *IWGIA Newsletter* 1/93.

Taussig, M. (1980) *The Devil and Commodity Fetishism in South America*. University of North Carolina Press, Chapel Hill.

Taylor, C. Nicholas; Bryan, C. Hobson and Goodrich, Colin G. (1990) *Social Assessment: Theory, Process and Techniques*. (Studies in Resource Management 7). Centre for Resource Management, Lincoln University, Canterbury.

Taylor, Peter (1995) *Caring for Country Strategy: an Indigenous Approach to Regional and Environmental Programmes*. Northern Land Council, Darwin.

Taylor, Peter J. (1982) A materialist framework for political geography. *Transactions of the Institute of British Geographers* 7: 15–34.

—— (1987) The paradox of geographical scale in Marx's politics. *Antipode* 19(3) 287–306.

—— (1993) *Political Geography: World-Economy, Nation-State and Locality*. 3rd edn. Longman, Harlow.

Teehan, Maureen (1994) Practising land rights: the Pitjantjatjara in the Northern Territory, South Australia and Western Australia, in Goot, Murray and Rowse, Tim (eds) *Make a Better Offer: The Politics of Mabo*.Pluto Press, Leichhardt: 34–54.

Teehan, Maureen (1997) Indigenous peoples, access to land and negotiated agreements: experiences and post-Mabo possibilities for environmental management. *Environmental and Planning Law Journal* April: 114–34.

Terebikhin, Nikolaj M (1993) Cultural geography and cosmography of the Saami. *Acta Borealia* 10(1), 3–17.

Thomas, Ian (1987) *Environmental Impact Assessment, Australia Perspectives and Practice*. Graduate School of Environmental Science, Monash University, Clayton.

Thomas, Pete (1975) *The Nymboida Story: the Work-ins that Saved a Coal Mine*. Miners Federation, Sydney.

—— (1979) *The Mine the Workers Ran: the 1975–1979 Success Story at Nymboida*. Miners Federation, Sydney.

Tjamiwa, Tony (1988) Tourism workshop, in Kean, John; Richardson, Garry and Trueman, Norma (eds) *Aboriginal Role in Nature Conservation: Emu Conference June 7–9, 1988*. South Australian National Parks and Wildlife Service, Adelaide, 3.3.

Torp, Eivind (1992) Sami rights in a political and social context, in Lyck, Lise. *Nordic Arctic Research on Contemporary Arctic Problems*. Proceedings from Nordic Arctic Research Symposium 1992, Copenhagen, Denmark, *Nordic Arctic Research Forum*: 85–92.

—— (1994) Sami Bill up in smoke: discourse strategies within Sami ethnopolitics in Sweden, in Greiffenberg, Tom (ed.) *Sustainability in the Arctic*. Nordic Arctic Research Forum, Aalborg: 127–43.

Tough, Frank (1993) The criminalization of Indian hunting in Ontario, Canada 1892–1930, in Cant, Garth; Overton, John and Pawson, Eric (eds) *Indigenous Land Rights in Commonwealth Countries*. Department of Geography, University of Canterbury and Ngai Tahu Maori Trust Board, Christchurch: 37.

Toyne, Philip (1994) *The Reluctant Nation: Environment, Law and Politics in Australia*. ABC Books, Sydney.

Toyne, Phillip and Vachon, Daniel (1984) *Growing Up the Country: The Pitjantjatjara Struggle for their Land*. McPhee Gribble/Penguin, Melbourne.

Trent, John E.; Young, Robert and Lachapelle, Guy (eds) (1996) *Québec-Canada: What is the Path Ahead?/Nouveaux sentiers vers l'avenir*. University of Ottawa Press, Ottawa.

Trigger, David (1996) Contesting ideologies of resource development in British Columbia, Canada. *Culture* 16(1) 55–69.

—— (1997a) Mining, landscape and the culture of development ideology in Australia. *Ecumene* 4(2) 161–80.

—— (1997b) Reflections on Century Mine: preliminary thoughts on the politics of indigenous responses, in Smith, Diane E. and Finlayson, Julie (eds) *Fighting Over Country: Anthropological Perspectives*. CAEPR Research Monographs vol. 12. Centre for Aboriginal Economic Policy Research, Australian National University, Canberra: 110–28.

Tropical Forests Task Force (1985) *Tropical Forests: a Call for Action*. Report of an International Task Force convened by the World Resources Institute, The World Bank and UNDP (3 vols). World Resources Institute, Washington.

Turner, Edith (1997) There are no peripheries to humanity: northern Alaska nuclear dumping and the Inupiat's search for redress. *Anthropology and Humanism* 22(1) 95–109.

Ury, W. (1991) *Getting Past No: Negotiating with Difficult People*. Business Books, London.

Varley, Pamela (1995) *Cultures in Collision: Battling Over the Environmental Review of Québec's Great Whale Project*. Kennedy School of Government Case Program C18-95-1277.0 and C18-95-1277.1, 475–507.

Vincent, Sylvie (1988) Hydroelectricity and its lessons, in Vincent, Sylvie and Bowers, Garry (eds) *Baie James et Nord Québécois: Dix Ans Après/James Bay and Northern Québec: Ten Years After*. Recherches amérindiennes au québec, Montréal: 245–50.

Vitebsky, Piers (1990) Gas, environmentalism and native anxieties in the Soviet Arctic: the case of the Yamal peninsula. *Polar Record* 26(156) 19–26.

Wagner, M.W. (1991) Footsteps along the road. *Alternatives* 18(2) 23–7.

Wallman, Sandra (ed.) (1977) *Perceptions of Development*. Cambridge University Press, Cambridge.

Wand, Paul (1996) CRA exploration and mine development in Australia after Mabo. Paper presented to 'Australia and the Mabo Judgement' conference, London, April 1996.

Ward, Michael Don (ed.) (1992) *The New Geopolitics*. Gordon & Breach, Philadelphia.

Watts, Charlie (1988) The stakes from an Inuit negotiator's point of view, in Vincent, Sylvie and Bowers, Garry (eds) *Baie James et Nord Québécois: Dix Ans Après/James Bay and Northern Québec: Ten Years After*. Recherches amérindiennes au québec, Montréal: 54–6.

WCED (1990) *Our Common Future*. Australian edn. Oxford University Press, Melbourne.

Wei Li; Huadong Wang and Liu Dongxia (1998) Progress of environmental impact assessment and its methods in China, in Alan Porter and John Fittipaldi (eds) *Environmental Methods Review: Retooling Impact Assessment for the New Century*. Army Environmental Policy Institute and International Association for Impact Assessment, Atlanta and Fargo: 50–7.

Weller, Geoffrey (1988) Self-government for Canada's Inuit: the Nunavut proposal. *American Review of Canadian Studies* 18(3) 341–57.

Wells, Edgar (1982) *Reward and Punishment in Arnhem Land 1962–1963*. Australian Institute of Aboriginal Studies, Canberra.

Wensing, Ed (1997) Native title and planning: some practical advice. *Australian Planner* 34(1) 4–5.

West, Cornell (1990) The new cultural politics of difference, in Ferguson, R.; Gever, M.; Minh-Ha, T. and West, C. (eds) *Out there: Marginalization and Contemporary Cultures*. New Museum of Contemporary Art, New York and MIT Press, Cambridge, MA: 19–36.

West, Mary Beth (1992) Natural resources development on Indian reservations: overview of tribal, state and federal jurisdiction. *American Indian Law Review* 17(1) 71–98.

Westcoat, J. L. (1991) Resource management: the long-term global trend. *Progress in Human Geography* 15(1) 81–93.

Westcoat, J. L. (1992) Resource management: oil resources and the Gulf conflict. *Progress in Human Geography* 16(2) 243–56.

Wharton, Geoffrey S. (1996) *The Day They Burned Mapoon: A Study of the Closure of a Queensland Presbyterian Mission*. BA(Hons) Thesis, University of Queensland, Brisbane.

Whitehead, Alfred N. ([1925] 1997) *Science and the Modern World*. Free Press, New York.

—— (1985) *Adventures of Ideas*. Free Press, New York.

Widders, Terry and Noble, Greg (1993) On the dreaming track to the republic: indigenous people and the ambivalence of citizenship. *Communal/plural* 2: 95–112.

Wildman, Paul (1985) Social impact analysis in Australia: policy issues for the 1980s. *Australian Journal of Social Issues* 20(2) 136–51.

Wilkins, David E. (1987) *Diné Bibeehaz'aanii: a Handbook of Navajo Government*. Navajo Community College Press, Tsaile, AZ.

Wilkinson, Charles F. (1993) *The Eagle Bird: Mapping a New West*. Vintage Books, New York.

Willems-Braun, Bruce (1997) Buried epistemologies: the politics of nature in (post)colonial British Columbia. *AAAG* 87(1) 3–31.

Williams, N. and Hunn, E. (eds) (1982) *Resource Managers: North American and Australia Hunter-Gatherers*. Australian Institute of Aboriginal Studies, Canberra.

Williams, Nancy (1986a) *The Yolngu and Their Land: A System of Land Tenure and the Fight for its Recognition.* AIAS, Canberra.

—— (1986b) Multi-disciplinary research: an anthropologist's prolegomenon. *Australian Aboriginal Studies* 1: 38–42.

—— (1998) Territory, land and property: milestones and signposts, in Academy of the Social Sciences in Australia (ed.) *Challenges for the Social Sciences and Australia* (vol. 2). Australian Government Publishing Service, Canberra: 83–113.

Williams, Nancy M. and Dillon, Michael (1985) East Kimberley feasibility study. *Australian Aboriginal Studies* 1985/2: 85.

Williams, Robert A. Jr (1990) *The American Indian in Western Legal Thought: The Discourses of Conquest.* Oxford University Press, New York.

Wilson, John (1984) Towards social impact assessment in Western Australia, as part of the environmental impact assessment process. *Anthropologica* 5(2), 189–299.

Wilson, Paul R. (1982) *Black Death White Hands.* Allen & Unwin, Sydney.

Woenne-Green, Susan; Johnston, Ross; Sultan, Ros and Wallis, Arnold (1994) *Competing Interests: Aboriginal Participation in National Parks and Conservations in Reserves in Australia – a Review.* Australian Conservation Foundation, Melbourne.

Wolf, C.P. (1980) Comments on the Indian social impact assessment workshop. *Newsline* (Rural Sociology Society) 8(6) 3–5.

Wolf, Eric (1982) *Europe and the People Without History.* University of California Press, Berkeley.

Wood, Christopher (1995) *Environmental Impact Assessment: a Comparative Review.* Longman, Harlow.

Wood, William B. (1990) Tropical deforestation: balancing regional development demands and global environmental concerns, *Global Environmental Change* 1: 23–41.

Woodley, Alison (1992) Cross-cultural interpretation planning in the Canadian Arctic. *Journal of Cultural Geography* 11(2) 45–62.

Woodward, J. (1974) *Aboriginal Land Rights Commission, Second Report.* AGPS, Canberra.

Yapp, G. (1989) Wilderness in Kakadu National Park: Aboriginal and other interests, *Natural Resources Journal* 29: 170–84.

York, Geoffrey and Pindera, Loreen (1992) *People of the Pines: the Warriors and the Legacy of Oka.* Little Brown, Boston.

Young, E. (1992) Hunter-gatherer concepts of land and its ownership in remote Australia and North America, in Anderson, K. and Gayle, F. (eds) *Inventing Places: Studies in Cultural Geography.* Longman, Melbourne: 255–72.

Young, Elspeth (1995) *Third World in the First: Development and Indigenous Peoples.* Routledge, London.

Young, Elspeth; Ross, Helen; Johnson, Judy and Kesteven, Jenny (1991) *Caring for Country: Aborigines and Land Management.* Australian National Parks and Wildlife Service, Canberra.

Young, Iris Marion (1990) *Justice and the Politics of Difference.* Princeton University Press, Princeton New Jersey.

—— (1997) Unruly categories: a critique of Nancy Fraser's dual systems theory. *New Left Review* 222: 147–60.

Yu, Peter (1997) Multilateral Agreements: a new accountability in Aboriginal affairs, in Yunupingu, Galawurruy (ed.) *Our Land Is Our Life: Land Rights – Past Present and Future*. University of Queensland Press, Brisbane: 168–80.

Yunupingu, Mandawuy (1994) Yothu Yindi – finding the balance, in *Voices from the Land: 1993 Boyer Lectures*. ABC Books, Sydney: 1–11.

Zimmerman, E. (1964) *Introduction to World Resources*. Harper & Row, New York (Hunker, H. (ed.), original edition published in 1933).

Index